# Lecture Notes in Computer Science 2655

Edited by G. Goos, J. Hartmanis, and J. van Leeuwen

Springer
*Berlin*
*Heidelberg*
*New York*
*Barcelona*
*Hong Kong*
*London*
*Milan*
*Paris*
*Tokyo*

Jean-Pierre Rosen   Alfred Strohmeier (Eds.)

# Reliable
# Software Technologies –
# Ada-Europe 2003

8th Ada-Europe International Conference
on Reliable Software Technologies
Toulouse, France, June 16-20, 2003
Proceedings

 Springer

Series Editors

Gerhard Goos, Karlsruhe University, Germany
Juris Hartmanis, Cornell University, NY, USA
Jan van Leeuwen, Utrecht University, The Netherlands

Volume Editors

Jean-Pierre Rosen
Adalog
19-21 rue du 8 mai 1945
94110 Arcueil, France
E-mail: rosen@adalog.fr

Alfred Strohmeier
Swiss Federal Institute of Technology Lausanne
Software Engineering Laboratory
1015 Lausanne EPFL, Switzerland
E-mail: alfred.strohmeier@epfl.ch

Cataloging-in-Publication Data applied for

A catalog record for this book is available from the Library of Congress.

Bibliographic information published by Die Deutsche Bibliothek
Die Deutsche Bibliothek lists this publication in the Deutsche Nationalbibliografie;
detailed bibliographic data is available in the Internet at <http://dnb.ddb.de>.

CR Subject Classification (1998): D.2, D.1.2-5, D.3, C.2-4, C.3, K.6

ISSN 0302-9743
ISBN 3-540-40376-0 Springer-Verlag Berlin Heidelberg New York

Springer-Verlag Berlin Heidelberg New York
a member of BertelsmannSpringer Science+Business Media GmbH

http://www.springer.de

© Springer-Verlag Berlin Heidelberg 2003
Printed in Germany

Typesetting: Camera-ready by author, data conversion by PTP-Berlin GmbH
Printed on acid-free paper      SPIN: 10936917      06/3142      5 4 3 2 1 0

# Preface

The 8th International Conference on Reliable Software Technologies, Ada-Europe 2003, took place in Toulouse, France, June 18–20, 2003. It was sponsored by Ada-Europe, the European federation of national Ada societies, and Ada-France, in cooperation with ACM SIGAda. It was organized by members of Adalog, CS, UPS/IRIT and ONERA.

Toulouse was certainly a very appropriate place for this conference. As the heart of the European aeronautic and space industry, it is a place where software development leaves no place for failure. In the end, reliability is a matter of human skills. But these skills build upon methods, tools, components and controlled practices. By exposing the latest advances in these areas, the conference contributed to fulfilling the needs of a very demanding industry.

As in past years, the conference comprised a three-day technical program, during which the papers contained in these proceedings were presented, along with vendor presentations. The technical program was bracketed by two tutorial days, when attendees had the opportunity to catch up on a variety of topics related to the field, at both introductory and advanced levels. On Friday, a workshop on "Quality of Service in Component-Based Software Engineering" was held. Further, the conference was accompanied by an exhibition where vendors presented their reliability-related products.

## Invited Speakers

The conference presented four distinguished speakers, who delivered state-of-the-art information on topics of great importance, for now and for the future of software engineering:

- An Invitation to Ada 2005
  *Pascal Leroy, Rational Software, France*
- Modules for Crosscutting Models
  *Mira Mezini and Klaus Ostermann, Darmstadt University of Technology, Germany*
- Software Fault Tolerance: An Overview
  *Jörg Kienzle, McGill University, Canada*

We would like to express our sincere gratitude to these distinguished speakers, well known to the community, for sharing their insights with the conference participants and for having written up their contributions for the proceedings.

## Submitted Papers

A large number of papers were submitted, from as many as 20 different countries. The program committee worked hard to review them, and the selection process proved to be difficult, since many papers had received excellent reviews. Finally, the program committee selected 29 papers for the conference. The final result was a truly international program with authors from Austria, Belgium, Finland, France, Germany, Hong Kong, India, Israel, Italy, Portugal, Spain, the United Kingdom, and the USA, covering a broad range of software technologies: fault tolerance, high integrity and the Ravenscar profile, real-time systems, distributed systems, formal specifications, performance evaluation and metrics, testing, tools, OO programming, software components, HOOD, UML, XML, language issues, and teaching.

## Tutorials

The conference also included an exciting selection of tutorials, featuring international experts who presented introductory and advanced material in the domain of the conference:

- The Personal Software Process$^{SM}$ for Ada, *Daniel Roy*
- Developing High Integrity Systems with GNAT/ORK, *Juan Antonio de la Puente and Juan Zamorano*
- Implementing Design Patterns in Ada 95, *Matthew Heaney*
- Principles of Physical Software Design in Ada 95, *Matthew Heaney*
- High Integrity Ravenscar Using SPARK, *Peter Amey*
- Architecture Centric Development Using Ada and the Avionics Architecture Description Language, *Bruce Lewis*
- A Semi-formal Approach to Software Systems Development, *William Bail*
- An Overview of Statistical-Based Testing, *William Bail*

## Workshop on Quality of Service in Component-Based Software Engineering

It is now widely recognized that what makes Component-Based Software Engineering (CBSE) hard is not the production of components but their composition. The workshop focused on methods and tools, but also on the difficulties involved in predicting the overall behavior of a composite from the properties of its components, including emerging properties, i.e. performance, reliability, safety, etc. Ideally, it should be possible for the software engineer to express the qualities of service required for each component, and a composition "calculus" would then predict the quality of service of the composite. The workshop brought together practitioners and academics currently working on these issues.

## Acknowledgements

Many people contributed to the success of the conference. The program committee, made up of international experts in the area of reliable software technologies, spent long hours carefully reviewing all the papers and tutorial proposals submitted to the conference. A subcommittee comprising Pierre Bazex, Agusti Canals, Dirk Craeynest, Michel Lemoine, Albert Llamosi, Thierry Millan, Laurent Pautet, Erhard Plödereder, Jean-Pierre Rosen and Alfred Strohmeier met in Toulouse to make the final paper selection. Some program committee members were assigned to shepherd some of the papers. We are grateful to all those who contributed to the technical program of the conference. Special thanks to Jean-Michel Bruel whose dedication was key to the success of the workshop.

We would also like to thank the members of the organizing committee, and especially the people at CS, UPS/IRIT and ONERA, for the work spent in the local organization. Agusti Canals managed the general organization of the conference. Pierre Bazex and Thierry Millan supervised the preparation of the attractive tutorial program. Frédéric Dumas was in charge of the conference exhibition. Dirk Craeynest and Michel Lemoine worked hard to make the conference prominently visible. Jean-Marie Rigaud dealt with all the details of the local organization, and Carole Bernon did a great job in preparing the web pages and all the Internet facilities. Voyages 31 had the important duty of taking care of the registration and financial management. Many thanks also to Laurent Pautet, who was in charge of the liaison with Ada-Europe.

A great help in organizing the submission process and the paper reviews was the START Conference Manager, provided graciously by Rich Gerber.

Special thanks are due to our sponsors, the French DGA, Mairie de Toulouse and Conseil Régional Midi-Pyrénées.

Last, but not least, we would like to express our appreciation to the authors of the papers submitted to the conference, and to all the participants who helped in achieving the goal of the conference, providing a forum for researchers and practitioners for the exchange of information and ideas about reliable software technologies. We hope they all enjoyed the technical program as well as the social events of the 8th International Conference on Reliable Software Technologies.


June 2003                                                    Jean-Pierre Rosen
                                                             Alfred Strohmeier

# Organizing Committee

**Conference Chair**

Agusti Canals, CS, France

**Program Co-chairs**

Jean-Pierre Rosen, Adalog, France
Alfred Strohmeier, Swiss Fed. Inst. of Technology, Lausanne, Switzerland

**Tutorial Co-chairs**

Pierre Bazex, UPS/IRIT, France
Thierry Millan, UPS/IRIT, France

**Exhibition Chair**

Frédéric Dumas, CS, France

**Publicity Co-chairs**

Dirk Craeynest, Offis, Belgium
Michel Lemoine, ONERA, France

**Local Organization Co-chairs**

Jean-Marie Rigaud, UPS/IRIT, France
Carole Bernon, UPS/IRIT, France

**Ada-Europe Conference Liaison**

Laurent Pautet, ENST, France

## Program Committee

Alejandro Alonso, ETSI Telecomunicacion, Spain
Ángel Álvarez, Technical University of Madrid, Spain
Lars Asplund, Uppsala University, Sweden
Neil Audsley, University of York, UK
Janet Barnes, Praxis Critical Systems Ltd., UK
Pierre Bazex, IRIT, France

Guillem Bernat, University of York, UK
Johann Blieberger, Technische Universität Wien, Austria
Maarten Boasson, University of Amsterdam, The Netherlands
Ben Brosgol, ACT, USA
Bernd Burgstaller, Technische Universität Wien, Austria
Agusti Canals, CS, France
Ulf Cederling, Vaxjo University, Sweden
Roderick Chapman, Praxis Critical Systems Ltd., UK
Dirk Craeynest, Offis nv/sa and KU Leuven, Belgium
Alfons Crespo, Universidad Politécnica de Valencia, Spain
Juan A. de la Puente, Universidad Politécnica de Madrid, Spain
Peter Dencker, Aonix GmbH, Germany
Raymond Devillers, Universite Libre de Bruxelles, Belgium
Wolfgang Gellerich, IBM, Germany
Jesús M. González-Barahona, Universidad Rey Juan Carlos, Spain
Michael González-Harbour, Universidad de Cantabria, Spain
Thomas Gruber, Austrian Research Centers, Seibersdorf, Austria
Helge Hagenauer, University Salzburg, Austria
Andrew Hately, Eurocontrol, Belgium
Hubert B. Keller, Institut für Angewandte Informatik, Germany
Yvon Kermarrec, ENST Bretagne, France
Jörg Kienzle, Swiss Federal Institute of Technology, Lausanne, Switzerland
Fabrice Kordon, UPMC, France
Michel Lemoine, ONERA, France
Albert Llamosi, Universitat de les Illes Balears, Spain
Kristina Lundqvist, Massachusetts Institute of Technology, USA
Franco Mazzanti, Istituto di Elaborazione della Informazione, Italy
John W. McCormick, University of Northern Iowa, USA
Thierry Millan, IRIT, France
Pierre Morere, Aonix, France
Pascal Obry, EdF, France
Laurent Pautet, ENST Paris, France
Erhard Plödereder, University of Stuttgart, Germany
Ceri Reid, CODA Technologies, UK
Jean-Marie Rigaud, Université Paul Sabatier, France
Alexander Romanovsky, University of Newcastle, UK
Jean-Pierre Rosen, Adalog, France
Bo I. Sandén, Colorado Technical University, USA
Bernhard Scholz, Technische Universität Wien, Austria
Edmond Schonberg, New York University and ACT, USA
Gerald Sonneck, ARC Seibersdorf research GmbH, Austria
Alfred Strohmeier, Swiss Fed. Inst. of Technology, Lausanne, Switzerland Tullio
Vardanega, University of Padova, Italy
Andy Wellings, University of York, UK
Jürgen Winkler, Friedrich-Schiller-Universität, Germany
Thomas Wolf, Paranor AG, Switzerland

# Table of Contents

## Distributed Information Systems

## Metrics

## Software Components

## Formal Specification

## Real-Time Kernel

## Testing

## Real-Time Systems Design

# An Invitation to Ada 2005

Pascal Leroy

Rational Software
1 place Charles de Gaulle
78067 St Quentin en Yvelines, France
pleroy@rational.com

**Abstract.** Starting in 2000, the ISO technical group in charge of maintaining the Ada language has been looking into possible changes for the next revision of the standard, around 2005. Based on the input from the Ada community, it was felt that the revision was a great opportunity for further enhancing Ada by integrating new programming practices, e.g., in the OOP area; by providing new capabilities for embedded and high-reliability applications; and by remedying annoyances encountered during many years of usage of Ada 95. This led to the decision to make a substantive revision rather than a minor one. This paper outlines the standardization process and schedule, and presents a number of key improvements that are currently under consideration for inclusion in Ada 2005.

## 1 Introduction

Every ISO standard is reviewed every five years to determine if it should be confirmed (kept as is), revised, or withdrawn. The international standard for Ada [1], known as ISO/IEC 8652 in ISO jargon, was published in 1995. A Technical Corrigendum, ISO/IEC 8652 Corr. 1, was published at the end of the first review period, in 2001. This document corrected a variety of minor errors or oversights in the original language definition, with the purpose of improving the safety and portability of programs. However, it did not add significant new capabilities.

With 7 years' experience with Ada 95, the ISO working group in charge of maintaining the language (known as JTC1/SC22/WG9) has come to the conclusion that more extensive changes were needed at the end of the second review period, in 2005. As a result, it was decided to develop an Amendment to integrate into the Ada language the result of more that 10 years of research and practice in programming languages. (In ISO terms, an Amendment is a much larger change than a Technical Corrigendum.)

WG9 has asked its technical committee, the Ada Rapporteur Group (ARG) to prepare proposals for additions of new capabilities to the language. Such capabilities could take the form of new core language features (including new syntax), new predefined units, new specialized needs annexes, or secondary standards, as appropriate. The changes may range from relatively minor to quite substantial.

J.-P. Rosen and A. Strohmeier (Eds.): Ada-Europe 2003, LNCS 2655, pp. 1–23, 2003.
© Springer-Verlag Berlin Heidelberg 2003

## 2    Revision Guidelines

As part of the instructions it gave to the ARG, WG9 has indicated that the main purpose of the Amendment is to address identified problems in Ada that are interfering with the usage or adoption of the language, especially in those application areas where it has traditionally had a strong presence: high-reliability, long-lived real-time and embedded applications, and very large complex systems.

In particular, the ARG was requested to pay particular attention to two categories of improvements:

- Improvements that will maintain or improve Ada's advantages, especially in those user domains where safety and criticality are prime concerns;
- Improvements that will remedy shortcomings of Ada.

Improvements that fall into the first category include: new real-time features, such as the Ravenscar profile, and features that improve static error detection. Improvements that fall into the second category include a solution to the problem of mutually dependent types across packages, and support of Java-like interfaces.

In selecting features for inclusion in the Amendment, it is very important to avoid gratuitous changes that could impose unnecessary disruption on the existing Ada community. Therefore, the ARG was asked to consider the following factors when evaluating proposals:

- Implementability: can the proposed feature be implemented at reasonable cost?
- Need: does the proposed feature fulfill an actual user need?
- Language stability: would the proposed feature appear disturbing to current users?
- Competition and popularity: does the proposed feature help improve the perception of Ada, and make it more competitive with other languages?
- Interoperability: does the proposed feature ease problems of interfacing with other languages and systems?
- Language consistency: is the provision of the feature syntactically and semantically consistent with the language's current structure and design philosophy?

In order to produce a technically superior result, it was also deemed acceptable to compromise strict backward compatibility when the impact on users is judged to be acceptable. It must be stressed, though, that the ARG has so far been extremely cautious on the topic of incompatibilities: incompatibilities that can be detected statically (e.g., because they cause compilation errors) might be acceptable if they are necessary to introduce a new feature that has numerous benefits; on the other hand, incompatibilities that would silently change the effect of a program are out of the question at this late stage.

In organizing its work, the ARG creates study documents named Ada Issues (AI) that cover all the detailed implications of a change. The Ada Issues may be consulted on-line at http://www.ada-auth.org/ais.html. The reader should keep in mind though that these are working documents that are constantly evolving, and that there is no guarantee that any particular AI will be included in the final Amendment.

# 3    Revision Schedule

At the time of this writing, WG9 targets the following schedule for the development of the Amendment:

- September 2003: receipt of the final proposals from groups other than WG9 or delegated bodies.
- December 2003: receipt of the final proposals from WG9 or delegated bodies.
- June 2004: approval of the scope of the Amendment, perhaps by approving individual AIs, perhaps by approving the entire Amendment document.
- Late 2004: informal circulation of the draft Amendment document, receipt of comments, and preparation of final text.
- Spring 2005: completion of proposed text of the Amendment.
- Mid 2005: WG9 email ballot.
- 3Q2005: SC22 ballot (SC22 is the parent body of WG9, and is in charge of the standardization of all programming languages).
- Late 2005: JTC1 ballot, final approval.

The elaboration of the Amendment being a software-related project, and a very complex one to boot, it would not be too surprising if the above schedule slipped by a couple of months. However, the work will essentially be schedule-driven, and the ARG will effectively stop developing new features at the beginning of 2004. The following two years will be spent writing and refining the Amendment document to ensure that its quality is on a par with that of the original Reference Manual.

# 4    Proposed New Features

The rest of this paper gives a technical overview of the most important AIs that are currently being considered for inclusion in the Amendment. Again, this is just a snapshot of work currently in progress, and there is no guarantee that all of these changes will be included in the Amendment. And even if they are, they might change significantly before their final incarnation.

For the convenience of the reader, the following table lists the proposals that are discussed below, in the order in which they appear:

| General-Purpose Capabilities |
| --- |
| Handling Mutually Dependent Type across Packages |
| Access to Private Units in the Private Part |
| Downward Closures for Access to Subprograms |
| Aggregates for Limited Types |
| Pragma Unsuppress |

| Real-Time, Safety and Criticality |
| --- |
| Ravenscar Profile for High Integrity Systems |
| Execution-Time Clocks |

Object-Oriented Programming
>     Abstract Interface to Provide Multiple Inheritance
>     Generalized Use of Anonymous Access Types
>     Accidental Overloading When Overriding

Programming By Contract
>     Pragma Assert, Pre-Conditions and Post-Conditions

Interfacing with Other Languages or Computing Environments
>     Unchecked Unions: Variant Records with No Run-Time Discriminant
>     Directory Operations

## 4.1   Handling Mutually Dependent Types across Packages (AI 217)

The impossibility of declaring mutually dependent types across package boundaries
has been identified very early after the standardization of Ada 95. This problem,
which existed all along since Ada 83, suddenly became much more prominent because the introduction of child units and of tagged types made it more natural to try
and build interdependent sets of related types declared in distinct library packages.

A simple example will illustrate the problem. Consider two classes, Department
and Employee, declared in two distinct library packages. These classes are coupled in
two ways. First, there are operations in each class that take parameters of the other
class. Second, the data structures used to implement each class contain references to
the other class. The following (incorrect) code shows how we could naively try to represent this in Ada:

```ada
with Employees;
package Departments is
   type Department is tagged private;
   procedure Choose_Manager
               (D        : in out Department;
                Manager : in out Employees.Employee);
private
   type Emp_Ptr is access all Employees.Employee;
   type Department is tagged
      record
         Manager : Emp_Ptr;
         ...
      end record;
end Departments;

with Departments;
package Employees is
   type Employee is tagged private;
   type Dept_Ptr is access all Departments.Department;
   procedure Assign_Employee
                 (E : in out Employee;
                  D : in out Departments.Department);
```

```
    function Current_Department (D : in Employee)
                                    return Dept_Ptr;
  private
      type Employee is tagged
        record
            Department : Dept_Ptr;

            ...

        end record;
  end Employees;
```

This example is patently illegal, as each specification has a with clause for the other one, so there is no valid compilation order.

The ARG has spent considerable effort on this issue, which is obviously a high-priority one for the Amendment. It is unfortunately very complex to solve, and at the time of this writing three proposals are being considered: type stubs, incomplete types completed in children, and limited with clauses. These proposals will be quickly presented in the sections below (in no particular order).

All proposals have in common that they somehow extend incomplete types to cross compilation units: incomplete types are the natural starting point for solving the problem of mutually dependent types, since they are used for that purpose (albeit in a single declarative part) in the existing language. When incomplete types are used across compilation units' boundaries, the normal limitations apply: in essence, it is possible to declare an access type designating an incomplete type, and not much else.

In addition, all proposals introduce the notion of tagged incomplete type. For a tagged incomplete type T, two additional capabilities are provided. First, it is possible to reference the class-wide type T'Class. Second, it is possible to declare subprograms with parameters of type T or T'Class. The reason why tagged incomplete types may be used as parameters is that they are always passed by-reference, so the compiler doesn't need to know their physical representation to generate parameter-passing code: it just needs to pass an address.

### 4.1.1 Type Stubs

A type stub declares an incomplete type that will be completed in another unit, and gives the name of the unit that will contain the completion:

```
type T1 is separate in P;
type T2 is tagged separate in P;
```

These declarations define two incomplete types, T1 (untagged) and T2 (tagged) and indicate that they will be completed in the library package P.

The rule that makes it possible to break circular dependencies is that, when the above type stubs are compiled, the package P doesn't have to be compiled. Later, when P is compiled, the compiler will check that it declares an untagged type named T1 and tagged type named T2.

For safety reasons, a unit that declares type stubs like the above must include a new kind of context clause to indicate that it contains a forward reference to package P. The syntax for doing this is:

```
abstract of P;
```

The package that provides the completion of the type stub, P, must have a with clause for the unit containing the stub. Moreover, each type may have at most one type stub (this is checked at post-compilation time).

With type stubs, the example presented above can be written as follows:

```
abstract of Departments;
package Departments_Interface is
   type Department is tagged separate in Departments;
end Departments_Interface;

abstract of Employees;
package Employees_Interface is
   type Employee is tagged separate in Employees;
end Employees_Interface;

with Departments_Interface, Employees_Interface;
package Departments is
   type Department is tagged private;
   procedure Choose_Manager
                (D        : in out Department;
                 Manager : in out
                     Employees_Interface.Employee);
private
   type Emp_Ptr is access all
                Employees_Interface.Employee;
   … -- As above.
end Departments;

with Departments_Interface, Employees_Interface;
package Employees is
   type Employee is tagged private;
   type Dept_Ptr is access all
                Departments_Interface.Department;
   procedure Assign_Employee
                (E : in out Employee;
                 D : in out
                     Departments_Interface.Department);
   function Current_Department (D : in Employee)
                               return Dept_Ptr;
private
   … -- As above.
end Employees;
```

Here we first declare two auxiliary packages, Departments_Interface and Employees_Interface, which contain the stubs. The context clauses make it clear that each package acts as the "abstract" of Departments and Employees, respectively. These packages are compiled first, and they declare two incomplete types, which will be completed in Departments and Employees.

Then Departments and Employees may be compiled, in any order, as they don't depend semantically on each other (although chances are that each body will depend semantically on the other package). Each package has a with clause for both Departments_Interface and Employees_Interface, and this requires an explanation. Take the case of Departments. Obviously it needs a with clause for Employees_Interface in order to declare the second parameter of Choose_Manager and the access type Emp_Ptr, using the tagged incomplete type Employees_Interface.Employee. But it also needs a with clause for Departments_Interface, as it completes a type stub declared there.

Note that in order to break the circularity it would be sufficient to declare only one type stub, say for type Department, for use by package Employees. Then package Departments could have a direct with clause for Employees. Here we have chosen to keep the example symmetrical, and to create two type stubs.

### 4.1.2  Incomplete Types Completed in Child Units

Another approach that is being considered is to allow incomplete types to be completed in child units. This can be thought of as a restricted version of type stubs, as the stub and the completion may not be in unrelated packages. The rationale for imposing this restriction is that types that are part of a cycle are tightly related, so it makes sense to require that they be part of the same subsystem. This somewhat simplifies the user model, as it doesn't require so much new syntax to declare the stub and tie it to its completion. Of course, this simplification comes at the expense of some flexibility, as all types that are part of a cycle have to be under the same root unit.

The syntax used to declare such an incomplete type is simply:

```
type C.T1;
type C.T2 is tagged;
```

These declarations define two incomplete types, T1 (untagged) and T2 (tagged) and indicate that they will be completed in unit C, a child of the current unit. As usual, the normal restrictions regarding the usages of incomplete types before their completion apply.

With incomplete types completed in child units, our example can be written as follows:

```
package Root is
    type Departments.Department;
    type Employees.Employee;
end Root;

package Root.Departments is
    type Department is tagged private;
    procedure Choose_Manager
                (D       : in out Department;
                 Manager : in out Employees.Employee);
private
    type Emp_Ptr is access all Employees.Employee;
    ... -- As above.
end Root.Departments;

package Root.Employees is
    type Employee is tagged private;
    type Dept_Ptr is access all Departments.Department;
    procedure Assign_Employee
```

```
                    (E  :  in out Employee;
                     D  :  in out Departments.Department);
      function Current_Department (D : in Employee)
                                  return Dept_Ptr;
   private
      … -- As above.
   end Root.Employees;
```

Here we first declare a root unit, Root, to hold the two incomplete type declarations. Then we can write the two child units in a way which is quite close to our original (illegal) example. Note however that in package Root.Departments the name Employees.Employee references the incomplete type declared in the parent, not the full type declared in the sibling package Root.Employees. Therefore, the two children don't depend semantically on each other, and the circularity is broken. However, the usages of Employees.Employee in Root.Departments are restricted to those that are legal for a (tagged) incomplete type.

### 4.1.3    Limited With Clauses

A different approach is to look again at the original (illegal) example, and see what it would take to make it legal. Obviously we cannot allow arbitrary circularities in with clauses, but when we look at the specifications of Departments and Employees, we see that they make a very restricted usage of the types Employees.Employee and Departments.Department, respectively: they only try to declare parameters and access types. If we introduce a new kind of with clause that strictly restricts the entities that are visible to clients and how they can be used, we can allow circularities among these new with clauses.

A limited with clause has the following syntax:

```
limited with P, Q.R;
```

A "limited with" clause gives visibility on a *limited* view of a unit. A limited view is one that contains only types and packages (excluding private parts), and where all types appear incomplete. Thus, a limited view of the original Departments package looks as follows:

```
package Departments is
   type Department is tagged;
end Departments;
```

The rule that makes it possible to break circularity is that a limited with clause doesn't introduce a semantic dependence on the units it mentions. Thus limited with clauses are allowed in situations where normal with clauses would be illegal because they would form cycles.

Using limited with clauses, our example can be written as follows:

```
limited with Employees;
package Departments is
   type Department is tagged private;
   procedure Choose_Manager
                    (D       : in out Department;
                     Manager : in out Employees.Employee);
private
   type Emp_Ptr is access all Employees.Employee;
   ... -- As above.
end Departments;

limited with Departments;
package Employees is
   type Employee is tagged private;
   type Dept_Ptr is access all Departments.Department;
   procedure Assign_Employee
                 (E : in out Employee;
                  D : in out Departments.Department);
   function Current_Department (D : in Employee)
                                      return Dept_Ptr;
private
   ... -- As above.
end Employees;
```

Here we have simply changed the normal with clauses to be limited with clauses. In package Employees, the limited with clause for Departments gives visibility on the limited view shown above, so the type Department may be named, but it is a tagged incomplete type, and its usages are restricted accordingly. The situation is symmetrical in package Departments.

Internally, the compiler first processes each unit to create the limited views, which contain incomplete type declarations for Department and Employee. During this first phase the with clauses (limited or not) are ignored, so any ordering is acceptable. Then the compiler analyzes each unit a second time and does the normal compilation processing, using the normal compilation ordering rules. During this second phase the limited with clauses give visibility on the limited views, and therefore on the incomplete types. This two-phase processing is what makes it possible to find a proper compilation ordering even in the presence of cycles among the limited with clauses.

## 4.2    Access to Private Units in the Private Part (AI 262)

The private part of a package includes part of the implementation of the package. For instance, the components of a private type may include handles and other low-level data structures.

Ada 95 provides private packages to organize the implementation of a subsystem. Unfortunately, these packages cannot be referenced in the private part of a public package—the context clause for the private package is illegal. This makes it difficult to use private packages to organize the implementation details of a subsystem.

AI 262 defines a new kind of with clause, the "private with" clause, the syntax of which is:

```
private with A, B.C, D;
```

This clause gives visibility on the entities declared in units A, B.C, and D, but only in the private part of the unit where the clause appears. Thus, a public unit may reference a private unit in a private with.

The following is an example derived from code in the CLAW library (which provides a high-level interface over the Microsoft® Windows® UI). The low-level interface for an image list package looks like:

```
private package Claw.Low_Level_Image_Lists is
   type HImage_List is new DWord;
   type IL_Flags is new UInt;
   ILR_DEFAULTCOLOR : constant IL_Flags := 16#0000#;
   ILR_MONOCHROME   : constant IL_Flags := 16#0001#;
   ...
end Claw.Low_Level_Image_Lists;
```

The high-level interface for the image list package looks like:

```
private with Claw.Low_Level_Image_Lists;
package Claw.Image_List is
   type Image_List_Type is tagged private;
   procedure Load_Image
                    (Image_List : in out Image_List_Type;
                     Image      : in String;
                     Monochrome : in Boolean := False);
   ...
private
   type Image_List_Type is tagged
     record
        Handle : Claw.Low_Level_Image_Lists.HImage_List;
        Flags : Claw.Low_Level_Image_Lists.IL_Flags;
        ...
     end record;
end Claw.Image_List;
```

Here the private part of the high-level package needs to reference declarations from the low-level package. Because the latter is a private child, this is only possible with a private with clause.


### 4.3   Downward Closures for Access to Subprogram Types (AI 254)

One very important improvement in Ada 95 was the introduction of access-to-subprogram types, which make it possible to parameterize an operation by a subprogram. However, this feature is somewhat limited by the accessibility rules: because Ada (unlike, say, C), has nested subprograms, the language rules must prevent the creation of dangling references, i.e., access-to-subprogram values that outlive the subprogram they designate.

Consider for example the case of a package doing numerical integration:

```
package Integrate is
   type Integrand is access function (X : Float)
                                  return Float;
   function Do_It (Fn : Integrand; Lo, Hi : Float)
                  return Float;
end Integrate;
```

This package works fine if the function to be integrated is declared at library level. For instance, to integrate the predefined square root function, one would write:

```
Result := Integrate.Do_It
          (Ada.Numerics.
              Elementary_Functions.Sqrt'Access,
           0.0, 1.0);
```

However, the package Integrate cannot be used with functions that are not declared at library level. One good practical example of this situation is double integration, where the function to be integrated is itself the result of an integration:

```
function G (X : Float) return Float is
   function F (Y : Float) return Float is
   begin
       … -- Returns some function of X and Y.
   end F;
begin
   return Integrate.Do_It (F'Access, -- Illegal.
                           0.0, 1.0);
end G;

Result : Float := Integrate.Do_It (G'Access, 0.0, 1.0);
```

The accessibility rules are unnecessarily pessimistic here, because the integration algorithm does not need to store values of the access-to-subprogram type Integrand, so there is no risk of creating dangling references.

AI 254 proposes to introduce anonymous access-to-subprogram types. Such types can only be used as subprogram parameter types. Because they cannot be used to declare components of data structures, they cannot be stored and therefore cannot be used to create dangling references. With this feature, the package Integrate would be rewritten as:

```
package Integrate is
   function Do_It
          (Fn : access function (X : Float)
                                  return Float;
           Lo, Hi : Float)
           return Float;
end Integrate;
```

Now, the double integration example above would be legal, as there are no accessibility issues with anonymous access types. Note that the syntax may look a bit heavy-weight, but it actually follows the one that was used in Pascal.

## 4.4   Aggregates for Limited Types (AI 287)

Limited types allow programmers to express the idea that copying values of a type does not make sense. This is a very useful capability; after all, the whole point of a compile-time type system is to allow programmers to formally express which operations do and do not make sense for each type.

Unfortunately, Ada places certain limitations on limited types that have nothing to do with the prevention of copying. The primary example is aggregates: the programmer is forced to choose between the benefits of aggregates (full coverage checking) and the benefits of limited types. These two features ought to be orthogonal, allowing programmers to get the benefits of both.

AI 287 proposes to allow aggregates to be of a limited type, and to allow such aggregates as the explicit initial value of objects (created by object declarations or by allocators). This change doesn't compromise the safety of limited types; in particular, assignment is still illegal, and the initialization expression for a limited object cannot be the name of another object; also, there is no such thing as an aggregate for a task or protected type, or for a limited private type.

Because a limited type may have components for which it is not possible to write a value (e.g., components of a task or protected type), AI 287 also allows a box "<>" in place of an expression in an aggregate. The box is an explicit request to use default initialization for that component.

The following is an example of an abstraction where copying makes no sense, so the type Object should be limited. Actually, this type has a component of a protected type, so it *has* to be limited.

```
package Dictionary is
   type Object is limited private;
   type Ptr is access Object;
   function New_Dictionary return Ptr;
   ...
private
   protected type Semaphore is ...;
   type Tree_Node;
   type Access_Tree_Node is access Tree_Node;
   type Object is limited
      record
         Sem: Semaphore;
         Size : Natural;
         Root : Access_Tree_Node;
      end record;
end Dictionary;

package body Dictionary is
   function New_Dictionary return Ptr is
```

```
begin
   return new T'(Sem => <>,
                 Size => 0,
                 Root => null);
end New_Dictionary;
...
end Dictionary;
```

With the introduction of aggregates for limited types, the allocator in the body of New_Dictionary is legal. Note that in Ada 95, one would have to first allocate the object, and then fill the components Size and Root. This is error-prone, as during maintenance components might be added and not initialized. With limited aggregates, the coverage rules ensure that if components are added, the compiler will flag the aggregates where new component associations must be provided.

### 4.5   Pragma Unsuppress (AI 224)

It is common in practice for some parts of an Ada program to depend on the presence of the canonical run-time checks defined by the language, while other parts need to suppress these checks for efficiency reasons. For example, consider a saturation arithmetic package. The multiply operation might look like:

```
function "*" (Left, Right : Saturation_Type)
                return Saturation_Type is
begin
   return Integer (Left) * Integer (Right);
exception
   when Constraint_Error =>
      if (Left > 0 and Right > 0) or
         (Left < 0 and Right < 0) then
         return Saturation_Type'Last;
      else
         return Saturation_Type'First;
      end if;
end "*";
```

This function will not work correctly without overflow checking. Ada 95 does not provide a way to indicate this to the compiler or to the programmer.

AI 224 introduces the configuration pragma Unsuppress, which has the same syntax as Suppress, except that it cannot be applied to a specific entity. For instance, to ensure correctness of the above function, one should add, in its declarative part, the pragma:

```
pragma Unsuppress (Overflow_Check);
```

The effect of this pragma is to revoke any suppression permission that may have been provided by a preceding or enclosing pragma Suppress. In other words, it ensures that the named check will be in effect regardless of what pragmas Suppress are added at outer levels.

## 4.6    Ravenscar Profile for High-Integrity Systems (AIs 249 and 305)

The Ravenscar profile [2] is a subset of Ada (also known as a tasking profile) which was initially defined by the 8th International Real-Time Ada Workshop (IRTAW), and later refined by the 9th and 10th IRTAW. This profile is intended to be used in high-integrity systems, and makes it possible to use a run-time kernel of reduced size and complexity, which can be implemented reliably. It also ensures that the scheduling properties of programs may be analyzed formally.

The Ravenscar profile has been implemented by a number of vendors and has become a *de facto* standard. The ARG plans to include it in the Amendment. However, in looking at the details of the restrictions imposed by Ravenscar, it was quickly realized that many of them would be generally useful to Ada users, quite independently of real-time or high-integrity issues. So the ARG decided to proceed in two phases: first, it added to the language a number of restrictions and a new configuration pragma; then it defined the Ravenscar profile uniquely in terms of language-defined restrictions and pragmas. This approach has three advantages:

– Current users of the Ravenscar profile are virtually unaffected.
– Users who need to restrict the usage of some Ada constructs without abiding by the full set of restrictions imposed by Ravenscar can take advantage of the newly defined restrictions.
– Implementers may define additional profiles that cater to the needs of particular users communities.

AI 305 defines new restrictions corresponding to the Ada constructs that are outlawed by the Ravenscar profile. To take only two examples, restrictions No_Calendar and No_Delay_Relative correspond to the fact that Ravenscar forbids usage of package Ada.Calendar and of relative delay statements. AI 305 also defines a new configuration pragma, Detect_Blocking, which indicates that the runtime system is required to detect the execution of potentially blocking operations in protected operations, and to raise Program_Error when this situation arises.

AI 249 then defines a new configuration pragma, Profile, which specifies as its argument either the predefined identifier Ravenscar or some implementation-defined identifier (this is to allow compiler vendors to support additional profiles). The Ravenscar profile is defined to be equivalent to a list of 19 restrictions, to the selection of policies Ceiling_Locking and FIFO_Within_Priorities, and to the detection of blocking operations effected by pragma Detect_Blocking.

## 4.7    Execution-Time Clocks (AI 307)

The 11th IRTAW identified as one of the most pressing needs of the real-time community (second only to the standardization of the Ravenscar profile) the addition of a capability to measure execution time.

AI 307 proposes to introduce a new predefined package, Ada.Real_Time.Execution_Time, exporting a private type named CPU_Time, which represents the processor time consumed by a task. Arithmetic and conversion operations similar to those in Ada.Real_Time are provided for this type. In addition, it is possible to query the CPU time consumed by a task by calling the function

CPU_Clock, passing the Task_ID for that task. This package also exports a protected type, Timer, which may be used to wait until some task has consumed a predetermined amount of CPU time.

Ada.Real_Time.Execution_Time makes it possible to implement CPU time budgeting, as shown by the following example, where a periodic task limits its own execution time to some predefined amount called WCET (worst-case execution time). If an execution time overrun is detected, the task aborts the remainder of its execution, until the next period:

```
with Ada.Task_Identification;
with Ada.Real_Time.Execution_Time;
   ...
   use Ada.Real_Time;
   use type Ada.Real_Time.Time;
   ...
   task body Periodic_Stopped is
      The_Timer : Execution_Time.Timer;
      Next_Start : Real_Time.Time := Real_Time.Clock;
      WCET : constant Duration := 1.0E-3;
      Period : constant Duration := 1.0E-2;
   begin
      loop
         The_Timer.Arm
             (To_Time_Span (WCET),
              Ada.Task_Identification.Current_Task);
         select
             -- Execution-time overrun detection.
             The_Timer.Time_Expired;
             Handle_The_Error;
         then abort
             Do_Useful_Work;
         end select;
         The_Timer.Disarm;
         Next_Start := Next_Start +
                         Real_Time.To_Time_Span (Period);
         delay until Next_Start;
      end loop;
   end Periodic_Stopped;
```

Here the call to Arm arms the timer so that it expires when the current task has consumed WCET units of CPU time. The loop then starts doing the useful work it's supposed to do. If that work is not completed before the timer expires, the entry call Time_Expired is accepted, the call to Do_Useful_Work is aborted, and the procedure Handle_The_Error is called.

#### 4.8    Abstract Interfaces to Provide Multiple Inheritance (AI 251)

At the time of the design of Ada 95, it was decided that multiple inheritance was a programming paradigm which imposes too heavy a distributed overhead to introduce in Ada, where performance concerns are prevalent.

Since then, an interesting form of multiple inheritance has become commonplace, pioneered notably by Java and COM. An "interface" defines what methods a class must implement, without supplying the implementation of these methods. A class may "implement" any number of interfaces, in which case it must provide implementations for all the methods inherited from these interfaces. Interfaces have the attractive characteristic that they are relatively inexpensive at run-time. In particular, unlike full-fledged multiple inheritance, they do not impose a distributed overhead on programs which do not use them.

The ARG is studying the possibility of introducing interfaces in Ada. This is a sizeable language change, which affects many areas of the language, so this proposal is somewhat less mature than most of the others discussed in this paper.

AI 251 proposes to add new syntax to declare interfaces:

```
type I1 is interface; -- A root interface.
type I2 is interface; -- Another root interface.

-- An interface obtained by composing I1 and I2:
type I3 is interface with I1 and I2;
```

An interface may have primitive subprograms, but no components. In many respects, it behaves like an abstract tagged type; in particular, constructs which would create objects of an interface type are illegal.

It is expected that most componentless abstract types in Ada 95 could be turned into interfaces with relatively little effort, and with the added flexibility of using them in complex inheritance lattices.

Note that, in order to preserve compatibility, the word "interface" is not a new keyword. It is a new kind of lexical element, dubbed "non-reserved keyword", which serves as a keyword in the above syntax, but is a normal identifier everywhere else.

A tagged type may implement one or more interfaces by using a new syntax for type derivation:

```
type I4 is interface …;
type T1 is tagged …;
type T2 is new T1 and I3 and I4 with
   record
      …
   end record;
```

In this instance, T2 is normally derived from T1 (and it inherits the primitive operations of T1) and implements interfaces I3 and I4. It must therefore override all the primitive operations that it inherits from I3 and I4.

Interfaces become really handy in conjunction with class-wide types. The Class attribute is available for interfaces. Thus, a parameter of type I'Class can at execution be of any specific tagged type that implements the interface I. Similarly, it is possible to declare an access type designating I'Class, and to build data structures that gather objects of various types which happen to implement I.

Because many reusable components in Ada are written as generics, and for consistency with the rest of the language, new kinds of generic formal types are added. They are useful to pass interface types to a generic or to indicate that a formal derived type implements some number of interfaces. Their syntax is similar to that of normal interface and derived type declarations.

In implementation terms, the only operations that may incur a significant performance penalty are membership tests and conversions from a class-wide interface. Say that X is a parameter of type I'Class (I being an interface) and T is a specific tagged type. Evaluating T (X) (a conversion) requires a check at execution time that X is actually in the derivation tree rooted at T. This check already exists for normal tagged types, but it is more complex in the presence of interfaces, because the derivation structure is an acyclic graph, not a tree. Also, as part of the conversion the dispatch table must be changed so as to ensure that dispatching executes the operations of T.

The following example shows a concrete case of usage of interfaces:

```ada
type Stack is interface;
procedure Append (S : in out Stack;
                  E : Element) is abstract;
procedure Remove_Last (S : in out Stack;
                       E : out Element) is abstract;
function Length (S : Stack) return Natural is abstract;

type Queue is interface;
procedure Append (Q : in out Queue;
                  E : Element) is abstract;
procedure Remove_First (Q : in out Queue;
                        E : out Element) is abstract;
function Length (Q : Queue) return Natural is abstract;

type Deque is interface with Queue and Stack;
```

Here Stack and Queue are two interfaces which declare the operations (or the contract) characteristic of stacks and queues. Deque is an interface obtained by mixing these two interfaces. It has the characteristic that elements can be removed at either end of the deque.

```ada
package … is
   type My_Deque is new Deque with private;
   procedure Append (D : in out My_Deque;
                     E : Element;
   procedure Remove_First (D : in out My_Queue;
                           E : out Element);
   procedure Remove_Last (D : in out My_Deque;
                          E : out Element);
   function Length (D : My_Deque) return Natural;
private
   type My_Deque is new Blob and Deque with
      record
         … -- Implementation details here.
```

```
        end record;
    end ...;

    procedure Print (S : Stack'Class) is
        E : Element;
    begin
        if Length (S) /= 0 then
            Remove_Last (S, E);
            Print (E);
            Print (S);
            Append (S, E);
        end if;
    end Print;
```

Here My_Deque is a concrete implementation of Deque, which declares record components as needed, and overrides the subprograms Append, Remove_First, Remove_Last and Length.

The procedure Prints recursively prints a stack by removing its first element, printing it, printing the rest of the stack, and re-appending the first element. It can be used with any type that implements the interface Stack, including Deque and My_Deque (but not with type Queue, which does not have a Remove_Last primitive operation).

## 4.9    Generalized Use of Anonymous Access Types (AI 230)

In most object-oriented languages, types that are references to a subclass are freely convertible to types that are references to its superclass. This rule, which is always safe, significantly reduces the need for explicit conversions in code that deals with complex inheritance hierarchies.

In Ada however, explicit conversions are generally required, even when going from a subclass to its superclass. This obscures the explicit conversions that really do need attention (because they may fail a run-time check), and makes the whole object-oriented coding style more cumbersome than in other languages.

Another problem that is somewhat related is that of "named access type proliferation". This commonly occurs when, for one reason or another, an access type is not defined at the point of the type declaration (for instance, a pure package can not declare an access type). Ultimately, if they need to create references, clients end up declaring their own "personal" access type, causing yet more need for unnecessary explicit conversions.

To address these problems, AI 230 proposes to generalize the use of anonymous access types to components and object renaming declarations. As an example, consider the following (toy) object hierarchy:

```
    type Animal is tagged ...;
    type Horse is new Animal with ...;
    type Pig is new Animal with ...;

    type Acc_Horse is access all Horse'Class;
    type Acc_Pig is access all Pig;
```

An anonymous access type may be used as a component subtype in the declaration of an array type or object:

```
Napoleon, Snowball : Acc_Pig := ...;
Boxer, Clover      : Acc_Horse := ...;
Animal_Farm : constant array (Positive range <>) of
                access Animal'Class := (Napoleon,
                                        Snowball,
                                        Boxer,
                                        Clover);
```

Note the use of an anonymous access type for the component type of Animal_Farm. As is customary with anonymous access types, a value of a named access type can be implicitly converted to an anonymous access type, provided that the conversion is safe (i.e., goes towards the root of the type hierarchy). That's why the various animal names can be directly used in the aggregate. Contrast this situation with Ada 95, where a named access type would have to be used for the array component, and explicit conversions would be required in the aggregate.

Anonymous access types may also be used in record components, with the same implicit convertibility rules:

```
type Noah_S_Arch is
   record
      Stallion, Mare : access Horse;
      Boar, Sow      : access Pig;
   end record;
```

For the purpose of the accessibility rules, such components have the same accessibility level as that of the enclosing type. This is necessary to prevent dangling references, and it means that from this perspective the anonymous access types behave as if they were implicitly declared immediately before the composite type.

Finally, it is also possible to use anonymous access types in object renaming declarations. This is useful to reference a value without changing its accessibility level. Consider the example:

```
procedure Feast (The_Arch : access Noah_S_Arch) is
   A_Pig : access Pig renames The_Arch.Sow;
begin
   Roast (A_Pig);
end Feast;
```

Here the accessibility level of The_Arch is that of the actual parameter. The renaming makes it possible to reference the component Sow in a manner that doesn't alter the accessibility level (in other words, the accessibility level of A_Pig is that of the actual parameter passed to Feast). Thus the accessibility level is preserved when passing A_Pig to Roast. Using a named access type for A_Pig would change the accessibility level, which may be problematic when executing the body of Roast.

## 4.10  Accidental Overloading When Overriding (AI 218)

It is possible in Ada 95 (and in other programming languages, e.g., Java, C++) for a typographic error to change overriding into overloading. When this happens, the dynamic behavior of a program won't be what the author intended, but the bug may be very hard to detect. Consider for instance:

```
with Ada.Finalization;
package Root is
   type Root_Type is new Ada.Finalization.Controlled
      with null record;
   -- Other primitive operations here.
   procedure Finalize (Object : in out Root_Type);
end Root;

with Root;
package Leaf is
   type Derived_Type is new Root.Root_Type
      with null record;
   -- Other primitive operations here.
   procedure Finalise (Object : in out Derived_Type);
end Leaf;
```

Here presumably the author of package Leaf intended to redefine the procedure Finalize to provide adequate finalization of objects of type Derived_Type. Unfortunately, she used the British spelling, so the declaration of Finalise (note the 's') does not override the inherited Finalize (with a 'z'). When objects of type Derived_Type are finalized, the code that is executed is that of Root.Finalize, not that of type Leaf.Finalise.

This is obviously a safety concern, and some programming languages (e.g., Eiffel, C#) provide mechanisms to ensure that a declaration is indeed an override. The ARG intends to provide a similar mechanism, although the details are still unclear. The approach which seems the most promising is that of introducing new syntax. An overriding declaration would have to include the word "overriding" (a non-reserved keyword) as in the following example:

```
overriding
procedure Finalise (Object : in out Derived_Type);
```

This declaration would actually result in an error, since Finalise doesn't override any subprogram (in particular, it doesn't override the parent type's Finalize), and the spelling error would be detected early, at compilation time.

There are a number of issues that are still unresolved, though, having to do with overriding in generics and with the interactions with private types, so a lot of work is still needed to consolidate this proposal.

## 4.11  Pragma Assert, Pre-conditions and Post-conditions (AIs 286 and 288)

Several Ada compilers support an Assert pragma, in largely the same form. As part of the Amendment work, the ARG intends to standardize this pragma, an associated package, and an associated configuration pragma for controlling the effect of the pragma on the generated code.

AI 286 defines pragma Assert as taking a boolean expression and optionally a message:

```
pragma Assert (Angle in 0.0 .. Pi / 2 or
               Angle in Pi .. 3 * Pi / 2,
               Message => "Angle out of range");
```

This pragma may appear anywhere, including in a declarative part. At execution, the boolean expression is evaluated, and if it returns False the exception Assertion_Error is raised with the indicated message. This exception is declared in a new predefined package, Ada.Assertions:

```
package Ada.Assertions is
   pragma Pure (Ada.Assertions);
   Assertion_Error : exception;
end Ada.Assertions;
```

In practice, it is also useful to be able to enable or disable assertion checking on an entire program or on a collection of units. In order to help with this usage model, AI 286 also defines a configuration pragma, Assertion_Policy. Thus, the pragma:

```
pragma Assertion_Policy (Ignore);
```

disables all assertion checking in the entire environment or in specific units, while the pragma:

```
pragma Assertion_Policy (Check);
```

enables normal assertion checking.

On a related topic, the ARG has studied the possibility of improving the support of the "programming by contract" model, in a manner similar to what Eiffel provides. This would be done by expressing pre- and post-conditions for subprograms, and invariants for types and packages. AI 288 defines a number of pragmas to that effect. However, this proposal is not yet mature, and its details are very much in flux.

## 4.12  Unchecked Unions: Variant Records with No Run-Time Discriminant (AI 216)

Ada does not include a mechanism for mapping C unions to Ada data structures. At the time of the design of Ada 95, it was thought that using Unchecked_Conversion to obtain the effect of unions was satisfactory. However, easy interfacing with C unions is important enough that several compilers have defined a method to support it. The ARG has decided to standardize this interfacing technique.

AI 216 defines a new representation pragma, Unchecked_Union, which may be applied to a discriminated record type with variants to indicate that the discriminant must not exist at run-time. For example, consider the following C type, which could represent an entry in the symbol table of a compiler, where a symbol could be either an object or a package:

```
struct sym {
    int id;
    char *name;
    union {
        struct {
            struct sym *obj_type;
            int obj_val_if_known;
        } obj;
        struct {
            struct sym *pkg_first_component;
            int pkg_num_components;
        } pkg;
    } u;
};
```

This data structure maps to the following unchecked union type in Ada:

```
type Sym;
type Sym_Ptr is access Sym;
type Sym_Kind_Enum is (Obj_Kind, Pkg_Kind);
type Sym (Kind : Sym_Kind_Enum :=
                    Sym_Kind_Enum'First) is
record
    Id : Integer := Assign_Id (Kind);
    Name : C_Ptr;
    case Kind is
        when Obj_Kind =>
            Obj_Type : Sym_Ptr;
            Obj_Val_If_Known : Integer := -1;
        when Pkg_Kind =>
            Pkg_First_Component : Sym_Ptr;
            Pkg_Num_Components  : Integer := 0;
    end case;
end record;
pragma Unchecked_Union (Sym);
```

Because of the presence of the pragma, the discriminant Kind does not exist at run-time. This means that the application has to be able to determine from the context if a symbol is a package or an object. Those operations which would normally need to access the discriminant at run-time (like membership test, stream attributes, conversion to non-unchecked union types, etc.) raise Program_Error. Note that representation clauses may be given for an unchecked union type, but it is obviously illegal to give a component clause for a discriminant.

## 4.13  Directory Operations (AI 248)

Most modern operating systems contain some sort of tree-structured file system. Many applications need to manage these file systems (by creating and removing directories, searching for files, and the like). Many Ada 95 compilers provide access to these operations, but their implementation-defined packages differ in many ways, making portable programs impossible.

The POSIX libraries provide operations for doing this, but these libraries usually are available only on POSIX systems, leaving out many popular operating systems including MS-DOS®, most flavors of Windows®, and even Linux.

Therefore, the ARG has decided to standardize a minimum set of capabilities to access directories and files. The purpose is not to provide interfaces to of all the features of the underlying file system, but rather to define a common set of interfaces that makes it possible to write programs that can easily be ported on different operating systems. This is similar in spirit to the definition of Ada.Command_Line in Ada 95.

AI 248 defines a new predefined package, Ada.Directories, with operations to:

- Query and set the current directory;
- Create and delete a directory;
- Delete a file or a directory or an entire directory tree;
- Copy a file and rename a file or a directory;
- Decompose file and directory paths into enclosing directory, simple name, and extension;
- Check the existence, and query the size and modification time of a file;
- Iterate over the files and directories contained in a given directory.

It is intended that implementations will add operating-system-specific children of Ada.Directories to give access to functionalities that are not covered by this unit (e.g., file protections, encryption, sparse files, etc.).

## 5    Conclusion

This paper has given an overview of the standardization process for the 2005 revision of Ada, and it has presented those proposals that are reasonably stable and mature.

The ARG has also studied (and is still studying) a large number of other topics including: improving formal packages; allowing access-to-constant parameters; improving visibility rules for dispatching operations; and supporting notification of task termination. Some of these ideas seem promising, other might be disruptive, and so it's too early to tell which ones will make it into the Amendment and which ones will not.

The ARG's goal is to make Ada more powerful, safer, more appealing to its users, without imposing unreasonable implementation or training costs. Whatever the details of the changes that ultimately make it into the Amendment, we anticipate that 2005 will be a good vintage.

## References

1. S.T. Taft, R.A. Duff, R.L. Brukardt, E. Ploedereder: Consolidated Ada Reference Manual, LNCS 2219, Springer-Verlag.
2. A. Burns: "The Ravenscar Profile", ACM Ada Letters, Vol. XIX, No. 4, pp. 49–52, December 1999.

# Modules for Crosscutting Models

Mira Mezini and Klaus Ostermann

Darmstadt University of Technology, D-64283 Darmstadt, Germany
{mezini,ostermann}@informatik.tu-darmstadt.de

**Abstract.** Traditional programming languages assume that real-world systems have "intuitive", mind-independent, preexisting concept hierarchies. However, our perception of the world depends heavily on the context from which it is viewed: Every software system can be viewed from multiple different perspectives, and each of these perspectives may imply a different decomposition of the concerns. The hierarchy which we choose to decompose our software system into modules is to a large degree arbitrary, although it has a big influence on the software engineering properties of the software. We identify this *arbitrariness of the decomposition hierarchy* as the main cause of 'code tangling' and present a new model called CAESAR[1], within which it is possible to have multiple different decompositions *simultaneously* and to add new decompositions on-demand.

## 1   Introduction

The criteria which we choose to decompose software systems into modules has significant impact on the software engineering properties of the software. In [14] Parnas observed that a data-centric decomposition eases changes in the representation of data structures and algorithms operating on them. Following on Parnas work, Garlan et al. [2] argue that function-centric decomposition on the other side better supports adding new features to the system, a change which they show to be difficult with the data-centric decomposition.

Software decomposition techniques so far, including object-oriented decomposition, are weak at supporting multi-view decomposition, i.e., the ability to simultaneously breakdown the system into inter-related units, whereby each breakdown is guided by independent criteria. What current decomposition technology does well is to allow us to view the system at different abstraction levels, resulting in several *hierarchical* models of it, with each model be a refined version of its predecessor in the abstraction levels.

By multi-view decomposition, we mean support for simultaneous *crosscutting* rather than *hierarchical* models. The key point is that our perception of the world depends heavily on the perspective from which we look at it: Each perspective may imply a different decomposition of the concerns. In general, these view-specific decompositions are equally reasonable, none of them being a

---

[1] Project homepage at
http://www.st.informatik.tu-darmstadt.de/pages/projects/caesar

J.-P. Rosen and A. Strohmeier (Eds.): Ada-Europe 2003, LNCS 2655, pp. 24–44, 2003.

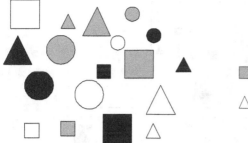

**Fig. 1.** Abstract concern space

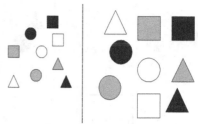

**Fig. 2.** Divide by size

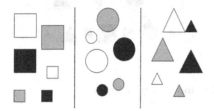

**Fig. 3.** Divide by shape

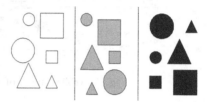

**Fig. 4.** Divide by color

sub-ordinate of the others, and the overall definition of the system results from a superimposition of them.

One of the key observations of the aspect-oriented software development is that a programming technique that does not support simultaneous decomposition of systems along different criteria suffers from what we call *arbitrariness of the decomposition hierarchy* problem, which manifests itself as tangling and scattering of code in the resulting software, with known impacts on maintainability and extendibility. In Fig. 1 to 5 we schematically give an idea of the problem at a non-technical level. Assuming the domain symbolically represented in Fig. 1 is to modeled, we can decompose according to three criteria, size, shape, and color, represented in Fig. 2, Fig. 3, and Fig. 4 respectively, whereby each of these decompositions is equally reasonable.

With a 'single-minded' decomposition technique that supports only hierarchical models, we have to choose one fixed classification sequence, as e.g., *color → shape → size* illustrated in Fig. 5. However, the problem is that with a fixed classification sequence, only the first element of the list is localized whereas all other concerns are tangled in the resulting hierarchical structure. Figure 5 illustrates this by the example of the concern 'circle', whose definition is scattered around several places in the hierarchical model. Only the color concern is cleanly separated into white, grey and black, but even this decomposition is not satis-

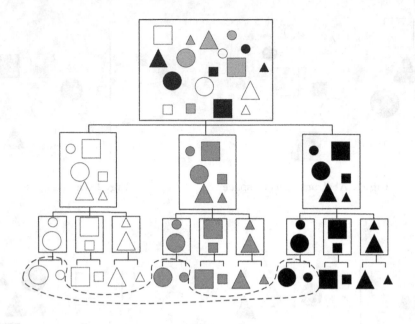

**Fig. 5.** Arbitrariness of the decomposition hierarchy

factory because the color concern is still tangled with other concerns: "white" is e.g., defined in terms of white circles, white rectangles and white triangles.

The problem is that models resulting from simultaneous decomposition of the system according to different criteria are in general "crosscutting" with respect to the execution of the system resulting from their composition. With the conceptual framework introduced so far, *crosscutting* can be defined as a relation between two models with respect to the abstract concern space. This relation if defined via *projections* of models (hierarchies). A projection of a model $M$ is a partition of the concern space into subsets $o_1, \ldots, o_n$ such that each subset $o_i$ corresponds to a leaf in the model. For illustration Fig. 6 shows a projection of the `color` model from Fig. 4 onto the abstract concern space of Fig. 1.

Now, two models, $M$ and $M'$, are said to be crosscutting, if there exist at least two sets $o$ and $o'$ from their respective projections, such that, $o \cap o'$, and neither $o \subseteq o'$, nor $o' \subseteq o1$. Figure 7 illustrates how the `black` concern of the `color` model (Fig. 4) crosscuts the `big` concern of the `size` model (Fig. 2). These two concerns have in common the big, black shapes, but neither is a subset of the other: the black module contains also small black shapes, while the size model contains also non-black big shapes.

On the contrary, a model $M$ is a *hierarchical refinement* of a model $M'$ if their projections $o_1, \ldots, o_n$ and $o'_1, \ldots, o'_m$ are in a subset relation to each other as follows: there is a mapping $p : \{1, \ldots, n\} \to \{1, \ldots, m\}$ such that $\forall i \in \{1, \ldots, n\} : o_i \subseteq o'_{p(i)}$. Crosscutting models are themselves not the problem, since they are inherent in the domains we model. The problem is that our languages

**Fig. 6.** Projection of the 'color' decomposition

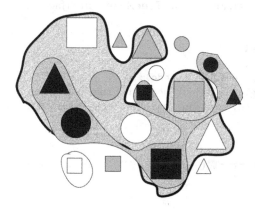

**Fig. 7.** Crosscutting hierarchies

and decomposition techniques do not (properly) support crosscutting modularity (see the discussion on decomposition arbitrariness above).

Motivated by these observations, we are working on a language model, called CAESAR, with explicit module support for crosscutting models. This paper puts special emphasis on the conceptual rationale and motivation behind CAESAR. It leaves out many technical details and features that are not of interest for the first-time reader. For more technical details we refer to two recent conference papers [11,12]. For instance, in [11] we describe how CIs, CI bindings, and CI implementations can be specified incrementally by an appropriate generalization of inheritance. In [12] we enrich CAESAR with means for describing *join points* (points in the execution of a base program to intercept) and *advices* (how to react at these points) and describe a new idea called *aspectual polymorphism*, with which components/aspects can be deployed polymorphically.

The remainder of this paper is structured as follows: In Sect. 2 we will discuss a concrete example that illustrates shortcomings of current language technology with respect to crosscutting hierarchies. In Sect. 3 we will shortly present our solution to cope with the identified problems. Section 4 is a wrap-up of what we have gained with the new technology. Section 5 discusses related work. Section 6 summarizes the paper.

## 2   Problem Statement

In this section, we identify shortcomings of current language technology with respect to supporting crosscutting models. We will use the Java programming language to illustrate shortcomings of current language technology, but the results are applicable to other module-based and object-oriented languages as well.

We introduce a simple example, involving two view-specific models of the same system: the GUI perspective versus the data types perspective. Figure 8 shows a simplified version of the TreeModel interface in Swing[2], Java's GUI

```
interface TreeModel {
  Object getRoot();
  Object[] getChildren(Object node);
  String getStringValue(Object node, boolean selected,
    boolean expanded, boolean leaf, int row, boolean focus);
  }
}

interface TreeGUIControl {
  display();
}

class SimpleTreeDisplay implements TreeGUIControl {
  TreeModel tm;
  display() {
    Object root = tm.getRoot();
    ... tm.getChildren(root) ...
    ...
    // prepare parameters for getStringValue
    ... tm.getStringtValue(...);
    ...
  }
}
```

**Fig. 8.** Simplified version of the Java Swing TreeModel interface

---

[2] Swing separates our interface into two interfaces, TreeModel and TreeCellRenderer. However, this is irrelevant for the reasoning in this model.

```
class Expression {
  Expression[] subExpressions;
  String description() { ... }
  Expression[] getSubExpressions() { ... }
}
class Plus extends Expression { ... }
```

**Fig. 9.** Expression trees

framework [4]. This interface provides a generic description of data abstractions that can be viewed and, hence, displayed as trees. Figure 8 also presents a interface for tree GUI controls in `TreeGUIControl`, as well as an implementation of this interface in `SimpleTreeDisplay` (the latter roughly corresponds to `JTree`).

In our terminology the code in Fig. 8 defines a GUI-specific model of any system that supports the display of arbitrary data structures that are organized as trees. When this model is used in a concrete context, e.g., in the context of a system that manipulates arithmetic expression object structures, as schematically represented in Fig. 9, it provides to this context the `display()` functionality. In turn, it expects a concrete implementation of `getChildren(...)` and `getStringValue(...)`.

Please note how the GUI-specific model crosscuts a model resulting from a data-centric decomposition of the system that manipulates expressions, and/or other aggregate, tree-structured data types. This is exemplified in Fig. 9, which shows sample code for a data-centric decomposition of arithmetic expressions. In a running system, resulting from a composition of the GUI-specific and data-specific models, the tree module from the GUI-specific model and the expression module from the data type model intersect, since there will be display-enabled expression objects. However, none of them is a subset of the other: There might be non-displayable expression objects, as well as displayable tree-structured data type objects that are not expressions.

## 2.1   Lack of Appropriate Module Constructs

The need for crosscutting modularity calls for module constructs that capture individual crosscutting models of the system by encapsulating the definition of multiple inter-dependent abstractions. In addition, mechanisms are needed to allow such modules to be self-contained (closed), while at the same time be opened to be composed with other crosscutting modules, while remaining decoupled from them. We argue that common module constructs in object-oriented languages and especially the common concept of interfaces as we know it, e.g., from Java, lack these two features.

**Lack of Support for Declaring a Set of Mutually Recursive Types.** Defining generic models involves in general several related abstractions. We claim that current technology falls short in providing appropriate means to express the different abstractions and their respective features and requirements that are

```
class ExpressionDisplay implements TreeModel {
  ExpressionDisplay(Expression r) { root = r; }
  Expression root;
  Object getRoot() { return root; }
  Object[] getChildren(Object node) {
      return ((Expression) node).getSubExpressions();
  }
  String getStringValue(Object node, boolean selected,
    boolean expanded, boolean leaf, int row, boolean focus){
    String s = ((Expression) node).description();
    if (focus) s ="<"+s+">";
      return s;
  }
}
```

**Fig. 10.** Using `TreeModel` to display expressions

involved in a particular collaboration. Let us analyze the model in Fig. 8 from this point of view. The first "bad smell" is the frequent occurrence of the type `Object`. We know that a tree abstraction is defined in terms of smaller tree node abstractions. However, this collaboration of the tree and tree node abstractions is not made explicit in the interface. Since the interface does not state anything about the definition of tree nodes, it has to use the type `Object` for nodes.

The disadvantages of using the most general type, `Object`, are twofold. First, it is conceptually questionable. If every abstraction that is involved in the GUI model definition is only known as `Object`, no messages, beside those defined in `Object`, can be directly called on those abstractions. Instead, a respective top-level interface method has to be defined, whose first parameter is the receiver in question. For example, the methods `getChildren(...)` and `get-StringValue(...)` conceptually belong to the interface of a tree node, rather than of a tree. Since the tree definition above does not include the declaration of a tree node, they are defined as top-level methods of the tree abstraction whose first argument is `node: Object`.

Second, we loose type safety, as illustrated in Fig. 10, where we 'compose' the expression and GUI views of the system by adapting expressions to fit in the conceptual world of a `TreeModel`. Since we use `Object` all the time, we cannot rely on the type checker to prove our code statically safe because type-casts are ubiquitous.

The question naturally raises here: Why didn't the Swing designers define an explicit interface for tree nodes as in Fig. 11? The problem is that than it becomes more difficult to decouple the two contracts, i.e., the data structures to be displayed from the display algorithm. The idea is that the wrapper classes around e.g., `Expression` would look like in Fig. 12. The problem with such kind of wrappers, as also indicated in [3], is that we create new wrappers every time we need a wrapper for an expression. This leads to the *identity hell*: we loose the state and identity of previously created wrappers for the same node.

```
interface TreeDisplay {
  TreeNode getRoot();
}
interface TreeNode {
  TreeNode[] getChildren();
  String getStringValue(boolean selected,
    boolean expanded, boolean leaf, int row, boolean focus);
}
```

**Fig. 11.** TreeDisplay interface with explicitly reified TreeNode interface

```
class ExprAsTreeNode
implements TreeNode {
  Expression expr;
  void getStringValue(...) {  /*as before*/  }
  TreeNode[] getChildren() {
    Expressions[] subExpr = expr.getSubExpressions();
    TreeNode[] children =
        new TreeNode[subExpr.length];
    for (i = 0; i<subExpr.length; i++) {
      children[i] = new ExprAsTreeNode(subExpr[i]));
    }
    return children;
  }
}
```

**Fig. 12.** Mapping TreeNode to Expression

The questionable alternative would be to use hash tables which is not only laborious but does also involve the definition and use of additional classes for maintaining these hashtables, thereby rendering the code more complex and less readable[3].

A final important point to make before leaving this branch of the discussion is that it is difficult and awkward to associate state with abstractions like our tree nodes. We might want to associate state with tree nodes in both the ExpressionDisplay class in Fig. 10 and also inside the tree GUI control. For example, we might want to cache the computed string value or children in Fig. 10, because the re-computation might be expensive. In the GUI control itself, we might want to associate state like whether a tree node is selected or not or its position on the screen with the respective tree node. The only means to associate state with tree nodes is to make extensive use of hash tables, which is laborious and awkward.

**Lack of Support for Bidirectional Communication.** Interfaces provide clients with a contract as what to expect from a server object that implements

---

[3] Actually, Swing offers a TreeNode interface similar to the one in Fig. 11. However, classes that define data structures to be displayed as tree nodes should anticipate this and explicitly implement the interface.

the interface. We say, they express the *provided* contract. In order to define generic partial, view-specific models which are decoupled from their potential contexts of use, expressing expectations that a server might have on potential contexts of use is as important. We use the term *expected* contract to denote these expectations. What is needed is support for a loose coupling of client and server, that is (a) decoupling them to facilitate reuse, while (b) enabling them to tightly communicate with each other as part of a whole.

In our example, `TreeGUIControl` corresponds roughly to what we called the provided contract, while `TreeModel` corresponds roughly to what we called the expected contract. The class `SimpleTreeDisplay` represents a sample implementation of the provided contract. It expects from the context a concrete implementation of `getChildren(...)` and `getStringValue(...)`. In our terminology, `ExpressionDisplay` in Fig. 10 represents an implementation of the *expected* contract.

The design in Fig. 8 does actually a good job in decoupling the expected and provided contracts. Different implementation of GUI controls can be written to the `TreeModel` interface and can therefore be reused with a variety of concrete implementations of it, i.e., with a variety of data structures. The other way around, any data structure to be displayed is decoupled from a specific tree GUI control (e.g., `JTree`), such that the data structure can be displayed with different GUI tree controls.

Now, consider the `getStringValue(...)` method in Fig. 8 and Fig. 10. This method has noticeable many parameters that might be of interest when computing a string representation of the node. *Might be.* The sample implementation in Fig. 10 uses only the `selected` parameter and ignores the others. That means, the tree GUI control, which calls this method on the `TreeModel` interface, has to perform expensive computations to obtain the parameter values for this method (see implementation of `SimpleTreeDisplay::display(...)` in Fig. 8), although they might be rarely all used.

This is a typical case where we would like to establish a bidirectional communication between the two contracts of the tree displaying component. Here we would like `ExpressionDisplay.getStringValue` to explicitly ask the tree GUI control to compute only relevant values for it, like `selected` or `hasFocus`, implying the GUI control interface provides respective operations. Recall that the GUI control interface corresponds to the provided interface of our generic component for displaying arbitrary data structures that can be viewed as trees in a GUI. As for now, the interfaces are completely separated (into `TreeModel` and `TreeGUIControl`), and there is nothing in the design that would suggest their tight relation as two faces of the same abstraction. As such, there is no build-in support for bidirectional communication between their respective implementations. Build-in means by the virtue of implementing two faces of the same abstraction, which serves as the implicit communication channel.

One can certainly achieve the desired communication by additional infrastructure (e.g., via cross-references) which has to be communicated to the respective programmers. However, we think that bidirectional communication is such a natural and frequent concept that the overhead that is necessary to enable bidirectional communication with conventional interfaces is too high. Please note that the additional `TreeNode` interface would also be of no help concerning the bidirectional communication problem exemplified by the `getStringValue()` method.

# 3    Caesar in a Nutshell

In this section, we will give an overview of the concepts that comprise our model by means of the `TreeDisplay` example from the previous section.

## 3.1    Collaboration Interfaces and Their Implementations and Bindings

In order to cope with the problems discussed in Sect. 2 we propose the notion of *collaboration interfaces* (*CI* for short), which differ from standard interfaces in two ways. First, CIs introduce the `provided` and `required` modifiers to annotate operations belonging to the provided and the expected contracts, respectively, hence supporting bidirectional interaction between clients and servers. Second, CIs exploit interface nesting in order to express the interplay between multiple abstractions participating in the definition of a generic component.

For illustration, the CI `TreeDisplay` that bundles the definition of the generic tree displaying functionality from Sect. 2 is shown in Fig. 13. As an example for the "reification" of provided and expected contract, consider the methods `TreeDisplay.display()` and `TreeDisplay.getRoot()` in Fig. 13. Any tree display object is able to display itself on the request of a client – hence the `provided` modifier for `TreeDisplay.display`. However, in order to do so, it expects a client specific way of how to access the root tree node. What the root

```
interface TreeDisplay {
  provided void display();
  expected TreeNode getRoot();

  interface TreeNode {
    expected TreeNode[] getChildren();
    expected String getStringValue();
    provided display();
    provided boolean isSelected(),
    provided boolean isExpanded();
    provided boolean isLeaf();
    provided int row();
    provided boolean hasFocus();
  }
}
```

**Fig. 13.** Collaboration interface for TreeDisplay

```
class SimpleTreeDisplay implements TreeDisplay {
  void onSelectionChange(TreeNode n, boolean selected) {
    n.setSelected(true);
  }
  void display() { getRoot().display(); }

  class TreeNode {
    boolean selected;
    ...
    boolean isSelected() { return selected; }
    // other provided methods similar to selected
    void setSelected(boolean s) { selected =s;}
    void display() {
      ... TreeNode c = getChildren()[i];
      ... paint(position, c.getStringValue());
      ...
    }
  }
}
```

**Fig. 14.** A sample implementation of `TreeDisplay`

of a displayable tree will be depends on (a) which modules in a concrete deployment context of `TreeDisplay` will be seen as tree nodes and, (b) which one of them will play the role of the root node. Hence, the declaration of `getRoot` with the expected modifier. `TreeDisplay` comes with its own definition of a tree node: The CI `TreeNode` is nested into the declaration of `TreeDisplay`. Please note that nesting of bidirectional interfaces in our approach has a much deeper semantics than usual nested classes and interfaces in Java: the nested interfaces are namely *virtual types* as in [1]. We will elaborate on that in Sect. 3.4.

The categorization of the operations into expected and provided comes with a new model of what it means to implement an interface. We explicitly distinguish between *implementing* an interface's provided contract and *binding* the same interface's expected contract. Two different keywords are used for this purpose: `implements`, respectively `binds`. In the following, we refer to classes that are declared with the `implements` keyword as *implementation classes*. Similarly, we refer to classes that are declared with the `binds` keyword as *binding classes*

An implementation class of a CI must (a) implement all `provided` methods of the CI and (b) provide an implementation class for each of the CI's nested interfaces. In doing so, it is free to use respective `expected` methods. In addition, an implementation class may or may not add additional methods and state to the CI's abstractions it implements. Fig. 14 shows a sample tree GUI control that implements `TreeDisplay`. The class `SimpleTreeDisplay` implements the only provided operation of `TreeDisplay`, `display()`, by forwarding to the result of calling the expected operation `getRoot()`. In addition to implementing `display()`, `SimpleTreeDisplay` must also provide a nested class that implements `TreeNode` - the only nested interface of `TreeDisplay`. The correspondence between a nested implementation class and its corresponding nested interface is based on name identity – `SimpleTreeDisplay` e.g., defines a class named `Tree-Node` which is the implementation of the nested interface with the same name in `TreeDisplay`. This nested class has to implement all `provided` methods of the `TreeNode` interface, e.g., `display()`. The declaration of the instance variable `boolean selected` and the corresponding query operation `isSelected` in `SimpleDisplay.TreeNode` are examples of new declarations added by an implementation class. Please note that just as nested interfaces, all nested implementation classes are virtual types (see Sect. 3.4).

A binding class of a CI must (a) implement all `expected` methods of the CI, and (b) provide zero or more binding classes for each of the CI's nested interfaces (we may have multiple bindings of the same interface, see subsequent discussion). Just as implementation classes can use their respective expected facets, the implementation of the expected methods of a CI and its nested interfaces can also call methods declared in the respective provided facets. The process of binding a CI instantiates its nested types for a concrete usage scenario of the generic functionality defined by the CI. Hence, it is natural that in addition to their provided facets, binding classes also use the interface of abstractions from that concrete usage scenario. We say that bindings wrap abstractions from the world of the concrete usage scenario and map them to abstractions from the generic component world.

For illustration, the class `ExpressionDisplay` in Fig. 15 shows an example of binding the generic `TreeDisplay` CI from Fig. 13 for the concrete usage scenario, in which `Expression` structures are to be viewed as the trees to display. First, `ExpressionDisplay` binds the nested type `TreeNode` as shown in the nested class `ExprTreeNode`. The latter implements all expected methods of `TreeNode` by using (a) the provided facet of `TreeNode`, and (b) the interface of the class `Expression` (via the instance variable `e`). Consider e.g., the implementation of the method `ExprTreeNode.getStringValue()`, which calls both `TreeNode.hasFocus()` as well as `Expression.getDescription()`.

In addition to binding `TreeNode`, `ExpressionDisplay` also implements the method `getRoot()` - the only method declared in the *expected* facet of `TreeDisplay`. Here is where the reference `root` to the `Expression` object to be seen as the root of the expression structure to display is transformed into a `TreeNode` by being wrapped into an `ExprTreeNode` object. Please note that this wrapping does not happen via an ordinary constructor call - `new ExprTreeNode(root)` in this case -, but rather by means of the *wrapper recycling* call `ExprTreeNode(root)`. We will elaborate on the concept of *wrapper recycling* in a moment.

Except of binding the interface `TreeNode`, `ExprTreeNode` is basically a usual class that, in this case, wraps an instance of `Expression`. Since wrapping of objects in these classes is a very common task, we add some syntactic sugar for the most common case, namely by a `wraps` clause.

The semantics of `wraps` is that

```
class ExprTreeNode binds TreeNode wraps Expression {...}
```

is equivalent to

```
class ExpressionDisplay binds TreeDisplay {
  Expression root;
  public ExpressionDisplay(Expression rootExpr) {
    root = rootExpr;
  }
  TreeNode getRoot() { return ExprTreeNode(root); }
  class ExprTreeNode binds TreeNode {
    Expression e;
    ExprTreeNode(Expression e) { this.e=e;}
    TreeNode[] getChildren() {
      return ExprTreeNode[](e.getSubExpressions());
    }
    String getStringValue() {
      String s = e.description();
      if (hasFocus()) s ="<"+s+">";
      return s;
    }
  }
}
```

**Fig. 15.** Binding of `TreeDisplay` for expressions

```
class ExpressionDisplay binds TreeDisplay {
  ...
  class ExprTreeNode binds TreeNode wraps Expression{
    TreeNode[] getChildren() {
      return ExprTreeNode[](wrappee.getSubExpressions());
    }
    String getStringValue() {
      String s = wrappee.description();
      if (hasFocus()) s ="<"+s+">";
      return s;
    }
  }
}
```

**Fig. 16.** Alternative encoding of `ExprTreeNode` using the `wraps` clause

```
class ExprTreeNode binds TreeNode {
  Expression wrappee;
  ExprTreeNode(Expression e) { wrappee = e;}
  ... }
```

Using **wraps**, the code in Fig. 15 can be rewritten as in Fig. 16. In the following code we will make frequent use of **wraps** but it is important to understand that it is just syntactic sugar and does not prevent us to create arbitrarily complex initialization procedures by using ordinary constructors.

Binding classes and their nested classes are "almost standard" classes – we do not use more declarative mapping constructs because the full computational power of a general-purpose programming language is needed to express arbitrarily complex mappings, and this is very hard to achieve with declarative means. "Almost standard", however, stands for two differences. First, nested binding classes are also virtual types (see Sect. 3.4). Second, they make use of the notion of *wrapper recycling*, which we discuss next.

## 3.2   Wrapper Recycling

*Wrapper recycling* is our mechanism to escape the wrapper identity hell mentioned in Sect. 2. It is a concept on how to create and maintain wrapper instances, and a way to navigate between abstractions of the component world and abstractions of the base world – the concrete usage scenario world –, ensuring that the same (identical) wrapper instance will always be retrieved for a set of constructor arguments. This way the state and the identity of the wrappers is preserved.

Syntactically, wrapper recycling refers to the fact that, instead of creating an instance of a wrapper W with a standard new W(args) constructor call, a wrapper is retrieved with the construct outerClassInstance.W(args). For illustration consider once again the expression return ExprTreeNode(root) in the method ExpressionDisplay.getRoot() in Fig. 15. We already mentioned in the previous section that the expression in the return statement is not a standard constructor call, but rather a wrapper recycling operator. We use the usual Java scoping rules, i.e., return ExprTreeNode(root) is just an abbreviation for return this.ExprTreeNode(root).

The idea is that we want to avoid creating a new ExprTreeNode wrapper each time the method getRoot() is called on an ExpressionDisplay. The call to the wrapper recycling operation ExprTreeNode(root) is equivalent to the corresponding constructor call, only if a wrapper for root does not already exist, ensuring that there is a unique ExprTreeNode wrapper for each expression within the context of the enclosing ExpressionDisplay instance. That is, two subsequent wrapper retrievals for an expression e yield the same wrapper instance - the identity and state of the wrapper are preserved.

This is due to the semantics of a wrapper recycling call, which is as follows: The outer class instance maintains a map mapW for each nested wrapper class W. An expression outerClassInstance.W(wrapperargs) corresponds to the following sequence of actions:

1. Create a compound key for the constructor arguments, lookup this key in mapW.
2. If the lookup for the key fails, create an instance of outerClassInstance.W with the annotated constructor arguments, store it in the hash table mapW, and return the new instance. Otherwise return the object already stored in mapW for the key.

The wrapper recycling call ExprTreeNode[](...) in the method ExprTreeNode.getChildren in Fig. 15 is an example for the syntactic sugar we use to express wrapper recycling of arrays, namely an automatic retrieval of an array of wrappers for an array of base objects.

## 3.3   Composing Bindings and Implementations

Both classes defined in Fig. 14 and 15 are not operational, i.e., cannot be instantiated, even if they are not annotated as abstract. These classes are indeed not abstract, since they are complete implementations of their respective contracts. The point is that the respective contracts are parts of a whole and make sense only within a whole. Operational classes that completely implement an interface are created by composing an implementation and a binding class, syntactically denoted as *aCollabIfc<aBinding,anImpl>*. This is illustrated by the class `SimpleExpressionDisplay` in Fig. 17, which declares itself as an extension of the composed class `TreeDisplay <SimpleTreeDisplay,ExpressionDisplay>`. Only such compound classes are allowed to be instantiated by the compiler. For instance, Fig. 17 also shows sample code that instantiates and uses the compound class `SimpleExpressionDisplay`.

Combining two classes as in Fig. 17 means that we create a new compound class within which the respective implementations of the **expected** and **provided** methods are combined. The same combination also takes place recursively for the nested classes: All nested classes with a **binds** declaration are combined with the corresponding implementation from the component class. The separation of the two contracts, their independent implementation, and the dedicated late composition, allows us to freely reuse implementations of the two contracts in arbitrary compositions. We could combine `SimpleTreeDisplay` with any other binding of `TreeDisplay`. Similarly, `ExpressionDisplay` could be combined with any other implementation of `TreeDisplay`.

Note that the overall definition of the nested type, e.g., `TreeNode`, depends on the concrete composition of implementation and binding types within which the type is used. This does not only affect the external clients, but also the internal references. For instance, references to `TreeNode` within `ExpressionDisplay` and `SimpleDisplay` are rebound to the composed definition of `TreeNode` in `SimpleExpressionDisplay`, as illustrated in Fig. 18. Their meaning would be different in another compound class, e.g., resulting from composing `SimpleDisplay` with another binding class, or `ExpressionDisplay` with another implementation class. This is a natural consequence of the fact that nested types introduced by the collaboration interfaces are virtual types, on which we will elaborate in the following.

```
class SimpleExpressionDisplay extends
   TreeDisplay<SimpleTreeDisplay,ExpressionDisplay> {}
...

Expression test = new Plus(new Times(5, 3), 9);
TreeDisplay t = new SimpleExpressionDisplay(test);
t.display();
```

**Fig. 17.** Creating and using compound classes

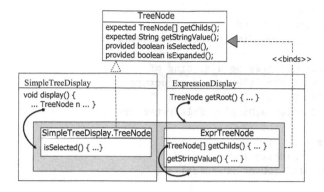

**Fig. 18.** Type rebinding in compound classes

## 3.4  Virtual Types

In CAESAR approach, all types that are declared as nested interfaces of a CI and all classes that implement or bind such interfaces (including classes that extend the latter) are *virtual types* and *virtual classes*, respectively [6]. In the context of this paper, we use the notion of virtual types of the family polymorphism approach [1]. This means: (a) similar to fields and methods, types also become properties of objects of the class in which they are defined, and consequently (b) their denotation can only be determined in the context of an instance of the enclosing class. Hence, the meaning of a virtual type is late bound depending on the receiver object that executes when the virtual type at hand is referenced.

Consequently, all type declarations, constructor calls, and wrapper recycling calls for virtual types/classes within a CI are actually always annotated with an instance of the enclosing class. That is, type declarations and constructor invocations are always of the form `enclInst.MyVirtual x`, respectively `enclInst.MyConstructor()`. Similarly, wrapper recycling calls are also always of the form `outerClassInstance.W(args)` and not simply `W(args)`. For the sake of simplification, we apply the scoping rules common for Java nested classes also to type declarations and constructor or wrapper recycling calls: A call `OuterClass.this.W(args)` can be shortened to `W(args)`, and the type declaration `OuterClass.this.W` can be shorted to `W` as long as there are no ambiguities. This scoping rule applies to all type declarations and wrapper recycling calls that have appeared so far in this chapter.

For instance, all references to `ExprTreeNode` in Fig. 15 should be read as `ExpressionDisplay.this.ExprTreeNode`. The implication is that the meaning of any reference to the type name `ExprTreeNode` within the code of `ExpressionDisplay` will be bound to the compound class that combines `ExpressionDisplay.ExprTreeNode` with the implementation class of `TreeNode` that is appropriate in the respective execution context. For example, in the context of a `SimpleExpressionDisplay` as in Fig. 17, `Expr-`

```
Expression e = ...;
final ExpressionDisplay ed =
                  new SimpleExpressionDisplay(e);
...
// let FileSystemDisplay be a binding of
// TreeDisplay to the file system structure
class SimpleFileSystemDisplay extends
 TreeDisplay<SimpleTreeDisplay,FileSystemDisplay> {};

FileSystem fs = ... ;
final FileSystemDisplay fsd =
      new SimpleFileSystemDisplay(fs);
...
ed.TreeNode t = ed.getRoot();
fsd.setRoot(t); // Type error detected by typechecker!
      // sd.TreeNode is not subtype of ed.TreeNode
```

**Fig. 19.** Type safety due to family polymorphism

TreeNode will be bound to the respective definition in the compound class
TreeDisplay<ExpressionDisplay,TreeNode>. The same references will be
bound differently if they occur in the execution context of an object of some
subclass of ExpressionDisplay or in the context of a different implementation
class. The same also applies to nested implementation and compound classes.

The rationale behind using virtual types lies in their power with respect to
supporting reuse and polymorphism, as argued in [1]. We will rather shortly
discuss how our specific use of virtual types (borrowed from [1]) does *not* suffer
from covariance problems usually associated with virtual types, as for example
the virtual type proposal in [18], which requires runtime type checks. If we
have a virtual type in a contravariant position, as for example the argument
type of setRoot in Fig. 19, type safety is still preserved, because subsumption
is disallowed if the enclosing instances are not identical. In order to make the
approach sound, all variables that are used as part of type declarations have to
be declared as final because otherwise the meaning of a type declaration might
change due to a field update. For illustration consider the declaration of the
variable ed in the sample code in Fig. 19. It is used as part of a type declaration
for the variable t and is therefore declared as final. For more details on typing
issues we refer to [1].

## 4   Evaluation

Before discussing the implications of our new model, let us at first compare the
new solution to the conventional solution discussed in Sect. 2.

- Other than the Swing interface in Fig. 8, we do not need to use Object;
  every item is well-typed and we do not need type casts. The methods that

are conceptually part of the interface of tree nodes, are expressed as methods of a dedicated nested interface.

- Due to bidirectional interfaces, we do not have the problem related to the getStringValue() parameters: The implementation of this method, as in Fig. 15, causes the computation of only those values about the state of displaying that are really needed by means of calling appropriate methods in the provided interface.
- It is easy to associate additional state with tree nodes. For example, the TreeNode implementation in Fig. 14 adds a selected field, and the TreeNode binding in Fig. 15 could as well have added extra state to ExprTreeNode.

Abstracting from the concrete example, what have we gained with the new language means proposed in this paper? The important point is the idea of encoding modules in terms of their own world-view, encoded in a collaboration interface, together with means to translate this world view into the world view of a particular application by means of CI bindings. Due to the independence of a binding from a particular CI implementation, CI bindings are universal, reusable "world-view translators".

## 5   Related Work

*Pluggable Composite Adapters (PCAs)* [13] and their predecessor, *Adaptive Plug and Play Components (APPCs)* [10], have been important starting points for our work. Both approaches offer different means for on-demand remodularization. The APPC model had a vague definition of required and provided interfaces. However, this feature was rather ad-hoc and not well integrated with the type system. Recognizing that the specification of the required and expected interfaces of components was rather ad-hoc in APPCs, PCAs even dropped this notion and reduced the declaration of the expected interface to a set of standard abstract methods. With the notion of collaboration interfaces, the approach presented here represents a qualitative improvement over PCA and APPC.

The Hyperspace model and its instantiation in Hyper/J [17] also target multiple co-existing hierarchies. However, despite the common goal, there are some important differences between these two approaches. In a nutshell, the functionality offered by Hyper/J can be summarized as *extracting concerns* and *composing concerns*.

*Extracting concerns* means that one can take a piece of existing software and tag parts of the software, e.g., method a() in class A and method b() in class B, by means of a so-called *concern mapping*. Later, this mapping can be used to extract a particular concern from this software and reuse it in a different context. This is similar to the old idea of retroactive generalization in inheritance hierarchies [15]. An important concept for extracting concerns is the notion of *declarative completeness*. Basically, this means that all methods that are used inside the tagged methods but are not tagged themselves are declared as *abstract* in the context of the extracted concern. Our model does not have any dedicated means for feature extraction.

However, we think that with respect to *composing concerns* our approach is in some important ways superior to Hyper/J. Composition in Hyper/J happens by means of a so-called *hypermodule specification*, which describes in a declarative sublanguage, how different concerns should be composed. In terms of our model, a hypermodule performs both the functionality of our binding classes and the actual composition with the + operator. Due to this mixing and due to the absense of an interface concept similar to our collaboration interface, Hyper/J has no polymorphism and reuse as in our approach, e.g., one cannot switch between different implementations and bindings, and one cannot use them polymorphically. Since the mapping sublanguage is declarative, it relies on similar signatures that can be mapped to each other, and transformations other than name transformations (e.g., type transformations), are very difficult. In addition, Hyper/J's sublanguage for mapping specifications from different hyperslices is fairly complex and not well integrated into the common OO framework.

The last important difference is that Hyper/J's approach is class-based: it is not possible to add the functionality defined in a hyperslice to individual objects, instead the objects have to be created as objects of the compound hypermodule from the very beginning. Therefore, multiple independent bindings that are added to individual objects at runtime are not possible.

Hölzle [3] analyses some problems that occur when combining independent components. Our proposal can be seen as an answer to the problems and challenges discussed in [3].

Our work is also related to *architecture description languages* (ADL) [16], for example Rapide [5], Darwin [7], C2 [9], and Jiazzi [8]. The building blocks of an architectural description are components, connectors, and architectural configurations. A component is a unit of computation or data store, a connector is an architectural building block used to model interactions among components and rules that govern those interactions, and an architectural configuration is a connected graph of components and connectors that describe architectural structure. In comparison with our approach, ADLs are less integrated into the common OO framework, and do not have a dedicated notion of on-demand remodularization in order to provide a new virtual interface to a system.

We think that collaboration interfaces might also prove very useful in the context of ADL. In ADL, components also describe their functionality and dependencies in the form of required and provided methods (so-called *ports*). The goal of these ports is to render the components reusable and independent from other components. However, although the components are syntactically independent, there is a very subtle semantic coupling between the components, because a component A that is to be connected with a component B has to provide the exact counterpart interface of B. The situation becomes even worse if we consider multiple components that refer to the same protocol. The problem is that there is no central specification of the communication protocol to which all components that use this protocol can refer to – in other words: we have no notion of a collaboration interface.

# 6 Summary

Traditional programming languages assume that real-world systems have "intuitive", mind-independent, preexisting concept hierarchies. We argued that this is in contrast to our perception of the world, which depends heavily on the context from which it is viewed. We identified the *arbitrariness of the decomposition hierarchy* as the main cause of 'code tangling' and presented a new model called CAESAR, within which it is possible to have multiple different decompositions *simultaneously* and to add new decompositions on-demand.

# References

1. E. Ernst. Family polymorphism. In *Proceedings of ECOOP '01*, LNCS 2072, pages 303–326. Springer, 2001.
2. D. Garlan, G.E. Kaiser, and D. Notkin. Using tool abstraction to compose systems. *Computer*, 25(6):30–38, 1992.
3. U. Hölzle. Integrating independently-developed components in object-oriented languages. In *Proceedings ECOOP '93, LNCS*, 1993.
4. Java Foundation Classes. http://java.sun.com/products/jfc/.
5. D.C. Luckham, J.L. Kenney, L.M. Augustin, J. Vera, D. Bryan, and W. Mann. Specification and analysis of system architecture using Rapide. *IEEE Transactions on Software Engineering*, 21(4):336–355, 1995.
6. O.L. Madsen and B.Møller-Pedersen. Virtual classes: A powerful mechanism in object-oriented programming. In *Proceedings of OOPSLA '89*. ACM SIGPLAN, 1989.
7. J. Magee and J. Kramer. Dynamic structure in software architecture. In *Proceedings of the ACM SIGSOFT'96 Symposium on Foundations of Software Engineering*, 1996.
8. S. McDirmid, M. Flatt, and W. Hsieh. Jiazzi: New age components for old fashioned Java. In *Proceedings of OOPSLA '01*, 2001.
9. N. Medvidovic, P. Oreizy, and R.N. Taylor. Reuse of off-the-shelf components in C2-style architectures. In *Proceedings of the 1997 international conference on Software engineering*, pages 692–700, 1997.
10. M. Mezini and K. Lieberherr. Adaptive plug-and-play components for evolutionary software development. In *Proceedings OOPSLA '98, ACM SIGPLAN Notices*, 1998.
11. M. Mezini and K. Ostermann. Integrating independent components with on-demand remodularization. In *Proceedings of OOPSLA '02, Seattle, USA*, 2002.
12. M. Mezini and K. Ostermann. Conquering aspects with Caesar. In *Proc. International Conference on Aspect-Oriented Software Development (AOSD '03), Boston, USA*, 2003.
13. M. Mezini, L. Seiter, and K. Lieberherr. Component integration with pluggable composite adapters. In M. Aksit, editor, *Software Architectures and Component Technology: The State of the Art in Research and Practice*. Kluwer, 2001. University of Twente, The Netherlands.
14. D.L. Parnas. On the criteria to be used in decomposing systems into modules. *Communications of the ACM*, 15(12):1053–1058, 1972.
15. C.H. Pedersen. Extending ordinary inheritance schemes to include generalization. In *OOPSLA '89 Proceedings*, 1989.

16. M. Shaw and D. Garlan. *Software Architecture: Perspectives on an Emerging Discipline*. PrenticeHall, 1996.
17. P. Tarr, H. Ossher, W. Harrison, and S.M. Sutton. N degrees of separation: Multi-dimensional separation of concerns. In *Proc. International Conference on Software Engineering (ICSE 99)*, 1999.
18. K.K. Thorup. Genericity in Java with virtual types. In *Proceedings ECOOP '97*, 1997.

# Software Fault Tolerance:
# An Overview

Jörg Kienzle

School of Computer Science, McGill University
Montreal, QC H3A 2A7, Canada
Joerg.Kienzle@mcgill.ca

**Abstract.** This paper presents an overview of the techniques that can be used by developers to produce software that can tolerate design faults and faults of the surrounding environment. After reviewing the basic terms and concepts of fault tolerance, the most well-known fault-tolerance techniques exploiting software-, information- and time redundancy are presented, classified according to the kind of concurrency they support.

**Keywords:** Software fault tolerance, failures, concurrency, exceptions.

## 1 Introduction

The scope, complexity, and pervasiveness of computer-based and controlled systems continue to increase dramatically, and hence the consequences of such systems failing can be considerable. Ideally, the processes by which the software controlling such systems is created, analyzed, designed, implemented and tested would have come to the point where software could be developed without errors [1].

Indeed, advances in *fault avoidance* (or fault prevention) techniques, such as rigorous specification of system requirements, structured design and good programming discipline, formal methods, and software reuse, have considerably improved the potential for high quality software development. Also, *fault removal* techniques, such as verification and validation methods, rigorous testing, formal inspection and formal proofs, can be used effectively to reduce the amount of faults that remain in a final software product when it is shipped. However, even if the best people, practices, and tools are used, it would be very risky to assume that the developed software is error-free.

*Fault forecasting* techniques concentrate on estimating the presence of faults in software based on software metrics and failure data obtained during system testing or system operation. Fault forecasting techniques can, for instance, justify the need for additional testing, especially in safety- and mission-critical applications, which must often guarantee a very low probability of failure.

Ultimately, preventing system failure in spite of faults remaining in the software can be achieved by applying *software fault tolerance* techniques. Some of these techniques have been available for close to 30 years now and are well understood. A survey of the most important techniques is presented in this paper. In general, software fault tolerance techniques are categorized into design diverse and data diverse techniques.

J.-P. Rosen and A. Strohmeier (Eds.): Ada-Europe 2003, LNCS 2655, pp. 45–67, 2003.
© Springer-Verlag Berlin Heidelberg 2003

This overview, however, presents software fault tolerance models based on the different forms of concurrency they support. The paper is structured as follows:

Section 2 presents the most important concepts and definitions in the field of fault tolerance. Section 3 looks at the base techniques that can be used in a sequential environment (recovery blocks, retry blocks). Section 4 presents how these techniques can be extended to loosely coupled concurrent systems (N-version programming, N-copy programming). Section 5 investigates techniques for addressing fault tolerance in competitive concurrent systems (transactions). Section 6 concentrates on techniques suited for handling fault tolerance in cooperative systems (atomic actions). Finally, Section 7 presents techniques that combine features of the previous ones, and therefore support more than one form of concurrency.

## 2   Fundamental Concepts

To discuss fault tolerance meaningfully, a definition of correct behavior of a program is needed — otherwise, how could one know that something went wrong? For the purposes of fault–tolerant computing, the *specification* of the program is considered to be the definition of correct program behavior: as long as the program meets its specification, it is considered correct.

A *failure* is the observation of an erroneous system state: an observable deviation from the specification is considered a failure. An *error* is that part of the system state that leads to a failure of the system. An error itself is caused by some defect in the system; those defects that cause observable errors are called *faults*. There may be defects in the system that remain undetected; only those that manifest themselves as errors are considered faults. Likewise, an error does not necessarily lead to a failure: it may be a *latent* error [2]. Only when the error in the system state causes the system to behave in a way that is contradictory to its specification, a failure occurs. This relationship is illustrated in Fig. 1.

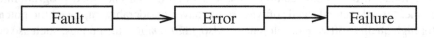

**Fig. 1.** Fault tolerance terminology

If we look at the occurrence of faults, errors and failures over time we can define the notions of *latency* and *inertia*, illustrated in Fig. 2.

*Latency* is the meantime between the fault occurrence and its initial activation as an error. In software systems, faults are generally introduced during the development process, but they are sometimes activated only by, for example, certain input values or situations. In this case, latency may vary according to the frequency of use of the module that contains the fault.

*Inertia* is the meantime of the duration between the occurrence of a failure, i.e. an observable deviation from the specification, and the beginning of external (irrecoverable) consequences to the environment.

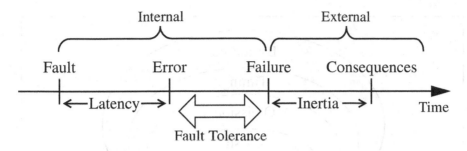

**Fig. 2.** Latency and inertia

The goal of fault tolerance is to avoid system failure in the presence of faults. Therefore, as soon as an error has been detected, it must be corrected to avoid a later potential failure: corrective actions have to be taken to restore a correct system state.

### 2.1 Fault Classification

Faults can be characterized in various ways. One can consider the temporal characteristics of a fault. A *transient* fault has a limited duration, e.g. a temporary malfunction of the system, or a fault due to external interference. If a transient fault occurs repeatedly, it is called an *intermittent* fault. In contrast, *permanent* faults persist, i.e. the faulty component of the system will not work correctly again unless it is replaced.

Another way to classify faults is to consider the software life-cycle phase in which they occur. Here, one can distinguish *design faults* (in particular software design faults) from *operational faults* occurring during the use of the system.

### 2.2 Failure Semantics

Failures, i.e. deviations from a program's specification, can manifest themselves in various ways [3]:

- A *crash* failure occurs when the system stops responding completely. One generally distinguishes *fail–silent* and *fail–stop* behavior: with the latter, the clients of the system have a means to detect that it has failed.
- *Omission* failures occur when the system does not respond to a request when it is expected to do so.
- *Timing* failures can occur in real–time systems if the system fails to respond within the specified time slice. Both early and late responses are considered timing failures; late timing failures are sometimes called *performance* failures.
- A system is said to exhibit *byzantine* failure semantics, if upon failure it behaves arbitrarily [4].

These failure semantics can be organized in a hierarchy: byzantine failures are the most general model, and subsume all others as shown in Fig. 3.

The algorithms used for achieving any kind of fault tolerance depend on the computational model, i.e. on what failure semantics we assume for the components in our system.

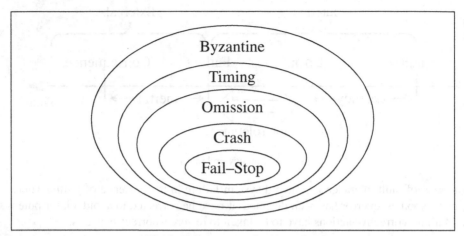

**Fig. 3.** Failure semantics hierarchy

## 2.3  Redundancy

The key supporting concept for fault tolerance is redundancy. In software, redundancy can take several forms: *functional redundancy, data redundancy* and *temporal redundancy.*

*Functional redundancy* aims at tolerating design faults. To the contrary of hardware fault tolerance, software design and implementation faults can not be detected simply by replication of identical software units, since the same fault will exist and manifest itself in all copies (provided that they all run with the same input). The idea is to introduce diversity into the software replicas, creating different *versions,* variants or alternates. These versions are functionally equivalent, i.e. based on the same specification, but internally use different designs, algorithms and implementation techniques.

Information or *data redundancy* includes the use of additional information that allows to check for integrity of important data, for example error-detecting or error-correcting codes. Diverse data, i.e. identical data represented in different formats, also fall into this category.

Finally, *temporal redundancy* involves the use of additional time to achieve fault tolerance. Temporal redundancy is an effective way of tolerating transient faults. Provided that the temporary circumstances causing the fault are absent at a later time, simple re-execution of the failed operation will result in success. In general, most software fault tolerance techniques add execution overhead to an application, and therefore use additional time compared to a non-fault-tolerant application.

## 2.4  Error Processing

Once an error has been detected in the system state, it should be corrected to avoid a potential system failure later on. Of course, the fault(s) causing the error also should be treated, which means that the reason for the error must be identified and then the defect be corrected in order to avoid that the fault causes more errors. As stated in the introduction, fault diagnosis and removal are very important steps when producing dependable software. In fault tolerance, however, we want to continue to provide system functionality according to the specification in spite of the presence of faults.

Once an error is detected, the damage to the system state must be assessed. Measures must be taken that keep the error from propagating to other parts of the system, preventing further damage. Once the error is "under control", error recovery is applied, i.e. the erroneous system state is substituted with an error-free one.

There are two base cases:

- *Forward error recovery* attempts to construct a coherent, error–free system state by applying corrective actions to the current, erroneous state.
- *Backward error recovery* replaces the erroneous system state with some previous, correct state.

Forward error recovery requires that a more or less accurate damage assessment be made. The error must be identified in order to apply corrective actions in a way that makes sense. This diagnosis for forward error recovery depends on the particular system. Exceptions are provided in programming languages to signal and at the same time identify the nature of an error. Forward error recovery can be achieved through exception handling.

Backward error recovery requires that a previous correct state exists: such systems periodically store a copy of a coherent state (sometimes called *recovery point, check point, savepoint* or *recovery line*, depending on the recovery technique), to which they can *roll back* in case of an error. Backward error recovery is a general method: because it re–installs a previous, hopefully correct system state, it does not depend on the nature of the error nor on the application's semantics. Its main drawback is that it incurs an overhead even in failure–free executions, because recovery points have to be established from time to time.

## 2.5 System Structuring for Fault Tolerance

Software systems and systems in general are not monolithic; they usually consist of several components or subsystems, and fault tolerance approaches must account for that. Different approaches may be applied to different components. The composite nature of systems also means that the classification of fault, error, and failure is not absolute: a given component may perceive the failure of a sub–component as a fault and have its own fault tolerance techniques in place to handle it.

This hierarchic model of a system gives rise to the notion of error confinement: the system is structured in regions beyond which the effects of a fault should not propagate undetected. This implies that a given component be accessible to other components only through a well–defined (and preferably narrow [5]) interface. Different error confinement regions may employ different means to achieve fault tolerance. The chosen technique depends upon the failure semantics the system component should adhere to according to its specification, as well as on the failure semantics of its sub–components.

[6, 7] advocate structuring system execution based on *idealized fault-tolerant components* (see Fig. 4). The component offers services that may return replies to the component that made a service request. If a request is malformed, the component signals

this by raising an *interface exception*, otherwise it executes the request and produces a reply. If an *internal exception* or *local exception* signaling an error occurs, error processing is activated in an attempt to handle the error. If it can be dealt with, normal processing in the component resumes; if not, the component itself signals its failure by a *failure exception*. It is immaterial whether exceptions are true exceptions in the sense of exceptions provided by programming languages or are indicated using exceptional replies to requests. It is even possible that some entity external to the system component observes its failure and initiates appropriate error processing in the users of the component.

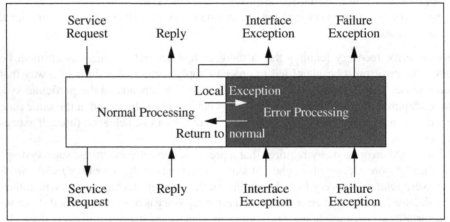

**Fig. 4.** Idealized fault-tolerant component

## 2.6   Classification of Concurrent Systems

The subsequent parts of the paper present and review the most commonly used software fault tolerance techniques, classified based on the different forms of concurrency they support: sequential techniques, independent or loosely-coupled concurrent techniques, competitive- and collaborative concurrent techniques, and hybrid techniques.

## 3   Sequential Techniques

### 3.1   Robust Software

Robust software is often not considered part of software fault tolerance, since it does not use any for of redundancy. However, it represents the base for achieving any form of dependability.

Robustness is defined as "the extent to which software can continue to operate correctly despite the introduction of invalid inputs" [8]. Invalid inputs must be defined in the specification. They include out of range inputs, inputs of the wrong type or format, corrupted inputs, wrong sequencing of input, and violations of pre-conditions.

Upon detection of such invalid input, several optional courses of action may be taken: requesting new input from the input source, using the last acceptable value, or using a pre-defined default value.

## 3.2   Recovery Blocks

*Recovery Blocks* [9, 10] have been introduced as a software structuring mechanism providing software fault tolerance based on the concept of design diversity and backward error recovery. A recovery block consists of one or more algorithms called alternatives that implement the same functionality, coupled with an acceptance test that determines whether an alternative has functioned correctly or not. Under normal conditions, only the first alternative is executed. Only if an error is detected by the acceptance test, other alternatives are tried successively. The typical syntax for expressing recovery blocks is as follows:

```
ensure Acceptance Test
by Primary Alternate
else by Second Alternate
...
else by N'th Alternate
else signal Failure
```

It is important to note that prior to execution of the primary alternate, a snapshot of the current state of the component is taken. When an alternative fails to pass the acceptance test, this state is restored, and the next alternative is tried. This is repeated until either an alternative succeeds, or there are no more alternatives available. In that case, the execution of the component is considered a failure, and the failure is signaled to the outside.

The overhead introduced by the recovery block scheme in fail-free mode is fairly low. It comprises establishing a checkpoint and running the acceptance test. However, every failure of an alternate requires restoring the checkpoint, executing another alternate and running the acceptance test again. Although unlikely, the potential overhead is huge. Also, "infinite loop" type of errors can not be detected easily.

In order to make it possible to use recovery blocks in real-time applications, the base scheme has been extended to include a watchdog timer that interrupts the recovery block execution after a specific time. The execution of the watchdog version of recovery blocks is illustrated in form of a UML state diagram in Fig. 5.

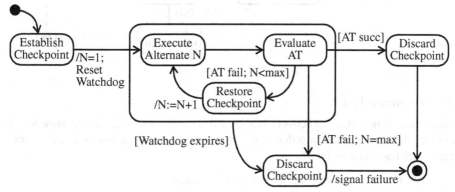

**Fig. 5.** Recovery block execution

### 3.3 Retry Blocks

*Retry Blocks* are the data diverse complement of recovery blocks [11]. They are based on the assumption that although an algorithm might fail on some specific input, it might succeed on related input, i.e. slightly different input that still produces acceptable results. Instead of relying on a secondary algorithm in case the primary one fails, retry blocks use a *data re-expression algorithm* (DRA) to slightly modify the input, and run the primary algorithm again. Apart from this, the retry blocks follow the same idea as recovery blocks, i.e. they rely on an acceptance test for determining validity of results, and on backward error recovery to restore a consistent application state in case of failure.

The syntax for expressing retry blocks is shown below:

```
ensure Acceptance Test
by Primary Algorithm (Original Input)
else by Primary Algorithm (Re-expressed Input)
...
...    [Deadline expires]
else by Backup Algorithm (Original Input)
else signal Failure
```

The execution is shown in the UML state diagram of Fig. 6.

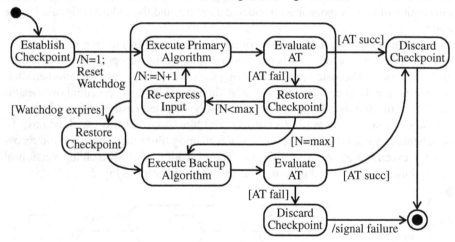

**Fig. 6.** Retry block execution

### 3.4 Acceptance Test

Central to both techniques, recovery blocks and retry blocks, is the *acceptance test*. It examines the system state in order to verify that the system's behavior is acceptable. Acceptance tests have to be:

- *Simple*, in order to keep run-time overhead reasonable,
- *Effective*, to ensure that anticipated faults are detected, and to ensure that non-faulty behavior is not incorrectly rejected,

- *Highly reliable*, to reduce the chance of introducing additional design faults.

Acceptance tests are difficult to develop. Comprehensive tests have a better chance of detecting errors, but create more run-time overhead and, in general, are more complex and hence more likely to introduce new faults into the system. Also, it is not always clear if an acceptance test should test for what a program should do or for what a program should not do.

Testing for satisfaction of the requirements stated in the specification often requires a computation as complex as the one performed by the algorithm to be tested. Moreover, if the acceptance test uses a similar algorithm than the one that is to be tested, chances of common-mode failures are increased. However, in certain situations, satisfaction of requirement test can be very efficient. For example, some mathematical operations, such as calculating the square root, can be verified by applying the inverse operation.

Often, testing for what a program should not do is simpler and provides a higher degree of independence between the acceptance test and the algorithm to be tested. Accounting tests, e.g. checksums, are well suited for data-manipulating applications with simple mathematical operations. Reasonableness tests or testing for violations of safety conditions are particularly suitable for process control systems. The tests can be based on physical constraints or rate change of values. Finally, run-time tests, i.e. testing for divide-by-zero, over- or underflow, end-of-file, and similar conditions, present the most cursory class of acceptance test.

## 4    Independent Concurrent Systems

If several processing nodes are available, executing multiple versions of an algorithm does not require additional time anymore, since the alternates can execute in parallel. Also, since each version has it's own copy of the state, there is no need to perform backward error recovery. The N-version programming and N-copy programming techniques exploit these facts.

### 4.1   N-Version Programming

The *N-Version Programming* scheme [12, 13] is a design diverse scheme similar to the recovery block scheme. The main difference is that the different versions execute concurrently (often in separate processes on separate processors). Once all versions have completed their execution, the results are compared, and the final result is determined using a decision mechanism.

The general syntax is given below:

```
run Version 1, Version 2, ... , Version n in parallel
if Decision Mechanism (Result 1, Result 2, .., Result n)
then
    return Determined Result
else signal Failure
```

The execution of the n-version programming scheme is illustrated in the UML state diagram shown in Fig. 7.

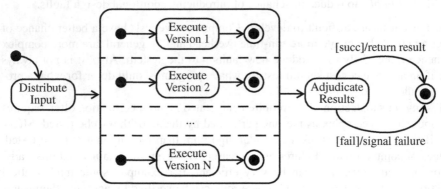

**Fig. 7.** N-version program execution

## 4.2 N-Copy Programming

*N-Copy Programming* [14] is the data diverse complement of n-version programming. In the n-copy scheme, the n processes all run the same algorithm, but each one using slightly different input.

The general syntax is given below:

```
run  Algorithm (DRA 1(original input)),
     Algorithm (DRA 2(original input)), ... ,
     Algorithm (DRA n(original input)) in parallel
if Decision Mechanism (Result 1, Result 2, .., Result n)
then
    return Determined Result
else signal Failure
```

The original input is distributed to the n processes, which each start by executing a different data re-expression algorithm to transform the input. Then, they all run the same algorithm concurrently. The results of each version are provided to the decision making mechanism, which determines if a correct result can be adjudicated. If yes, then the result is returned, if not, a failure is signaled to the outside.

## 4.3 Decision Mechanism

In both techniques, N-version programming and N-copy programming, multiple results are calculated concurrently. The task of the decision mechanism is to determine if one of the results can be considered correct. This is actually not that easy.

First of all, there might be *multiple correct results* (MCR). For instance, if the problem is calculating a root of an n-th order equation, then there are n different correct answers. Then there is the problem of *limited floating point arithmetic precision* (LFPA). Diverse algorithms for solving a given floating point problem might calculate results that vary slightly, but are within a given tolerance. This issue, combined with different execution paths taken by different variants results in a more fundamental problem called the *consistent comparison problem* (CCP). The difficulty is that, if n versions operate independently, then whenever the specification requires that they perform a comparison, it is not possible to guarantee that the versions will all make the same decision, i.e. make comparisons that are consistent [15]. It has been shown that, without communication among the variants, there is no solution for the consistent comparison problem.

Different decision algorithms have been developed, all having their strengths and weaknesses. A straightforward voter is the *exact majority voter* [16], which considers a result correct if a majority of versions agree. It is best used in situations where the results are discrete values. Unfortunately, it is very vulnerable to MCR and LFPA, and cannot be used in data diverse schemes that use approximate data re-expression algorithms. Other similar voters include the *mean voter* and the *consensus voter* [17].

Other voters have been developed, such as the *median voter*, which selects the median of the results provided by the versions. This can, or course, only be done with ordered values.

Finally, tolerance values can be added to voters in order to address LFPA. In this case, results that are within a pre-defined tolerance value $\varepsilon$ are considered identical.

## 5  Competitive Concurrent Systems

*Competitive concurrency* exists when two or more processes are designed separately, are not aware of each other, but share some resources. Programmers of such processes would like to live in an artificial world in which they do not have to care about other concurrent activities. They want to access objects as if they had them at their exclusive disposal.

Moreover, a fault of one process should not affect other processes running in the system. The ability to isolate the processes from each other is crucial for achieving reliability in such systems. Since the processes have been designed separately, they obviously can not anticipate any faults that might originate from others. Systems in which erroneous state from one process can propagate to other concurrent running processes might lead to disastrous results.

### 5.1  Transactions

*Transactions* [18] are the base technique for structuring the execution of competitive concurrent systems. A transaction groups an arbitrary number of operations on one or more data objects together, making the whole appear indivisible with respect to other concurrent transactions. Transactions guarantee the so-called ACID properties: *Atomicity, Consistency, Isolation* and *Durability* [18].

Transactions rely on backward error recovery to provide fault tolerance. The boundaries of a transaction are marked with two of the three standard operations: *begin, commit* and *abort*. After beginning a new transaction, all update operations on data objects are made on behalf of that transaction. At any time during the execution of the transaction it can *abort*, which means that the state of the system is restored to the state at the beginning of the transaction. An abort can be explicitly requested by the application, or automatically triggered due to some failure in the system (e.g. a remote object on some other machine can not be contacted due to a network failure). By *committing* a transaction, the application states that the updates to the system state are completed and the resulting state is error-free. From that point on, the state changes made on behalf of the transaction become permanent and are made visible to the outside.

*Flat transactions*, illustrated in Fig. 8, represent the simplest type of transaction. They are called *flat*, because there is only one layer of control available to the application programmer. Every statement inside the transaction is at the same level; that is, the transaction will either survive together with all modifications made to data objects on behalf of the transaction, or it will be rolled back, which means that all changes made to transactional objects will be undone.

The example in Fig. 8 represents a transaction that performs a money transfer from bank account A to bank account B. Both bank accounts are shared data objects that might be concurrently accessed by other processes running in the system. After starting the transaction, the amount of money to be transferred is first withdrawn from account A, then the amount is deposited on account B. If no problems are encountered, the transaction commits.

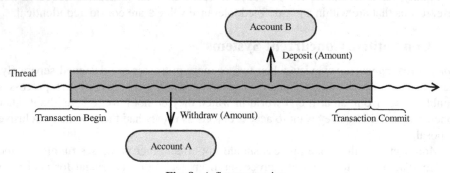

**Fig. 8.** A flat transaction

Banking systems extensively use transactions, and their importance is nicely illustrated by the transfer example. Without an enclosing transaction, a failure occurring after the withdraw operation has been completed on account A, but before the deposit operation has begun on account B, may result in the loss of the amount of money being transferred. Such a situation is not acceptable.

## 5.2   Transaction Model Extensions

Flat transactions have been extended in many ways, e.g. in order to provide more fine-grained rollback control. Extended models include flat transactions with savepoints, chained transactions, split and joint transactions [19], recoverable communicating actions [20], and Sagas [21].

One of the most important extension from the point of view of fault tolerance is the *nested transaction* model [22]. In the nested transaction model, a transaction is allowed to start subtransactions, thereby creating a hierarchy of transactions in form of a tree. The transaction at the root of the tree is called the top-level transaction. The transactions at the leaf level are flat transactions.

It is important to note that leaf-level subtransactions are not fully equivalent to classical flat transactions. The key point is that the properties of such transactions are valid only within the confines of the surrounding parent transaction. Leaf-level subtransactions are atomic from the perspective of the parent transaction, they *preserve consistency* with respect to the local function they implement; they are *isolated* from all other activities inside and outside the parent transaction. An immediate consequence of the commit rules is that subtransactions are *not durable*, since their changes are only made persistent when the top-level transaction commits.

## 5.3 Transactions and Software Fault Tolerance

The notion of transaction has first been introduced in database systems in order to correctly handle concurrent updates of data and to provide fault tolerance with respect to hardware failures [18]. However, transactions can also be applied in the context of software fault tolerance.

To begin with, transactions provide backward error recovery thanks to the atomicity property and the possibility of aborting voluntarily. Next, the isolation property hides state changes made on behalf of a transaction from other processes until the transaction commits. Therefore, potential errors are also confined within the transaction boundaries, and cannot propagate to the outside and spread to other processes. Finally, the durability property ensures that committed results remain available in the future in spite of any subsequent failures.

According to the rules, a transaction must be written to preserve consistency. Each transaction expects a consistent state when it starts, and recreates that consistency after making its modifications, provided it runs to completion. In a sense, the transaction is responsible for making sure that there are no errors in the state of the application that it has modified before committing the transaction. This can be achieved by using an acceptance test (see Section 3.4) at the end of every transaction. If an error is detected, the application can either choose to perform manual forward error recovery, or it can recover the initial consistent state by aborting the transaction.

Transactions combined with structured exception handling can be even more powerful. Exceptions signal abnormal situations during an application execution, i.e. potential erroneous state that must be addressed in order to guarantee consistency. The idea is to make transactions exception handling contexts. As long as an exception can be dealt with inside a transaction there is no danger. However, if an exception crosses a transaction boundary unhandled, then potential erroneous state might exist. To be on the safe side, such unhandled exceptions should be treated as an abort vote, i.e. a command to the transaction support to rollback all changes and re-establish the consistent state as it was at the beginning of the transaction.

Integrating transactions and exceptions also makes it possible to design and implement so-called *self-checking transactional objects* [23]. For such objects, methods are decorated with pre- and post-conditions. When an invariant, a pre- or a post-condition is violated by the execution of a method, an exception is propagated to the caller. The caller must then handle this exception in order to address a potential inconsistency. If handling fails, i.e. the exception propagates outside of the transaction context, the transaction is aborted, and all the changes made to transactional objects on behalf of the transaction are undone.

To make transactions a general technique for achieving software fault tolerance, the nested transaction model should be used. Nested transactions combined with exception handling make it possible to recursively structure the execution of an application into error confinement regions, following the idealized fault tolerant component approach described in Section 2.5.

# 6  Cooperative Concurrent Systems

*Cooperative concurrency* exists when several processes cooperate, i.e. do some job together and are aware of this. They can communicate by resource sharing or explicitly. They have been designed together. They cooperate to achieve their joint goal and use each other's help and results.

## 6.1  Conversations

The concept of a *Conversation* has been introduced by [24] to structure the execution of a set of collaborating processes. A fixed number of processes enter a conversation asynchronously; a recovery point is established in each of them. They freely exchange information within the conversation but cannot communicate with any outside process (violations of this rule are called *information smuggling*). When all processes participating in the conversation have come to the end of the conversation, their acceptance tests are to be checked. If all tests have been satisfied, the processes leave the conversation together. Otherwise, they restore their states from the recovery points and may try and execute a different *alternate*. By providing different alternates based on different algorithms that produce the same result, conversations can tolerate software design faults just as recovery blocks.

The occurrence of an error in a process inside a conversation requires the rollback of all (and only) the processes in the conversation to the checkpoint established upon entering the conversation. Conversations may be nested freely, meaning that any subset of the processes involved in a conversation at nesting level $i$ may enter a conversation at nesting level $i + 1$ [25].

## 6.2  Atomic Actions

Later on, the conversation scheme has been augmented with additional support for forward error recovery and exception resolution. The resulting model is named *Atomic Actions* [26, 6].

The structure of an atomic action is represented in Fig. 9. A fixed number of participants (threads, processes or active objects) enter an action and cooperate inside it to achieve joint goals. They share work and explicitly exchange information in order to complete the action successfully.

Atomic actions structure dynamic system behavior. To guarantee action atomicity, no information is allowed to cross the action border. Actions can be nested, meaning that a subset of the participants of the containing action can enter a nested action. The number of participants of an atomic action is fixed in advance, and hence no dynamic

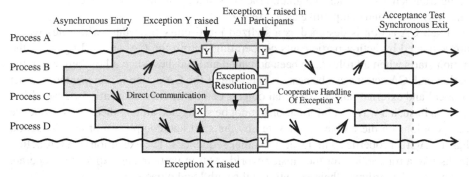

**Fig. 9.** An atomic action with coordinated exception handling

creation of threads is allowed. Participants leave the action together when all of them have completed their job.

Atomic actions have very detailed exception handling rules. A set of internal and external exceptions is associated with each atomic action, and these exceptions are clearly separated. The model is recursive, and all external exceptions of an action are viewed as internal ones of the containing action. Each participant of the action has a set of handlers for all internal exceptions. In this approach, action participants cooperate not only when they execute program functions (i.e. during normal activity) but also when they handle abnormal events. This is mainly due to the fact that when an atomic action is executed, an error can spread to all participants, and the system can be returned into a consistent state only if all participants are involved in handling. This is why, when an exception is raised in any participant, appropriate handlers are initiated in all of them. An action can be completed either normally (without raising any internal exceptions or after a successful cooperative handling of such exceptions) or by signaling an external exception to the context of the containing action. Concurrent internal exceptions are resolved using a resolution graph, and handlers for the resulting exception are called in all participants as illustrated in Fig. 9.

# 7 Hybrid Systems

This section covers some of the most important techniques that can handle different forms of concurrency. The next two subsections present coordinated atomic actions and open multithreaded transactions, two models that support competitive and cooperative concurrency. The remaining sections cover other advanced models, mainly extensions of recovery blocks or n-version programming, that combine the ideaa of sequential and loosely coupled concurrent models.

## 7.1 Coordinated Atomic Actions

The developers of the *Coordinated Atomic Action* (CA action) concept [27] have defined a model that fully integrates cooperative and competitive concurrency. They have extended the atomic action concept by allowing participants of an atomic action to access external objects. Atomic actions are used to control cooperative concurrency and to implement coordinated error recovery, whilst external objects are accessed using transactions in order to maintain the consistency of shared resources in the presence of failures and competitive concurrency.

Each CA action is designed as a stylized multi-entry procedure with roles which are activated by action participants cooperating within the CA action. Logically, the action starts when all roles have been activated and finishes when all of them reach the action end. CA actions can be nested. The state of the CA action is represented by a set of local and external objects. External objects can be used concurrently by several CA actions in such a way that information cannot be smuggled among them. Any sequence of operations on these objects bracketed by the start and completion of the CA action has the ACID properties with respect to other sequences. The execution of a CA action looks like a transaction for the outside world. Action participants explicitly cooperate (interact and coordinate their executions) through local objects.

All participants are involved in recovery if an error is detected inside a CA action. Conceptually it makes no difference which of them detects the error. The whole CA action represents the recovery region. Exception handling in CA actions is very similar to the one found in atomic actions: all action participants are involved in cooperative handling of any internal exception, internal exceptions raised concurrently are resolved and external exceptions are explicitly propagated by action participants. Exception handling in CA actions explicitly deals with local and transactional objects [28].

The CA action interface can contain one or more abort exceptions, a predefined failure exception and a number of exceptions corresponding to partial (committed and consistent) results which the action can provide. In the latter case, it uses external exceptions to inform the containing action of the fact that it has not been able to produce a complete required result and, indirectly, of the state in which objects have been left and of the available partial results. If one of the participants signals an abort interface exception, the CA action is aborted; all modifications made to transactional objects are undone and all local objects destroyed. Note that to improve performance, local objects can simply be re-initialized if software diversity or retry are used for recovery. A failure interface exception is signalled by the support when some serious problems are encountered. This might happen if, for example, the support cannot abort or commit the states of transactional objects. When an external exception corresponding to a partial result is signalled to the outside of an action, the state of all transactional objects is committed before raising this exception in the containing context. In all these cases signaling an interface exception means that the responsibility for dealing with the abnormal event is passed to a higher level in the system structure. At this level, detailed information about the current system state and the reasons for the exception occurrence can be determined based on the identity of the exception, optional output parameters and the post-conditions associated with the exception.

## 7.2  Open Multithreaded Transactions

The *Open Multithreaded Transaction* model [29] is a transaction model that provides features for controlling and structuring not only accesses to objects, as usual in transaction systems, but also threads taking part in transactions. The model allows several threads or processes, here called *participants*, to enter the same transaction in order to perform a joint activity. It provides a flexible way of manipulating threads executing inside a transaction by allowing them to be forked and terminated, but it restricts their behavior in order to guarantee correctness of transaction nesting and isolation among transactions.

Within an open multithreaded transaction, threads can access a set of transactional objects. Although individual threads evolve independently inside an open multi-threaded transaction, they are allowed to collaborate with other threads of the transaction by accessing the same transactional objects. The transactional objects therefore must preserve data consistency despite concurrent accesses from within a transaction, and at the same time provide isolation among concurrent accesses from different transactions.

The open multithreaded transaction model incorporates disciplined exception handling adapted to nested transactions. It allows individual threads to perform *forward error recovery* by handling an abnormal situation locally. If local handling fails, the transaction support applies *backward error recovery* and reverses the system to its "initial" state.

The model distinguishes *internal* and *external* exceptions. The set of internal exceptions for each participant consists of all exceptions that might occur during its execution. There are three sources of exceptions inside an open multithreaded transaction:

- An internal exception can be raised explicitly by a participant;
- An external exception raised inside a nested transaction is raised as an internal exception in the parent transaction;
- Transactional objects accessed by a participant of a transaction can raise an exception to signal a situation that violates the consistency of the state of the transactional object (see *self-checking transactional objects* in Section 5.3).

All these situations give rise to a possibly inconsistent application state. If a participant does not handle such a situation, the application's correct behavior can not be guaranteed.

A participant must therefore provide handlers for all internal exceptions. If such a handler is not able to deal with the situation, it can signal an external exception. If a participant "forgets" to handle an internal exception, the external exception `Transaction_Abort` is signaled, and the application consistency is restored by aborting the transaction.

If any participant of a transaction signals an external exception, the transaction is aborted, the exception is propagated to the containing context, and the exception `Transaction_Abort` is signalled to all other participants. If several participants signal an external exception, each of them propagates its own exception to its own context.

## 7.3  N-Version Programming Variants

*Acceptance Voting* [30] is an example of a minor extension of n-version programming. In this scheme, only results that pass an initial acceptance test are voted upon.

The *N-Version Programming with Tie Breaker and Acceptance Test* [31] technique strives for better performance by just comparing the results of the two fastest versions.

If they match, then the result is immediately output. However, if they do not agree, then all the results are voted upon. The final result of the voter is tested by means of an acceptance test.

## 7.4 Distributed Recovery Blocks

*Distributed Recovery Blocks* [32, 33] provide hardware and software fault tolerance specifically targeted to real-time applications. A distributed recovery block executes simultaneously on two processing nodes that are structured as a primary-shadow pair. The general syntax is given below:

```
ensure Acceptance Test on Node 1 or Node 2
by Primary on Node 1 or Alternate on Node 2
else by Alternate on Node 1 or Primary on Node 2
else signal Failure
```

First the input data is distributed to both nodes. Then one node starts executing the primary algorithm, whereas the other node starts with the alternate algorithm. If the primary result fails the acceptance test, then the alternate result is tested (forward error recovery). Only if both fail, then backward error recovery rolls back the state, and the roles are inverted, e.g. the first node executes the alternate, while the second one now tries the primary. Each node also contains two watchdogs: one that monitors the execution of the local algorithm, and one that monitors the execution of the alternate algorithm on the other node.

## 7.5 Consensus Recovery Blocks

The *Consensus Recovery Block* technique [34] combines recovery blocks and n-version programming ideas. It uses n variants that are ranked in order of their service and reliability. The n variants are first run in n-version programming fashion, and the results are checked by a voter. If, however, the voter can not determine a correct result, then the result of the highest ranked variant is submitted to the acceptance test. If this fails, then the next highest ranked variant's result is tested, and so on.

It is claimed that the consensus recovery block reduces the importance of the acceptance test used in recovery blocks and is able to handle cases where n-version programming would not be appropriate because of multiple correct results (see Section 4.3).

## 7.6 Two-Pass Adjudicators

The *Two-Pass Adjudicators* technique [35] combines data and design diverse software fault tolerance. The first pass uses n-version programming, meaning that the different alternates are executed concurrently and their results are passed to the voter. If, however, the voter can not determine a result, then the input data is re-expressed and the algorithms are executed again with the new input. The two-pass adjudicator technique can, in some situations, handle multiple correct results.

## 7.7  Self-Configuring Optimal Programming

*Self-Configuring Optimal Programming* [36] attempts to reduce the cost of fault-tolerant software in terms of space and time redundancy by providing a flexible architecture. The trade-off between dependability and efficiency can be dynamically adjusted at run-time.

In short, self-configuring optimal programming selects a set of variants to be run in the first phase according to how many results are needed to make a decision, and according to how much processing power is available. If no result can be determined during the first phase, then additional new variants are executed. This goes one until either a satisfying result has been determined, or until all versions have been unsuccessfully tried.

# 8  Conclusion and Discussion

This paper presented an overview of the most important techniques that can be used by developers to produce software that can tolerate design faults and faults of the surrounding environment. The different techniques have been classified with respect to the kind of concurrency they support.

Although some of the techniques presented here have been available for close to 30 years, software fault tolerance is still not used extensively in current software development, except in mission- and safety-critical systems. There are several reasons for this.

As we have seen thoughout the examples in this paper, software fault tolerance requires additional resources, such as processing power, memory, time, etc. Design diversity requires additional design and implementation efforts. In short, addressing fault tolerance in software increases the development cost.

This goes together with the fact that fault tolerance is often considered a non-functional requirement during the development of an application. As a result, it is not addressed properly during the analysis and design phases. This leads to poor integration of fault tolerance, since it is too late to make use of structured fault tolerance models once the implementation phase has begun. There still is a lot research and educational work to be done in order to make fault tolerance an integral part of software development processes.

Finally, modern programming languages do not provide any direct software fault tolerance support. Programmers have to manually implement the techniques, using the basic building blocks offered by programming languages, such as exceptions, threads, synchronization means, object-orientation, information hiding, controlled copying, and serialization. The interaction among these features, however, can be very subtle, which makes the development of software fault-tolerant applications complicated and error-prone.

One way to avoid this problem is to develop middleware that provides fault tolerance support. This can be done successfully in certain situations, as demonstrated by transactional middleware such as the CORBA Object Transaction Service [37] or Enterprise Java Beans [38]. However, if tight interaction between the application and the fault tolerance infrastructure is required, such a separation is artificial, results in complex interfaces, and might result in poor performance.

Ada [39] is a language that has been used extensively in safety- and mission-critical systems. It has a reputation of being highly reliable, and it is therefore not surprising that there have been many Ada-based implementations of software fault tolerance schemes, such as checkpointing and recovery cache support [40], shared recoverable objects [41], atomic actions [42, 43], coordinated atomic actions [44], and open multithreaded transactions [45]. All schemes follow a somehow similar approach. They provide a library or framework of reusable components with well-defined interfaces to the application programmer, and define rules and programming conventions of how to use them in a safe way. The elegance of the interfaces, however, depends heavily on the features offered by the programming language. Unfortunately, some characteristics of Ada, such as rules for limited types, access discriminants, and controlled types, lead to rather complicated interfaces.

Programming languages that provide powerful features such as reflection [46] make it possible to build nicer interfaces, as shown for instance in [47]. Also, approaches using the upcoming aspect-oriented programming paradigm [48] show promising results [49].

# References

[1]    Pullum, L.L.: *Software Fault Tolerance - Techniques and Implementation*, Artech House, Boston, 2001.

[2]    Laprie, J.-C.: "Dependable Computing and Fault Tolerance : Concepts and Terminology", in *Proceedings of the 15th International Symposium on Fault–Tolerant Computing Systems (FTCS–15)*, pp. 2–11, Ann Arbour, MI, USA, June 1985.

[3]    Cristian, F.: "Understanding Fault–Tolerant Distributed Systems", *Communications of the ACM 34(2)*, February 1991, pp. 56–78.

[4]    Lamport, L.; Shostak, R.; Pease, M.: "The Byzantine Generals Problem", *ACM Transactions on Programming Languages and Systems* 4(3), pp. 382–401, 1982.

[5]    Kopetz, H.: *Real–Time Systems – Design Principles for Distributed Embedded Applications*, Kluwer Academic Publishers, 1997.

[6]    Lee, P.A.; Anderson, T.: "Fault Tolerance – Principles and Practice", in *Dependable Computing and Fault-Tolerant Systems*, Springer Verlag, 2nd ed., 1990.

[7]    Randell, B.; Xu, J.: *The Evolution of the Recovery Block Concept*, chapter 1, pp. 1–21, in Lyu, M.R. (Ed.): *Software Fault Tolerance*, John Wiley & Sons, 1995.

[8]    IEEE Standard 729-1982: "IEEE Glossary of Software Engineering Terminology", 1982.

[9]    Horning, J.J, et al.: "A Program Strucure for Error Detection and Recovery", in E. Gelenbe and C. Kaiser (eds.), *Lecture Notes in Computer Science* 16, pp. 171–187, Springer, 1974.

[10]    Randell, B.: "System Structure for Software Fault Tolerance", *IEEE Transactions on Software Engineering* SE-1(2), pp. 220–232, 1975.

[11]    Ammann, P.E.; Knight, J.C.: "Data Diversity: An Approach to Software Fault Tolerance", *Proceedings of the 17th International Symposium on Fault–Tolerant Computing Systems (FTCS–17)*, Pittsburgh, PA, pp. 122–126, 1987.

[12]    Elmendorf, W.R.: "Fault Tolerant Programming", *Proceedings of the 2nd International Symposium on Fault–Tolerant Computing Systems (FTCS–2)*, Newton, MA, pp. 79–83, 1972.

[13]    Chen, L. and Avizienis, A.: "N-Version Programming: A Fault Tolerance Approach to Reliability of Software Operation", *Proceedings of the 8th International Symposium on Fault–Tolerant Computing Systems (FTCS–8)*, Toulouse, France, pp. 3–9, 1978.

[14]    Ammann, P.E.; Knight, J.C.: "Data Diversity: An Approach to Software Fault Tolerance", *IEEE Transactions on Computers* **37**(4), pp. 418–425, 1988.

[15]    Brilliant, S.S.; Knight, J.C.; Leveson, N.G.: "The Consistent Comparison Problem in N-Version Software", *IEEE Transactions on Software Engineering* **15**(11), pp. 1481–1485, 1989.

[16]    Avizienis, A.: "The N-Version Approach to Fault-Tolerant Software", *IEEE Transactions on Software Engineering* **SE-11**(12), pp. 1491–1501, 1985.

[17]    Vouk, M.A. et al.: "An Empirical Evaluation of Consensus Voting and Consensus Recovery Block Reliability in the Presence of Failure Correlation", *Journal of Computer and Software Engineering* **1**(4), pp. 367–388, 1993.

[18]    Gray, J.; Reuter, A.: *Transaction Processing: Concepts and Techniques*. Morgan Kaufmann Publishers, San Mateo, California, 1993.

[19]    Pu, C.; Kaiser, G.E.; Hutchinson, N.C.: "Split-Transactions for Open-Ended Activities", in *14th International Conference on Very Large Data Bases*, pp. 26–37, Los Angeles, California, Morgan Kaufmann, 1988.

[20]    Vinter, S.; Ramamritham, K.; Stemple, D.: "Recoverable Actions in Gutenberg", in *Proceedings of the 6th International Conference on Distributed Computing Systems*, pp. 242–249, Los Angeles, Ca., USA, IEEE Computer Society Press, 1986.

[21]    Garcia-Molina, H.; Salem, K.: "SAGAS", in *Proceedings of the SIGMod 1987 Annual Conference*, pp. 249–259, San Francisco, CA, ACM Press, May 1987.

[22]    Moss, J. E. B.: *Nested Transactions, An Approach to Reliable Computing*. PhD Thesis, MIT, Cambridge, April 1981.

[23]    Kienzle, J.; Strohmeier, A.; Romanovsky, A.: "Auction System Design Using Open Multithreaded Transactions". In *Proceedings of the 7th IEEE International Worshop on Object-Oriented Real-Time Dependable Systems (WORDS'02), San Diego, CA, USA, January 7th–9th, 2002*, pp. 95–104, IEEE Computer Society Press, Los Alamitos, California, USA, 2002.

[24]    Randell, B.: "System Structure for Software Fault Tolerance", *IEEE Transactions on Software Engineering* **1**(2), pp. 220–232, 1975.

[25]    Strigini, L.; Giandomenico, F.D.; Romanovsky, A.: "Coordinated Backward Recovery between Client Processes and Data Servers", *IEEE Proceedings – Software Engineering* **144**(2), pp. 134–146, April 1997.

[26]    Campbell, R.H.; Randell, B.: "Error Recovery in Asynchronous Systems", *IEEE Transactions on Software Engineering* **SE-12**(8), pp. 811–826, August 1986.

[27]    Xu, J.; Randell, B.; Romanovsky, A.; Rubira, C.M.F.; Stroud, R.J.; Wu, Z.: "Fault Tolerance in Concurrent Object-Oriented Software through Coordinated Error Recovery", in *Proceedings of the 25th International Symposium on Fault–Tolerant Computing Systems (FTCS–25)*, pp. 499–509, Pasadena, California, 1995.

[28]    Xu, J.; Romanovsky, A.; Randell, B.: "Concurrent Exception Handling and Resolution in Distributed Object Systems", *IEEE Transactions on Parallel and Distributed Systems* **11(11)**, pp. 1019–1032, November 2000.

[29]    Kienzle, J.; Romanovsky, A.; Strohmeier, A.: "Open Multithreaded Transactions: Keeping Threads and Exceptions under Control". In *Proceedings of the 6th International Worshop on Object-Oriented Real-Time Dependable Systems, Universita di Roma La Sapienza, Roma, Italy, January 8th–10th, 2001*, pp. 197–205, IEEE Computer Society Press, Los Alamitos, California, USA, 2001.

[30]   Athavale, A.: "Performance Evaluation of Hybrid Voting Schemes", M.S. thesis, North Carolina State University, Department of Computer Science, 1989.

[31]   Tai, A.T.; Meyer, J.F.; Aviziensis, A.: "Performability Enhancement of Fault-Tolerant Software", *IEEE Transactions on Reliability* **42**(2), pp. 227–237, 1993.

[32]   Kim, K.H.: "Distributed Execution of Recovery Blocks: An Approach to Uniform Treatment of Hardware and Software Faults", *Proceedings of the Fourth International Conference on Distributed Computing Systems*, pp. 526–532, 1984.

[33]   Kim, K.H.: "The Distributed Recovery Block Scheme", in M.R. Lyu (ed.), *Software Fault Tolerance*, New York, John Wiley & Sons, pp. 189–209, 1995.

[34]   Scott, R.K.; Gault, J.W.; Mc Allister, D.F.: "The Consensus Recovery Block", *Proceedings of the Total Systems Reliability Symposium*, Gaithersburg, MD, pp. 95–104, 1983.

[35]   Pullum, L.L.: "Fault-Tolerant Software Decision-Making Under the Occurrence of Multiple Correct Results", Ph.D. thesis, Southeastern Institute of Technology, 1992.

[36]   Bondavelli, A.; Di Giandomenico, F.; Xu, J.: "Cost-Effective and Flexible Scheme for Software Fault Tolerance", *Journal of Computer System Science & Engineering* **8**(4), pp. 234–244, 1993.

[37]   Object Management Group, Inc.: *Object Transaction Service, Version 1.1*, May 2000.

[38]   Shannon, B.; Hapner, M.; Matena, V.; Davidson, J.; Pelegri-Llopart, E.; Cable, L.: *Java 2 Platform Enterprise Edition: Platform and Component Specification*. The Java Series, Addison Wesley, Reading, MA, USA, 2000.

[39]   ISO: *International Standard ISO/IEC 8652:1995(E): Ada Reference Manual*, Lecture Notes in Computer Science **1246**, Springer Verlag, 1997; ISO, 1995.

[40]   Rodgers, P.; Wellings, A.J.: "An Incremental Recovery Cache Supporting Software Fault Tolerance", in *Reliable Software Technologies - Ada-Europe'99, Santander, Spain, June 7–11, 1999, Lecture Notes in Computer Science* **1622**, pp. 385–396, 1999.

[41]   Kienzle, J.; Strohmeier, A.: "Shared Recoverable Objects", in *Reliable Software Technologies - Ada-Europe'99, Santander, Spain, June 7–11, 1999, Lecture Notes in Computer Science* **1622**, pp. 397–411, 1999.

[42]   Romanovsky, A.; Mitchell, S.E.; Wellings, A.J.: "On Programming Atomic Actions in Ada 95", Ada Europe'97, London, *Lecture Notes in Computer Science* **1251**, pp. 254–265, 1997.

[43]   Mitchell, S.E.; Wellings, A.J.; Romanovsky, A.: "Distributed Atomic Actions in Ada 95", *The Computer Journal* **41**(7), pp. 486–502, 1998.

[44]   Romanovsky, A.; Randell, B.; Stroud, R.; Xu, J.; Zorzo, A.: "Implementation of Blocking Coordinated Atomic Actions Based on Forward Error Recovery", *Journal of System Architecture* (Special Issue on Dependable Parallel Computing Systems) **43**(10), pp. 687–699, September, 1997.

[45]   Kienzle, J.; Jiménez-Peris, R.; Romanovsky, A.; Patiño-Martinez, M.: "Transaction Support for Ada". In *Reliable Software Technologies – Ada-Europe'2001, Leuven, Belgium, May 14–18, 2001*, pp. 290 – 304, *Lecture Notes in Computer Science* **2043**, Springer Verlag, 2001.

[46]   Maes, P.: "Concepts and Experiments in Computational Reflection", *ACM SIGPLAN Notices 22(12)*, December 1987, pp. 147–155.

[47]   Xu, J.; Randell, B.; Zorzo, A. F.: "Implementing Software-Fault Tolerance in C++ and Open C++: An Object-Oriented and Reflective Approach", *Proc. Int. Workshop on Computer-Aided Design, Test, and Evaluation for Dependability* (CADTED96), Beijing, China, pp. 224–229, Int. Academic Publ., 1996.

[48]   Elrad, T.; Aksit, M.; Kiczales, G.; Lieberherr, K.; Ossher, H.: "Discussing Aspects of AOP". *Communications of the ACM* **44**(10), pp. 33–38, October 2001.

[49]    Kienzle, J.; Guerraoui, R.: "AOP – Does it make sense? The case of concurrency and failures". In *Proceedings of the 16th European Conference on Object-Oriented Programming (ECOOP 2002)*, pp. 37–54, Malaga, Spain, June 2002, *Lecture Notes in Computer Science* **2374**, Springer Verlag, 2002.

# High Integrity Ravenscar

Peter Amey and Brian Dobbing

Praxis Critical Systems, 20, Manvers St., Bath BA1 1PX, UK
peter.amey@praxis-cs.co.uk
brian.dobbing@praxis-cs.co.uk

**Abstract.** The Ravenscar Profile is an exciting development for the Ada community since it provides, for the first time in the history of our industry, support for deterministic, multi-tasking programming as an integral part of a standardized language. Despite its many advantages, the profile leaves several areas where behaviour is implementation defined and can result in run-time errors; this is unfortunate in a profile aimed clearly at the critical systems market. The SPARK language is a well-established sequential Ada subset that avoids ambiguity and allows all language rule violations to be detected prior to execution. The authors show how the principles of SPARK have been successfully extended to encompass the Ravencar Profile thereby statically eliminating the profile's problematic areas. The result should allow concurrent Ada programs to be constructed with the same degree of rigour that is now possible using sequential SPARK.

## 1   Introduction

The Ravenscar Profile (the definition of which can be found in Sect. 3 of [1]) is now established as providing the basic building blocks for constructing high integrity Ada95 concurrent programs. The profile has been accepted for inclusion in the next revision of the Ada language standard and is already supported by the major Ada product vendors. In addition, specialized versions of Ada runtime systems that implement the profile's concurrency model, whilst excluding the other concurrency features of Ada95 that are restricted by the profile definition, have been developed. In some cases, these Ada runtime systems are supported by certification evidence for rigorous standards such as RTCA/DO-178B level A [2].

This support for the profile forms an acceptable baseline for meeting the requirements of the dynamic semantics of high integrity systems. However, this basic level of support needs to be supplemented by additional rules and by static analysis techniques in order to be able to show the same level of proof of correctness and absence of run-time errors that is currently achievable in sequential programs using tools such as the SPARK Examiner [3,4]. The kinds of problems that need to be addressed and the role of static analysis techniques, are described in Sect. 6.2 of [1]. Essentially this identifies two roles for analysis:

1. ensuring that a program is well-formed; and free from run-time errors and erroneous behaviour; and
2. providing evidence for its overall correct behaviour.

J.-P. Rosen and A. Strohmeier (Eds.): Ada-Europe 2003, LNCS 2655, pp. 68–79, 2003.

The paper is in three main sections: Sect. 2 briefly restates the problems of wellformedness introduced by the Ravenscar Profile. Section 3 shows how extensions to the SPARK language and SPARK Examiner are sufficient to detect all such problems, statically, before execution. Finally, Sect. 4 outlines further forms of analysis contributing to the second of the two roles enumerated above.

## 2   Problems Associated with the Ravenscar Profile

Although it represents an enormous step forward in its support for reliable concurrent programming, the Profile identifies a number of error conditions which may give rise to run-time exceptions, erroneous behaviour or implementation-defined behaviour. Section 6 of [1] provides a detailed explanation of these problem areas and outlines theoretical approaches that could be taken to detect and eliminate them statically. We briefly restate the problems here before describing our practical analysis system for their elimination.

### 2.1   Errors Leading to Run-Time Exceptions

**Effect of Unexpected Exceptions.** The general concern within high integrity systems of the occurrence of unhandled exceptions is not addressed by the Ravenscar Profile since exceptions relate to the sequential, rather than the concurrent, part of the language. Nevertheless, whereas an unhandled exception will cause a sequential program to terminate, and hence offer an immediate opportunity for some program level control to invoke recovery actions, an unhandled exception during the execution phase of a concurrent program is not so readily detected. In particular, an unhandled exception can cause any of the following effects:

1. silent abandonment of the execution of an interrupt handler;
2. silent termination of a task;
3. premature exit from a protected action, possibly leaving it in an inconsistent state.

Appropriate static analysis techniques already exist to show proof of absence of run-time errors in sequential code due to language-defined exceptions, see for example [5]. The same techniques can also be applied to the sequential code within each task and protected object, leaving only the possibility of the exceptions that relate directly to the concurrency behaviour. These techniques can therefore be used to show statically that all three of the above effects of an unhandled exception due to check failure cannot occur. Since the elimination of run-time errors in this way is not specific to the Ravenscar Profile, it is not considered further in this paper.

**Exceptions Due to Concurrency.** Of the concurrency checks defined by Ada95, there are only two that apply to a Ravenscar Profile program:

1. detection of priority ceiling violation as defined by the Ceiling_Locking policy — calls from a task to protected operations must follow a non-decreasing priority chain;

2. detection of violation of not more than one task waiting concurrently on a suspension object (via the Suspend_Until_True operation).

In addition, two further concurrency checks are introduced by the Ravenscar Profile definition:

1. the maximum number of calls that are queued concurrently on a protected entry shall not exceed one;
2. a potentially blocking operation shall not be executed by a protected action.

### 2.2 Errors Leading to Erroneous Behaviour

**Use of Unprotected Shared Variables.** If two tasks share an unprotected variable the resulting program may be erroneous. In principle this does not prevent the sharing of such variables; however this would require a demonstration that the temporal properties of the program prevented concurrent access to it. More robust protection requires a check that unprotected data *can never* be shared.

**Race Conditions During Elaboration.** Within a sequential program, detection of access before elaboration errors is generally straightforward during program development due to the repeatable nature of the elaboration order, and the raising of Program_Error exception at the point of failure, causing the program to terminate. Undesired elaboration order variation can be prevented by explicit use of elaboration order pragmas. SPARK is even stronger in this respect since it eliminates all dependency on elaboration ordering.

Section 6.2.2 of [1] provides a detailed explanation of why the situation is less clear cut in the case of concurrent programs and, in particular, the risk of race conditions during elaboration. These problems can be eliminated by the use of a new Partition_Elaboration_Policy pragma which has been agreed for addition to the next revision of the Ada language standard; however, this pragma is *not* part of the Ravenscar Profile and not yet part of the Ada language. SPARK takes a different approach to elaboration order problems as described in Sect. 3.4.

**Program Incompleteness.** It is not easy to ensure that all the tasks and protected objects required to provide a program's intended behaviour have actually been included in its executable image. Unlike sequential code, the failure to "with" an active component does not make the program illegal; it simply becomes a legal but different program performing a subset of its intended action.

### 2.3 Errors Leading to Implementation-Defined Behaviour

**Task Termination.** The Ravenscar Profile includes the Restrictions pragma No_Task_Termination, but the dynamic effect of such termination is implementation-defined. Attempts at adding a formal task termination handling mechanism to the next revision of the Ada language standard are at an immature state of development. A means of showing that tasks will *not* terminate is therefore essential if implementation-defined behaviour is to be avoided.

# 3  Static Elimination of Ravenscar Errors

The problems outlined in Sect. 2 can be addressed and eliminated by static analysis. In the case of SPARK we have extended the SPARK95 language and implemented additional checks within its supporting SPARK Examiner tool.

## 3.1  Background

The rationale for SPARK, which is fully described in [3], places great emphasis on two things:

**Precision.** SPARK is not about the detection and elimination of *some* errors, it is concerned with the exact representation of programs and the elimination of all ambiguous interpretations of them.

**A Constructive Approach.** We believe in the maxim "Correctness by Construction". It should be possible to check continuously, throughout development, that a program is progressing towards its planned goal.

These goals remain unchanged in the context of the current work and we have therefore had to devise language rules whose violation can be detected *under all circumstances* and to ensure that analysis can take place on incomplete programs, especially in the absence of all *bodies* of program units. The design approach follows the pattern used for the design of the sequential SPARK language. The required language properties are obtained by the combination of two tactics:

1. additional language restrictions policed by wellformation checks; and
2. the use of annotations to clarify the programmer's intentions and to support the analysis of incomplete programs by asserting properties of units whose bodies may not yet be available for analysis.

## 3.2  Additional Language Rules

Here we are concerned with the compilable core language of SPARK rather than with the definition of its annotations or language checks that make use of those annotations. Very few rules additional to those either already required by sequential SPARK or imposed by the Ravenscar Profile are required. The principal restrictions are as follows:

1. All task and protected *types* must be declared in package specifications. This is to facilitate the analysis of incomplete programs during development. Note that SPARK does not require task and protected *objects* to placed in specifications although, in accordance with the Ravenscar rules, they must be declared at library level.
2. Discriminants of task and protected types must be static and, currently, may not include access discriminants. (We believe the latter restriction may be removable provided the former remains).
3. Protected elements must be initialized at declaration.
4. Each task must have a plain loop with no exits as its final statement. If the Environment Task is the only task, in a program that contains other concurrency features, e.g. an interrupt handler, then it must end with such a loop.

5. The attribute 'Count is not supported, nor is pragma Atomic_Components.

In addition, there are some naming restrictions for protected operations to prevent the need for overload resolution which is excluded from SPARK.

## 3.3   Additional Annotations

SPARK has an annotation at the package level which indicates that a package contains "state variables" and allows the effect of the package's operations on that state to be described. This *own variable* annotation has been extended to allow the state to be identified as **protected** or as a **task**. It can also be followed by a *property list* which indicates such things as: priority, whether interrupts are involved and whether a task may suspend and on what variable(s). The property list is deliberately extensible and uses identifiers rather than new reserved words; this makes it feasible to extend the annotation system to support third party tools such as timing analysers and model checkers.

The property list may also be used as part of a procedure or task type annotation where it can be used to indicate, for example, that the procedure may delay and must be considered a potentially blocking operation.

Some properties take arguments or list of arguments in the form of named aggregates. The current list of properties and their meanings is as follows:

| Property | Location | Argument | Meaning |
|---|---|---|---|
| **Priority** | own protected | expression | announces the priority that the object *will* have when it is declared |
| **Suspendable** | own protected | none | indicates that the object is a *predefined suspension object* or a protected object with an *entry* |
| **Interrupt** | own protected | optional, see example | indicates that the object contains one or more interrupt handlers and gives (optionally) a user-defined name for them |
| **Protects** | own protected | identifier list | shows that the unprotected variables in the identifier list are used solely by the protected object and can be treated exactly like protected elements |
| **Suspends** | unprotected procedure or task type | identifier list | indicates that the operation suspends on the identifiers given (by calling an entry in a named protected object, for example) |
| **Delay** | unprotected procedure | none | marks a procedure as potentially executing a delay statement (or other blocking action) |

The validity of these claimed properties is, of course, checked when the body of the unit concerned is analysed. It is, for example, an error for a subprogram to execute a delay statement unless its specification annotation includes the delay property.

The manner in which SPARK eliminates the Ravenscar errors described above, and the additional analysis that can be achieved, is illustrated using a small example program which we describe now. The program provides a simple stopwatch: three user buttons allow the stopwatch to be started, stopped and reset; these are achieved by interrupt routines, attached to each button, that set or reset a suspension object that controls the main timing loop. The timing loop is a periodic task; when released, this cycles at 1 second intervals and calls a protected object which is responsible for maintaining the current count of seconds and passing it to an out port that causes it to be displayed. The reset button clears the time count to zero but does not start or stop the timing loop.

In our illustrative example, we have three packages: *User* which provides the control buttons; *Timer* which contains the task providing the main timing loop; and *Display* which maintains the second count and copies it to the display port each time it changes. Applying the above annotations to these package specifications we have:

```
package User
--# own protected Buttons : PT (Interrupt =>
--#                              (StartClock => StartButton,
--#                               StopClock  => StopButton,
--#                               ResetClock => ResetButton),
--#                     Priority => 10);
is
    protected type PT is
        pragma Interrupt_Priority (10);

        procedure StartClock;
        --# global in out Timer.Operate;
        --# derives Timer.Operate from Timer.Operate;
        pragma Attach_Handler (StartClock, 1);

        procedure StopClock;
        --# global in out Timer.Operate;
        --# derives Timer.Operate from Timer.Operate;
        pragma Attach_Handler (StopClock, 2);

        procedure ResetClock;
        --# global in out Display.State;
        --# derives Display.State from Display.State;
        pragma Attach_Handler (ResetClock, 3);
    end PT;
end User;
```

This tells us that the package will contain a protected own variable called Buttons, of type PT. This object provides interrupt handling and, optionally in this case, we have chosen to associate each of 3 interrupt handling procedures with a programmer-selected name that will be use in the partition wide flow analysis (see Sect. 4.1 for an explanation of how these names are used). Finally, the priority of the object is announced. The SPARK Examiner can be made aware of the definition of the implementation-dependent subtypes for priority ranges and will make appropriate range checks on the values used in the annotation.

```
package Display
--#  own out     Port;
```

```
--#        protected State : PT (priority => 10, protects => Port);
is

    procedure Initialize;
    --# global in out State;
    --# derives State from State;

    procedure AddSecond;
    --# global in out State;
    --# derives State from State;

    protected type PT is
        pragma Priority (10);

        procedure Increment;  -- add 1 second to stored time and send it to port
        --# global in out PT;  --note use of type name here means "myself" or "this"
        --# derives PT from PT;

        procedure Reset;  -- clear time to 0 and send it to port
        --# global in out PT;
        --# derives PT from PT;
    private
        Counter : Natural := 0;
    end PT;
end Display;
```

This tells us that the package contains an own variable called Port which is an out port (this is a conventional, sequential SPARK annotation) and a protected own variable State of priority 10 which *owns and controls* the port. The latter information is very useful; protected elements cannot included external objects such as those with associated address clauses or pragma Import, so great care is needed to ensure that concurrent access to a shared port cannot occur. The *protects* property may only be used to provide protection for an otherwise unprotected own variable declared in the same package as the protected object that will protect it; the SPARK Examiner then ensures by static semantic checks that the port is never accessed from outside the protected object that protects it. Finally, we consider the main timing package:

```
package Timer
--# own protected Operate (suspendable);
--#      task TimingLoop : TT;
is

    -- These two procedures simply toggle suspension object Operate
    procedure StartClock;
    --# global in out Operate;
    --# derives Operate from Operate;

    procedure StopClock;
    --# global in out Operate;
    --# derives Operate from Operate;

    task type TT
```

```
--#  global in out Operate, Display.State;
--#       in       Ada.Real_Time.ClockTime;
--#  derives Operate        from Operate &
--#           Display.State from Display.State &
--#           null          from Ada.Real_Time.ClockTime;
--#  declare suspends => Operate;
     is
        pragma Priority (10);
     end TT;
end Timer;
```

Note that the task type includes a *declare* annotation which states that tasks of type TT may perform a suspension operation on object Operate.

The combination of the information provided in these annotations and the SPARK rule requiring task and protected *types* to be declared in package specifications means that, without access to package bodies, we have sufficient information to eliminate the Ravenscar errors outlined in Sect. 2. Some errors are detected during the examination of individual program units such as a package body and others when the main program (more correctly *Environment Task*) is analysed. Note that the latter class of check requires access only to the package specifications "withed" by the main program; at no stage is a complete, linkable closure of the entire program required.

### 3.4   Static Elimination of Ravenscar Errors – A Practical Implementation

**Priority Ceiling Violation.** In order to ensure that the priority ceiling check cannot fail, we identify all protected objects that may be called directly or indirectly from each task (including the Environment Task) and each interrupt handler, via the *global* annotations. The transitivity rule for this annotation ensures that the identity of indirectly-called protected objects propagates to the root of the call tree. The priority of each protected object is available at specification level via its *property* annotation. The priority of each calling task is known from its subtype. Since all priorities in SPARK must be static (including via actual values of discriminants), we can ensure that each protected object call chain does not cause a priority ceiling violation.

**Protected Entry and Suspension Object Queue Violation.** We enforce the protected entry and suspension object queue length check by ensuring statically that there can be at most one caller that can suspend on each object. Although this is more restrictive than is strictly required since queue length violations could be avoided by the temporal properties of the program, detection of timing behaviour is beyond the scope of the SPARK toolset. The restriction may be relaxed in the future if additional capability such as model checking is integrated. The implementation is based on the *suspends* property that identifies the objects that the caller may suspend on. This property is transitive, and so the list of all objects that each task may suspend on appears at the root of the call tree. The check cannot fail if the intersection of these lists is null.

**Execution of Potentially-Blocking Operation Violation.** Detection of calls to potentially-blocking operations from a protected action is supported in SPARK using the transitive *delay* or *suspends* property of a procedure. If a protected body may execute a delay statement or protected entry call directly, or may call a procedure that has at least one of the *delay* or *suspends* properties, then an error is raised. As above, it may be possible to relax this rule in the presence of integrated model checking if this can show that no state exists for which the protected action does call the blocking operation.

**Use of Unprotected Shared Variables.** In SPARK, we avoid the possibility of shared use of unprotected variables by allowing at most one task to access each unprotected global variable. We also prohibit protected objects from accessing unprotected state. These rules are enforced via the *global* annotations that identify the global variables that are referenced by each task, and *own* variable annotations that supply the protected property information for these variables. Note that a protected own variable includes a suspension object, an object of pragma Atomic type and objects protected via the *protects* property, in addition to Ada protected objects. As above, the rule may be relaxed in the future if model checking can show that concurrent access to an unprotected shared variable is not possible.

**Race Conditions during Elaboration.** To avoid race conditions during elaboration we need to ensure that no task or interrupt handler is dependent, for its correct behaviour, on the earlier execution of package body elaboration code. Several SPARK rules and annotations combine to ensure that this is the case. We require that all global variables marked in their own variable annotation as *protected* are statically initialized:

- at declaration (for protected elements, protected variables of pragma Atomic types and non-external own variables protected via the *protects* property;
- automatically (as is the case for predefined suspension objects); or
- by the external environment (as is the case for a variable marked as an *in port* which is protected using the *protects* property.

We can therefore be sure that no protected state initialisation depends on body elaboration code. This leaves one further case to consider: the use of unprotected state by a single task. The model adopted is to ensure that the unprotected state is initialised only by the sole-user task, and not by library package elaboration code. This is enforced by examination of the task type's *global* and *derives* annotations to determine its use of unprotected variables, together with the *initializes* annotation for each of these variables, which defines whether the variable is initialised by elaboration code or not. Note that these rules do not apply to the Environment Task because the main subprogram is guaranteed not to run until all library package elaboration is complete; this is particularly useful since it provides full upward compatibility for existing legal sequential SPARK programs that may use initialised unprotected state.

We anticipate being able to relax the initialisation rules concerning use of unprotected state by tasks once pragma Partition_Elaboration_Policy is incorporated into the Ada language standard.

**Task Termination.** Task termination is prevented by the language rule requiring each task to end with a plain loop with no exit statements and by the elimination of run-time exceptions from the program.

# 4  Program-Wide Static Analysis

In addition to showing absence of run-time errors, it is also highly desirable to apply existing static analysis techniques for sequential code to an entire program that includes tasks, protected objects and interrupt handlers. Current technology supports data flow analysis, information flow analysis and proof that includes the use of pre and post conditions and assertions. For concurrent programs we also wish to support these analyses and to accommodate schedulability analysis and model checking.

## 4.1  Partition-Wide Information Flow Analysis

**Thread Level Analysis.** Sequential SPARK's data and information flow analysis is described in [6]. At the thread level, where we are concerned with an individual task or subprogram, this is largely unaffected by the addition of concurrency constructs. In particular, we do not concern ourselves with temporal aspects or with the effect of task suspension. Therefore delaying or waiting on an entry or suspension object does not affect task-level flow analysis. In summary, flow analysis of tasks and interrupt handlers is performed on the basis that they *will be* activated at some stage.

The only real change to sequential flow analysis is that references to potentially shareable protected variables must be considered volatile at all times because the value read may be generated by another program thread at any time. Updating a potentially shared protected variable does not mean that the value written will still be there when the object is next referenced. For example, if we foolishly try to exchange the values of variables X and Y using protected variable P as a temporary store:

```
P := X; X := Y; Y := P; -- dangerous, P may no longer contain X
```

then we find that instead of being described by the following flow relation (as it would be if P was not shareable):

```
--# derives Y from X    & X from Y & P from X;
```

we must write:

```
--# derives Y from X, P & X from Y & P from X;
```

which indicates that the final value of Y may depend not only on X but on any value of P that may be stored in the protected variable, by another thread, during its execution. This addition to the flow relation provides the necessary hook to allow us to track inter-task communication at the partition level.

**Partition-Wide Information Flow Analysis.** The SPARK annotation system is extended to include an additional *global* and *derives* annotation that describes the intended flow relation for the entire program partition. This intended behaviour is compared with a flow relation constructed as follows:

**For each task** (identified by the *own task* annotation of each package "withed" by the Environment Task), we add any object that the task suspends on (identified by the task's property annotation) as an import that influences the exports of that task.

**For each interrupt handler** (identified by the property list of the *own protected* annotation of each "withed" package), we add, as an import, the name of the source from which the interrupt is deemed to come. By default this is the name of the protected object that contains the interrupt handler but the property annotation allows a more descriptive, user-selected name to be used instead (the use of names of *system* signficance in annotations is encouraged, see [7]). The ability to name the source of interrupts is especially useful if a protected type declares more than one handler (as in our example) or if there are multiple objects of the same type.

**For the enriched annotations generated above** we take their union so as to establish connections between values generated by one task and referenced by another.

**Finally** we take the transitive closure of this union because each task runs continuously and the effects of inter-task information flows will eventually propagate to all tasks that share information. For example if Task 1 derives B from A and Task 2 derives C from B and Task 3 derives D from C then taking the closure ensures that the influence of A on D is detected.

It is important to note that none of the above steps require access to the bodies of the packages "withed" by the Environment Task. The information in the own variable annotations, associated property lists and the type declarations themselves is all that is needed. We can therefore check that our program is properly constructed before getting involved in implementation details. The veracity of the annotations is, of course, checked when the bodies are written.

We can obtain the partition flow relation for our stopwatch example by applying the above rules. The three interrupt handlers have their user-supplied names (from the *interrupt* property list) added as imports. The task has the suspension object it suspends on (from its *suspends* property list) added as an import. The relations are then unioned and closed giving:

```
--# derives Timer.Operate from Timer.Operate
--#            User.StartButton, User.StopButton &
--#         Display.State from Display.State
--#            Timer.Operate, User.StartButton,
--#            User.StopButton, User.ResetButton &
--#         null           from
--#            Ada.Real_Time.ClockTime;
```

which provides the useful information that all three user buttons affect the displayed time but the reset button does *not* affect suspension object Timer.Operate which controls whether the clock is running.

As well as providing a useful description of overall system behaviour, the partition flow analysis provides protection from the particularly nasty error of program incompleteness (see Sect. 2.2). With SPARK this error cannot occur because if an active program component is omitted then its effects will not be included in the calculated partition flow relation which is unlikely to agree with the claimed partition-level *global* and *derives* annotation. The check ensures that the program elements actually included in the program closure are those that provide the claimed and expected information flow.

## 4.2 Timing Analysis and Model Checking

The work described in this paper and incorporated in the SPARK language and toolset dovetails nicely with other research work on the Ravenscar Profile. The steps described above ensure that a Ravenscar program is "well formed" and will not violate the specified rules of the profile; however, they are not sufficient to ensure that the program will meet its timing deadlines. Further work on the model checking and response time analysis aspects of the concurrency model is on-going and is the subject of other research papers that are being developed by the University of York. Eventually it is hoped that these two analysis streams will supplement one another (for example, by extending SPARK's *property* annotation to provide information needed by timing analysis tools) to provide evidence of both statically and dynamically appropriate behaviour.

## 5   Conclusions

Ada remains unique in its comprehensive language-level support for multi-tasking. The Ravenscar Profile provides a framework for constructing dependable tasking programs with deterministic and analyseable timing properties. The extension of SPARK to encompass Ravenscar provides a way of statically showing such a program to be well-formed and free from run-time errors. The combination of these techniques with suitable, certifiable, run-time support must represent the most rigorous environment for producing high-integrity tasking programs.

## References

1.  Burns; Dobbing; Vardanega: *Guide for the use of the Ada Ravenscar Profile in high integrity systems*. University of York technical report YCS 348 (2003)
2.  *RTCA-EUROCAE: Software Considerations in Airborne Systems and Equipment Certification*. DO-178B/ED-12B. 1992.
3.  Finnie, Gavin et al: *SPARK 95 – The SPADE Ada 95 Kernel – Edition 3.1*. 2002, Praxis Critical Systems[1].
4.  Barnes, John: *High Integrity Software – the SPARK Approach to Safety and Security*. Addison Wesley Longman, ISBN 0-321-13616-0. 2003
5.  Chapman, Rod; Amey, Peter: *Industrial Strength Exception Freedom*. Proceedings of ACM SIGAda 2002[2]
6.  Bergeretti and Carré: *Information-flow and data-flow analysis of while-programs*. ACM Transactions on Programming Languages and Systems 1985[1], pp. 37–61.
7.  Amey, Peter: *A Language for Systems not Just Software*. Proceedings of ACM SIGAda 2001[2].

---

[1] Also available from Praxis Critical Systems

[2] Also downloadable from http://www.sparkada.com

# Adding Temporal Annotations and Associated Verification to the Ravenscar Profile

Alan Burns and Tse-Min Lin

Real-Time Systems Research Group
Department of Computer Science
The University of York
Heslington
York  YO10 5DD, UK
{burns,lin}@cs.york.ac.uk

**Abstract.** This paper presents a proposal for extending the Ravenscar Tasking Profile with annotations that can be used to express temporal properties. An approach using model checking for the verification of compliance to the annotations is also presented. An extended example is used to illustrate the application of the proposed approach.

**Keywords:** Ravenscar Profile, Model Checking, UPPAAL, SPARK.

## 1  Introduction

The Ravenscar profile [1,4,7,6,25] was defined as result of the 8th International Real-Time Ada Workshop (IRTAW) with the goal of restricting the tasking facilities of Ada95 to remove implementation dependencies and unspecified behaviour, making possible the verification of the concurrency aspects of an application. The Ravenscar Profile, as presented in [7], defines which features are allowed, which are disallowed, what dynamic semantics are required and how the restrictions can be represented.

One of the motivations for Ravenscar is to support the development of real-time software. But Ravenscar does not in itself directly represent all the primitives necessary to fully specify real-time behaviour. Deadlines are a common requirement of real-time systems, yet very few programming languages (for example, DPS [15] and TCEL [9])[1] allow the specification of deadlines in the code itself. This means that in the vast majority of programming languages used to implement real-time systems it is the responsibility of the programmer to ensure that the deadlines are always met.

In this paper we introduce a collection of annotations that can be used to represent deadlines and other temporal characteristics such as freshness and periodicity. We also show how temporal requirements can be verified – using model checking if necessary for complex algorithms. The overall motivation for the work presented here is as follows:

---

[1] Their work is restricted to a sequential programming language, and deadlines are only associated to a block of code.

J.-P. Rosen and A. Strohmeier (Eds.): Ada-Europe 2003, LNCS 2655, pp. 80–91, 2003.

1. The addition of an annotation for deadline;
2. The addition of annotations for concurrent task interaction via protected objects;
3. To facilitate the verification of each task's temporal annotations;
4. To facilitate the verification of end-to-end timing requirements through many tasks and protected objects;
5. The representation (and validation) of common structures such as periodic tasks;
6. The representation (and validation) of other static relationships such as those between the deadlines of periodic tasks and priority.

The work presented in this paper is focused on the first three objectives. A full description of all the annotations been proposed is available via a project report [17]. The annotations presented here are independent of other annotations that might be used on Ravenscar programs. They are represented as comments starting with the three characters --@.

This paper is organised as follows. The next section presents the deadline annotation. Section 3 introduces the rely and guarantee annotations. Section 4 illustrates, through a non-trivial example, how annotated programs can be verified. Finally, the last section presents the conclusions and a discuss of further work including the integration of these annotations and work being developed for SPARK (which is intending to support Ravenscar in a future release).

## 2   Deadline Annotation

Fidge et al. [8] presented the deadline command as a language primitive for real-time programming. The reason for its introduction is due to the lack of explicit support to express deadlines in programming languages that support the implementation of real-time systems (e.g. Ada95 [24] and Time-C [16]).

Fidge et al. also suggested the use of logical constants and assertions on time. Unlike the standard notion of declaration of constants, their proposal does not *fix* the value of a logical constant, but instead its value is chosen in an *angelic* manner so it can be constrained later by an assertion.

We present below a simple example using their proposal:[2]

```
1  procedure Use_of_Deadline       7   ...
2  ...                              8    --@ assert t = Spark_Time.Now;
3  --@ t: con Spark_Time.Time;      9   ...
4  is                              10    --@ deadline t+13.0;
5  ...                             11   ...
6  begin                          12  end Use_of_Deadline;
```

The *timed* annotation in line 8 indicates that the value of the logical constant t is equal to the current time obtained from the special variable Spark_Time.Now of type Spark_Time.Time. Spark_Time.Now denotes the "real-time", instead of the (approximated) computer's real-time available through the

---

[2] Note that their language is sequential.

function `Spark_Time.Clock`. Because of that, `Spark_Time.Now` can only be used in real-time assertions. The annotation in line 10 indicates that when execution reaches this point, the current time must not have passed `t+13.0`.[3] If the current time has already passed, then the procedure `Use_of_Deadline` must be rejected by the compiler (or associated tool) and this procedure must be reviewed by the programmer.

The package `Spark_Time` provides an interface between SPARK and the real-time features of Ada95 available through the package `Ada.Real_Time`. Although the implementation of `Spark_Time` was not presented in [8], it was suggested that this could be done following [3, §8.3].

While logical constants are helpful to support reasoning and timing constraint specification and analysis, they are of limited use because after having their value fixed they cannot be modified later. A more interesting approach would be to use *auxiliary variables*, as proposed in [10]. Like logical constants and the deadline command, auxiliary variables do not generate code but are more useful than logical constants in expressing and analysing timing constraints.

In our proposal we add **--@ aux** as a new annotation to introduce auxiliary variables. Auxiliary variables can only be used in annotations, therefore they do not generate code. We illustrate below how the example presented in lines 1–12 could be modified to use `t` as an auxiliary variable instead of a logical constant:

```
13  procedure Use_of_Deadline          19  ...
14  ...                                 20  --@ t := Ada.Real_Time.Clock;
15  --@ aux t: Ada.Real_Time.Time;      21  ...
16  is                                  22  --@ deadline t+13.0;
17  ...                                 23  ...
18  begin                               24  end Use_of_Deadline;
```

Another modification, in relation to [8], is that the deadline command can also be used as an annotation in the specification part of a task. An example of use is presented in Sect. 4.

Coupled with the deadline command is the delay statement. Together they allow a wide range of real-time requirements to be specified. The deadline command ensures (if the implementation satisfies the constraint) that the system is progressing "fast enough"; the delay statement (which is easy to satisfy) ensures the system does not proceed "too fast". We shall give a more detailed example of the combined use of deadline and delay in Sect. 4. Note, as delay until is part of Ravenscar it will use real program variables.

## 3   Rely and Guarantee Annotations

In a concurrent system, tasks usually cooperate to define the behaviour of the system. As such, the dependence relation among tasks should be made explicit.

---

[3] In another words, the execution of the code in line 9 must not take more than 13 seconds. The deadline annotation used in line 10 is equivalent to the timed assertion "`--@ assert Spark_Time.Now <= t+13.0;`".

For example, in a producer-consumer system, tasks can be used to implement the producer and the consumer, and a protected object can be used to encapsulate data and allow sharing under mutual exclusion.

The producer task might or might not require the presence of a consumer task, and the consumer task might require the presence of a producer task. A task might use some subprograms of a protected object while requiring some other task to use other subprograms of the same protected object.

While some relations might sound obvious, clearly a cooperative/distributed system that models a producer-consumer system is wrong if, for example, a producer task is missing. Unless there is some way to make it explicit that a consumer task depends on the existence of a producer task, such properties cannot be statically verified.

Therefore, it is proposed to introduce two annotations, --**@ guarantee** and --**@ rely**, to specify the relations mentioned above. As an example of use, a producer-consumer system could be written as follow:

```
25  package Example_Cooperative_System      36      protected type Shared_Data_PO
26  ...                                      37      ...
27  is                                       38      is
28    task Producer_Task;                    39         procedure Read (...);
29    --@ guarantee shared_data              40         procedure Write (...);
30    --@            uses Write;             41      ...
31    task Consumer_Task;                    42      end Shared_Data_PO;
32    --@ guarantee shared_data              43      ...
33    --@            uses Read;              44      shared_data: Shared_Data_PO;
34    --@ rely shared_data                   45  end Example_Cooperative_System;
35    --@       uses Write;
```

In this example, the producer task guarantees that it will use the subprogram Write of shared_data, while the consumer task guarantees that it will use the subprogram Read of shared_data and it also relies on another task to use the subprogram Write of shared_data. That is, the consumer task relies on the presence of a producer task, whereas the producer task does not require the existence of a consumer task.

The use of the words *guarantee* and *rely* are associated with Whysall's use of rely and guarantee conditions used in the ZERO approach presented in his PhD thesis [26], rather than Jones' more generic use of rely/guarantee to express interference presented in his PhD thesis [11] and subsequent work.

Using the ZERO approach, we say that a task can choose to *guarantee* some properties about the shared objects it uses – basically it guarantees which subprograms of a protected object it will use. In the same manner, this task can also choose to *rely* on some properties about the shared objects it uses – basically it will be able to rely on the fact that a protected object is not modified, or modified in predictable ways: which subprograms of a protected object can be used by other tasks.

Although Jones approach is more powerful, it is also more difficult to verify that the properties specified do indeed hold. For example, Woodcock reported in [27] some difficulties in the verification of an implementation that seems to be correct. Later, Jones reviewed his approach in [12,13]. Stølen also presented in his PhD thesis [23] another approach based on Jones early work. In [22], Stark presented a proof technique that can be used to break the rely/guarantee conditions into a finite collection of smaller rely/guarantee conditions. However, Stark acknowledged that the proof technique proposed cannot be fully automated. A review of some existing work on the rely–guarantee approach for concurrency is presented in [28].

It is clear that we do not allow the explicit specification of properties as in Jones, Stølen or Stark's approaches. Properties are *derived* from the behaviour of the subprograms of a shared object. We do however extend the notion to include temporal statements. An optional key **with_period** is introduced with the annotations rely and guarantee to specify the expected periodicity of use of a subprogram of a protected object. For example, the annotation:

46   −−@ **guarantee** buffer **uses** Write **with_period** 100;

specifies that a task uses the subprogram `Write` of a protected object `buffer` with a (upper bound) period of 100 milliseconds. The annotation:

47   −−@ **rely** buffer **uses** Reads **with_period** 110;

specifies that a task relies on another task to use the subprogram `Read` of a protected object `buffer` with a (upper bound) period of 110 milliseconds.

To allow for the specification of age (or temporal validity) another optional key, named **with_freshness**, can be used instead of **with_period** in a rely/guarantee annotation. For example:

48   −−@ **rely** buffer **uses** Write **with_freshness** 300;

If the annotation above is used in a consumer task, then it indicates that a producer task is expected to produce an item every 300 milliseconds – it can produce sooner, but not later.

## 4   Verification of Annotated Programs

The combined use of delays, deadlines and freshness allows a wide range of requirements to be specified. Of course a program cannot be fully verified until the final implementation has been completed and timing analysis (or measurement) undertaken. In this work we identify a clear two stage process for verification.

(1) Before implementation, check that the annotations are consistent and are sufficient to satisfy the overall timing requirements of the system.
(2) After implementation, check that all deadlines will be satisfied.

Under (1), we assume that the system will be fast enough to meet its deadlines – but check that delays and deadlines are compatible and that tasks satisfy the **guarantee** annotations on protected objects. A simple (obvious) example of an incompatibility is a task loop that has a deadline of 20ms but contains within one feasible path a delay of 30ms. Verification of these properties may be straightforward but may require the power of model checking [2] for some more complex properties.

Under (2), once execution times are known then standard timing and scheduling analysis can be undertaken to determine if deadlines will be met. We do not advocate the use of model checking to verify (1) and (2) together. Model checking tools are not powerful enough to undertake the full analysis, and scheduling analysis is sufficiently mature to accomplish the second verification step.

To illustrate how the first step of verification can be undertaken consider a relatively simple (but not obvious) problem. A task, `Producer`, reads data from the environment via either of two devices it has access to. A primary requirement is that it generates data in protected object, `Input_Data`, with a freshness of 300ms. The two devices have different temporal properties:

- device A is "smart", it provides a reading 30ms after it is enabled;
- device B is "simple", it must be read 10 times to gain a single reliable value – each reading must be taken between 2ms and 4ms after it is enabled and subsequent readings must be at least 10ms apart.

This specification thus has two free variables: the overall period $P$ of `Producer`; and the enforced delay $p$ of the inner loop for device B (which must be greater than 10ms). There is also the freedom to place extra deadline and delay until statements in the code. From a schedulability point of view $P$ and $p$ should be as large as possible. We assume at this level of description that `Producer` makes a non-deterministic choice between the two devices. We also assume overall period equal to deadline for the producer task.

Parameterisation (and subsequent verification) could now be undertaken in a number of ways. If there were only free variables then an algebraic formulation is possible. If there are trade offs to be made (e.g. between $P$ and $p$) then it may be possible to set up a linear programming solution. However the solution may not always be found in a systematic way. The freedom to place extra delay and deadline statements/annotations in the code means there is considerable design freedom still available. We contend here that the use of a time automaton model (and model checking) allows that freedom to be explored and verification undertaken. To illustrate this let us fix $P$ to be 150ms and $p$ to be 15ms. First we give the annotated code:[4]

---

[4] All programs in Ada were type checked using GNAT version 3.14p.

```
49  package Example                          82      delay until Current+
50  ...                                      83               Milliseconds (30);
51  is                                       84      -- read from device and
52    task Producer;                         85      -- construct data
53    --@ period 150;                        86      Input_Data.Write (Data);
54    --@ deadline 150;                      87    else
55    --@ guarantee Input_Data               88      for Count in 1..10 loop
56    --@        uses Write;                 89        -- enable device
57    --@        with_freshness 300;         90        Current := Clock;
58    -- The above annotation is clearly     91        delay until Current+
59    -- consistent as long as Producer      92                 Milliseconds (2);
60    -- calls Write on each iteration.      93        -- read from device
61    ...                                    94        --@ deadline Current+
62  end Example;                             95        --@   Milliseconds (4);
                                             96        if Count = 10 then
63  with Ada.Real_Time;                      97          -- construct Data
64  use Ada.Real_Time;                       98          Input_Data.Write (Data);
65  package body Example is                  99        else
66    task body Producer is                 100          Current := Clock;
67      type Device_Type is (smart,         101          delay until Current+
68                    simple);              102                   Milliseconds (15);
69      Device : Device_Type;               103          --@ deadline Current+
70      Next   : Time;                      104          --@   Milliseconds (15);
71      Current: Time;                      105        end if;
72      Data   : Some_Type;                 106      end loop;
73    begin                                 107    end if;
74      Next := Clock;                      108    --@ deadline Next+
75      -- other initialisations            109    --@       Milliseconds (150);
76      loop                                110    Next := Next+
77        Device := ...                     111            Milliseconds (150);
78        -- assumed non-deterministic      112    delay until Next;
79        if Device = smart then            113  end loop;
80          -- enable device                114  end Producer;
81          Current := Clock;               115 end Example;
```

Next we give a timed automaton, in the UPPAAL[5] framework [14,21]. We will develop the automaton in stages to illustrate the procedure. A key driver here is only to include the relevant features that refer to the temporal behaviour of the code. First consider only the "smart" branch. In UPPAAL two clocks are needed, one for the outer period (`clk1`) and one for the delay statement (`clk2`).

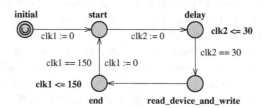

The automaton has five states. Each state has a name and a set of invariant that must always be true. Between the states are transitions that have boolean guards (must be true to be fired) and actions (assignment) that occur when the transition fires. For example, state **end** has an invariant that clock `clk1` must

---

[5] All models in UPPAAL were verified using UPPAAL2k version 3.2.13.

be less than (or equal to) 150; the transition out of this state requires clk1 to equal 150. These two properties together force the transition to occur when, and only when, clk1 is exactly 150. Hence a period of 150 is ensured. As a result of move to state start, clk1 is reset to 0. Note that within the formalism time is considered to be a continuous variable.

We will add the details of read_device_and_write shortly, but first we must ensure that the automaton makes progress. To do this we use the technique of [5] to bring the driving deadline back to all previous states. However, the forced delay in state delay means that start must be left after 120 (a value of 121, for example, leads to deadlock). The automaton then becomes:

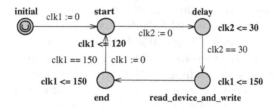

The protected object, Input_Data, is represented by another automaton. Communication with the Producer automaton is via a channel (call) .

The read_device_and_write location of Producer is changed to incorporate the channel call:

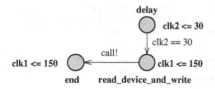

The freshness constraint on Input_Data can now be checked via model checking by proving that idclk never takes the value 300 or more. In the syntax of UPPAAL, "A[] Input_Data.idclk < 300". When applied to the model above this property is found to be true. If the deadline and period are increased however eventually it becomes false. The tool then illustrates a counter-example to show why the constraint is not met.

Now consider, again in isolation, the other branch (the simple sensor):

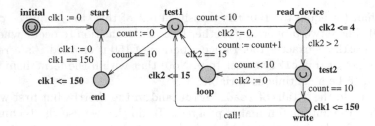

The locations with the symbol "∪" are urgent locations,[6] the model never stops in these states as at least one output transaction can always be taken (essentially they represent branch instructions – if statement, etc).

Verification of the above model fails due to deadlock. "A[] not deadlock" is found to be false because it is possible for location `write` to be entered with a value of `clk1` greater than 150 – this causes the automaton to deadlock. Changing the period and deadline to 200 solves this problem as long as the `start` location is given the constraint `clk1 <= 25`. Unfortunately the first path now fails; it must have a tighter deadline of 129 imposed on its cycle:

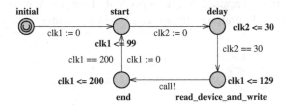

The final step is to combine the two paths:

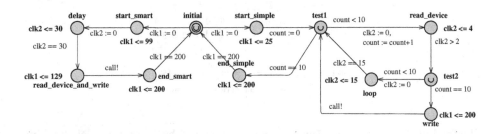

Although both paths on their own are correct (in their temporal properties) the combination is not. An initial execution down the smart sensor (finishing as early as possible) followed at the next iteration by the simple sensor route (finishing

---

[6] In this example, these locations could also be introduced as committed location without changing the behaviour of the model.

as late as allowed by the deadline statements) breaks the freshness constraint. There are a number of ways this could be fixed. One is to add an extra delay statement to the beginning of the smart loop:

Verification is now achieved:

```
A[] not deadlock
A[] Input_Data.idclk < 300
```

The properties above are found to be true. The modifications required in the annotated code presented in the beginning of this section are:

- change the period and deadline to 200 milliseconds (lines 53, 54, 109, and 111;
- add a delay of 71 milliseconds (**delay until** Next+Milliseconds(71)) after the statement in line 79;
- add a deadline of 129 milliseconds (**deadline** Next+Milliseconds(129)) after the statement in line 86.

To generalise, from this example, the annotations introduced in this paper allow:

- the consistent use of delay and deadline within each task to be checked;
- the temporal guarantee annotations to be checked against the task making that guarantee;
- the rely and guarantee annotation together to be checked against system level (or end-to-end) requirements.

This latter property has not been illustrated in the paper, but it is possible to check global properties using model checking with a set of timed automata. We have investigated these properties elsewhere [7] but not in the context of Ravenscar.

Although we ague that the process developed here increases the confidence with which real-time Ada programs can be developed, the approach is not formal. In particular there is, currently, no provable equivalence between the annotated Ada code and the automaton model. This continues to be studied (see work of [19,20]) and should be tractable in the sense that a tool should be able to automatically generate the outline Ada code from the automaton model.

## 5   Conclusion

The Ravenscar profile represents an important aspect of Ada technology. Its use emphasises the need for predictability and allows high integrity applications to be developed. For example, Lundqvist and Asplund [18] present a complete formal model of a Ravenscar-compliant run-time kernel in UPPAAL, and they show that the model behaves according to the semantics of Ada95 (e.g. a protected object is only updated under mutual exclusion and a task suspended due to a `delay until` statement is not released before the current time has reached that specified in the statement).

For simple systems with periodic tasks for example, confidence in the timing statements may be easy to obtain – although effective implementation may still be difficult to ensure. But Ravenscar is not just for such simple systems. We argue here for additional information to be added to the profile via annotation that capture important requirements such as deadlines and data freshness.

Although the approach is not yet completely formal, the use of timed automata models and model checking does enable a high level of confidence in the temporal characteristics of applications to be gained. The annotations presented here are concerned only with the temporal properties of Ravenscar programs. Other annotations will be needed to ensure, i.e. statically verify, that a full program does indeed conform to the Ravenscar profile. An example of the necessary annotations for this type of verification are the ones proposed in SPARK. There a technology that has been confined to sequential programs is being extended to concurrency by including the Ravenscar features. An extension of the work presented in this paper is to integrate the two sets of annotations. That is, show how our temporal annotations can be embedded within the broader set of SPARK annotations.

**Acknowledgements.** We would like to thank the anonymous referees for their helpful comments. This work is undertaken as part of the EPSRC funded RavenSPARK project (GR/N05963).

# References

1. L. Asplund abd B. Johnson and K. Lundqvist. Session summary: The Ravenscar profile and implementation issues. *Ada Letters*, XIX(2):12–14, 1999.
2. R. Alur and D. Dill. Automata for modeling real-time systems. In *ICALP 1990*, volume 443 of *LNCS*, pages 322–335. Springer-Verlag, 1990.
3. J.G.P. Barnes. *High Integrity Ada: The* SPARK *Approach*. Addison-Wesley, 1997.
4. B. Brosgol. Session summary: Future of the Ada language and language changes such as the Ravenscar profile. *Ada Letters*, XXII(4):113–119, 2002.
5. A. Burns. How to verify a safe real-time system: The application of model checking and timed automata to the production cell case study. *Real-Time Systems*, 24(2):135–151, 2003.
6. A. Burns, B. Dobbing, and G. Romanski. The Ravenscar tasking profile for high integrity real-time programs. In *Ada-Europe 98*, volume 1411 of *LNCS*, pages 263–275. Springer-Verlag, 1998.
7. A. Burns, B. Dobbing, and T. Vardanega. Guide for the use of the Ada Ravenscar profile in high integrity systems. Technical Report YCS 348, Department of Computer Science, The University of York, 2003.

8. C.J. Fidge, I.J. Hayes, and G. Watson. The deadline command. *IEEE Software – Special Issue on Real-Time Systems*, 146(2):104–111, 1999.

9. R. Gerber and S. Hong. Compiling real-time programs with timing constraint refinement and structure code motion. *IEEE Transactions on Software Engineering*, 21(5):389–404, May 1995.

10. I.J. Hayes. Real-time program refinement using auxiliary variables. In *FTRTFT2000*, volume 1926 of *LNCS*, pages 170–184. Springer-Verlag, 2000.

11. C.B. Jones. *Development Methods for Computer Programs Including a Notion of Interference*. PhD thesis, Programming Research Group, The University of Oxford, 1981.

12. C.B. Jones. Interference resumed. Technical Report UMCS-91-5-1, Department of Computer Science, The University of Manchester, May 1991.

13. C.B. Jones. Reasoning about interference in an object-based design method. In *FME'93*, volume 670 of *LNCS*, pages 1–18. Springer-Verlag, 1993.

14. K.G. Larsen, P. Pettersson, and W. Yi. UPPAAL in a nutshell. *Software Tools for Technology Transfer*, 1(1+2):134–152, 1997.

15. I. Lee and V. Gehlot. Language constructs for distributed real-time programming. In *IEEE Real-Time Systems Symposium*, pages 57–66. IEEE, 1985.

16. A. Leung, K. Palem, and A. Pnueli. TimeC: A time constraint language for ILP processor compilation. In *The 5th Australian Conference on Parallel and Real Time Systems*, pages 57–71. Springer-Verlag, 1998.

17. T.-M. Lin and A. Burns. Annotations for RavenSPARK. Technical report, Department of Computer Science, The University of York, 2002.

18. K. Lundqvist and L. Asplund. A Ravenscar-compliant run-time kernel for safety-critical systems. *Real-Time Systems*, 24(1):29–54, 2003.

19. D. Naydich and D. Guaspari. Timing analysis by model checking. In *LFM2000*, pages 71–82, June 2000.

20. Timing analysis for model checking (working document). Technical report, Odyssey Research Associates, June 2001.

21. P. Pettersson. *Modelling and Verification of Real-Time Systems Using Timed Automata:Theory and Practice*. PhD thesis, Department of Computer Systems, Uppsala University, February 1999.

22. E.W. Stark. A proof technique for rely/guarantee properties. In *FST & TCS 85*, volume 206 of *LNCS*, pages 369–391, New Dehli, 1985. Springer-Verlag.

23. K. Stølen. *Development of Parallel Programs on Shared Data-Structures*. PhD thesis, Department of Computer Science, The University of Manchester, 1990.

24. S.T. Taft and R.A. Duff, editors. *Ada 95 Reference Manual: Language and Standard Libraries*, volume 1246 of *LNCS*. Springer-Verlag, 1997.

25. A.J. Wellings. Session summary: Status and future of the Ravenscar profile. *Ada Letters*, XXI(1):5–8, 2001.

26. P.J. Whysall and J.A. McDermid. Object oriented specification and refinement. In *4th Refinement Workshop*, Workshops in Computing, pages 150–184. Springer-Verlag, 1991.

27. J.C.P. Woodcock and B. Dickinson. Using VDM with rely and guarantee conditions: Experiences of a real project. In *VDM'88*, volume 328 of *LNCS*, pages 434–458. Springer-Verlag, 1988.

28. Q. Xu, W-P. Roever, and J. He. The rely–guarantee method for verifying shared variable concurrent programs. *Formal Aspects of Computing*, 9(2):149–174, 1997.

# Impact of a Restricted Tasking Profile:
# The Case of the GOCE Platform Application Software

Niklas Holsti and Thomas Långbacka

Space Systems Finland Ltd., Kappelitie 6, 02200 Espoo, Finland

{Niklas.Holsti,Thomas.Langbacka}@ssf.fi
http://www.ssf.fi

**Abstract.** We are designing the on-board software for the GOCE satellite platform so that static analysis can verify its real-time requirements. Our aim is to analyse the binary code to derive bounds on the worst-case execution time for each task, followed by a schedulability analysis using preemptive fixed-priority scheduling. The code-analysis step constrains the sequential coding and the schedulability-analysis step constrains the concurrency design (tasking). We use the Ravenscar tasking profile further limited to a single suspension point per task and a single call of each protected entry. We discuss how this profile matches the real-time requirements and its impact on the designers, the software architecture and the resource usage. The main issue is that the single-suspension-point rule prevents procedural abstraction of operations that involve input-output. Some design patterns emerge from this example.

## 1 Introduction

The real-time performance of an embedded program is invariably verified by *measuring* execution times and response times when the program is *tested*. However, often testing can be done only in the final phase of a project, when the software and hardware can be integrated. Performance problems may be discovered late, making them expensive to correct. Furthermore, it may be hard to find the worst-case scenarios and test them. If the measured times are not the worst case they may be exceeded when the program is in use, perhaps making the system fail.

*Static analysis* is another approach in which the program is designed and coded in such a way that its real-time performance can be determined mathematically for all possible scenarios. Usually, this approach divides the program into a set of tasks or threads, maps the real-time requirements to task deadlines, and determines the overall schedulability using a suitable task scheduling model and the worst-case execution time of each task as computed from the code. This analysis can often be done before testing, at least in a preliminary way, and can be automated and repeated as the program evolves. Worst-case performance problems can be detected early.

Static analysis should be foreseen in the design and coding of the program. The algorithms must be designed with worst-case analysis in mind, and the tasking architecture must follow the rules of the computational model.

In this paper we discuss a particular set of tasking rules as used in the design of a medium-sized, real-time, embedded program: the Platform Application Software

J.-P. Rosen and A. Strohmeier (Eds.): Ada-Europe 2003, LNCS 2655, pp. 92–101, 2003.
© Springer-Verlag Berlin Heidelberg 2003

(PASW) of the Gravity and Steady-state Ocean Circulation Explorer satellite (GOCE), a scientific satellite under construction for the European Space Agency (ESA). This example points out some general design problems and solution patterns.

The rest of the paper is organized as follows. Section 2 summarizes the real-time requirements and development constraints of the PASW. In section 3 we describe and motivate the tasking rules which are a subset of the Ada Ravenscar profile. The impact of these constraints on the PASW design is discussed in section 4 with the support of examples. Section 5 summarizes the problems, traces them to the Ravenscar rules or to our stronger rules, and discusses the impact on the overall static analysis.

## 2    The GOCE Satellite and Its Platform Application Software

GOCE is a scientific Earth observation satellite being constructed for the European Space Agency by an industrial team led by the prime contractor Alenia Spazio of Turin, Italy. Space Systems Finland (SSF) is implementing the GOCE on-board Platform Application Software (PASW) for the avionics subcontractor, Astrium Space GmbH of Friedrichshafen, Germany.

The main instrument in GOCE is a highly sensitive gravity gradiometer that contains freely flying "proof masses" whose positions are measured and controlled electrostatically. The instrument is used both to measure the gradient of the Earth's gravitational field and to measure the non-gravitational acceleration and rotation of the GOCE satellite due to the Earth's atmosphere and magnetic field. These non-gravitational effects are large, because GOCE will use a very low orbit of about 250 km altitude, but they are compensated by continuously controlled ion thrusters, making GOCE fly a "drag-free" trajectory and maximizing the sensitivity of the instrument.

The PASW runs in the GOCE Central Data Management Unit (CDMU, implemented by LABEN, Italy) on a 24 MHz ERC32 (SPARC V7) processor and is written in Ada using the XGC compiler [1], a variant of GNU Ada. The PASW has two main functions: attitude and orbit control, which keeps GOCE on a drag-free orbit, and the so-called data-handling function, which means the routing of commands, housekeeping data, and scientific data between the ground and various functions in the PASW and in other subsystems on GOCE. This includes sensor data collection, mass-memory storage and replay, monitoring data for anomalies, autonomous response to anomalies, and time-tagged command execution. There is also a facility to upload and execute interpreted control procedures, using the On-board Command Language software from the IDA institute at the University of Braunschweig, Germany.

For the purpose of this paper, the main design problems arise from the input-output channels and the data-flows. Figure 1, below, illustrates them.

The four principal PASW real-time activities are: (1) the attitude and orbit control, which runs at 10 Hz and uses the MIL-bus for sensor and actuator data; (2) the TM down-link from mass-memory, which uses the fast serial lines and should fill all of the TM bandwidth of about 1.2 Mbs; (3) the periodic data collection, which samples the analog and digital lines and other data at 10 Hz; and (4) the routing of TC and TM packets, over the MIL-bus, between the CDMU and the remote terminals. The TC/TM traffic is partly periodic and partly sporadic and is extensively configurable by ground commands. The PASW manages a total of 15 interrupt sources.

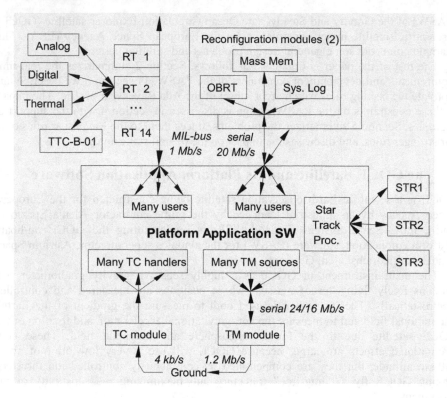

**Fig. 1.** The GOCE CDMU input-output channels and peripherals used by the PASW

Most I/O channels multiplex data and messages of several types to and from different devices and with different real-time requirements. The I/O channels must be shared by concurrent tasks, sometimes in a prioritized way. Moreover, the PASW and the peripheral units have finite buffering resources, and the real-time activities must be paced and synchronized to avoid buffer overflow.

We want to design the PASW to ensure predictable hard real-time performance under a bounded maximum load but to also provide soft-real-time qualities such as improved response times under light load and smoothly degraded performance under overload. These requirements on concurrency and timing may be hard to implement using a strongly restricted subset of Ada tasking. We aim in this paper to use the PASW as a test case for the subset described in the next section.

## 3  The Tasking Profile

The Ravenscar profile (here abbreviated RP) is a restriction of the Ada tasking features that aims to make the real-time behaviour of an Ada program statically analysable. A second aim is to allow a simpler, faster and more easily verifiable kernel (run-time system). The RP was defined at a sequence of IRTAW meetings (see [1] and references therein) and is proposed for ISO standardisation.

Briefly, an RP program contains a statically fixed number of Ada tasks. All tasks start at program boot, never terminate or abort, and communicate only through protected objects. Task entries, rendez-vous and asynchronous transfer of control are excluded. A protected object can have at most one entry, the barrier of the entry must be a single boolean variable local to the object, and the number of tasks waiting on the entry must never exceed one. The "requeue" statement is excluded. Interrupts are statically bound to protected operation handlers.

In the RP, a task can be suspended (wait) only when it calls a protected entry, when it executes a "delay until" statement, or by the explicit suspend and release operations in the predefined package `Synchronous_Task_Control`. The static analysis of an RP program is further simplified if we add two stricter rules:

- A task has a single suspension point.
- A protected entry is called by a single task.

The single-suspension-point rule means that tasks are clearly divided into cyclic or time-triggered tasks that suspend on a delay statement, and sporadic or event-triggered tasks that suspend on an entry call.

The second rule ensures in a static way that no more than one task can wait on an entry. It also implies that each interrupt is statically connected to one task, the task that calls the entry of the protected object that contains the interrupt handler procedure.

We have chosen this stricter profile to make the PASW suitable for the ESA schedulability analysis tools that follow the "HRT" model [3] and for our WCET analysis tool [4][5] without any manual translation from the Ada design to a simpler analysis model. However, the profile may make the design harder to create and understand. Furthermore, the schedulability analysis depends on the activation intervals, offsets and deadlines of the tasks, and these may now be harder to derive from the high-level real-time requirements because the task set is less directly related to the high-level functions. This trade-off is our main subject.

## 4   Design Problems and Patterns

### 4.1  Introduction

In the PASW case, we have found that our profile imposes design constraints and a design style that were not apparent to us from the start. The designer must take a new approach to some common real-time problems. In this section, we try to describe these issues from the point of view of the designer rather than as a list of included or excluded language features. The main issues are the following:

- Abstracting operations that include input/output or timing.
- Reifying input/output actions that use the same channel.
- Pacing data-flow from task to task.
- Queues with release controls.

Below, we will discuss how the tasking design is altered and how the scheduling attributes of the tasks are affected.

## 4.2  Procedural Abstraction Difficulties

The subprogram concept, or *procedural abstraction*, is without doubt the major tool of the software designer, and is itself necessary for the next most important tool, that of subsystems or *modules* as realized in Ada packages.

Traditionally, reading and writing data from or to peripheral devices is done by subprograms, such as `Ada.Text_IO.Put`, or the Unix system services `read()` and `write()`. For example, one GOCE hardware unit contains a real-time clock giving the On-Board Reference Time (OBRT) that the PASW must read to time-stamp telemetry packets and for other purposes. Normally, the designer would define a package `OBRT`, say, and a function `OBRT.Clock` to read and return the time. Our profile forbids this because the clock must be read using an I/O channel, which means that the reading task must wait for an interrupt to signal that the clock read-out is available, and this would introduce a new suspension point in the task.

The program's modular structure must now be based on the flow of data from the input channels, through the processing stages and to the output channels. No procedure can encapsulate a hidden suspension point. A procedure that contains a suspension point must be used carefully and only to be the single suspension point of a task.

This is a very strong constraint on the design. Moreover, it makes the whole design depend strongly on the hardware architecture, specifically on whether a peripheral unit is memory-mapped and so can be used without suspension, or connected to an I/O channel that uses interrupts or delays which imply suspension. If the hardware design changes in this respect, the change cannot be hidden in the lower-level procedures, but propagates to all tasks that use this part of the hardware.

For the PASW we designed a "proxy" clock that can be read without I/O. A cyclic task reads the physical OBRT clock (using I/O) and records the value together with the corresponding value of `Ada.Real_Time.Clock`. The "proxy" `OBRT.Clock` function uses `Ada.Real_Time.Clock` and converts the value to the OBRT scale. This works for the OBRT, but would not work for less predictable values.

## 4.3  Starter Tasks and Continuer Tasks

Consider the PASW task, referred to above, that cyclically reads the OBRT clock using an I/O channel. The design of this task seems simple: first use a delay statement to wait for the next cycle, then command the channel coupler to start the I/O, wait until the I/O is complete, and read the data from the channel buffer. The problem is that this design involves two suspension points: the delay statement and the wait for I/O completion which would usually be implemented as an entry call in the following way:

```
loop
   Compute Next_Time;
   delay until Next_Time;
   IO_Channel.Start (reading the OBRT);
   IO_Channel.Wait_For_Completion (Result);
   Check and use Result;
end loop;
```

Here we assume that `IO_Channel.Start` starts the reading of the OBRT clock and `IO_Channel.Wait` waits for it to finish (wait for interrupt). This code is not allowed by our profile because it contains two suspension points: `delay until` and `IO_Channel.Wait_For_Completion`.

Our design splits the task into two tasks: the "starter" task that waits for the next cycle and starts the I/O, and the "continuer" task that waits for the I/O to complete and then checks and uses the OBRT value read. The "starter" task body is:

```
loop
    Compute Next_Time;
    delay until Next_Time;
    IO_Channel.Start (reading the OBRT);
end loop;
```

The "continuer" task body is:

```
loop
    IO_Channel.Wait_For_Completion (Result);
    Check and use Result;
end loop;
```

Both tasks satisfy our profile, but the logical flow of the process is now harder to understand. Moreover, usually the "starter" must pass some data to the "continuer" to describe the I/O operation and what to do with the result. These data may be passed in a global variable, or via the IO_Channel object if protected access is needed.

This division into "starter" and "continuer" resembles I/O drivers in operating systems. Drivers are usually split into a subprogram that is called from the user process to start an I/O, and an interrupt handler that is executed later in the system context.

The PASW design contains several such groups of closely cooperating tasks. We call these groups *processes*. This splitting of tasks creates a problem for the static analysis because the high-level real-time requirements are usually expressed as a deadline on the whole process, while our schedulability analysis only ensures deadlines for each task. The analyst must split the process deadline into task deadlines and I/O deadlines, which is not trivial to do. Furthermore, the splitting usually creates a set of tasks with strong but implicit precedence relationships and timing offsets and if these are not taken into account the schedulability analysis may be too pessimistic. For example, in a typical PASW process it is impossible for both the "starter" and "continuer" tasks to be ready (schedulable) at the same time, so they cannot interfere.

## 4.4   Reifying I/O Actions

The I/O channel that accesses the OBRT is also the channel for several other HW registers, the mass memory and the "safeguard" memory which holds configuration data and a log of system events such as anomaly reports. The channel is thus used by other processes. Each process needs to start various kinds of I/O operations and wait for them to finish. The channel can only run I/O operations serially, one at a time, so mutual exclusion is needed from start to finish of the I/O operation.

In full Ada, we would define a protected object with "start I/O" entries for starting each kind of I/O operation, with entry barriers that ensure that the channel is idle. The object would also contain an interrupt-handling procedure and a "wait for I/O to stop" entry which waits for the I/O to end and then releases the exclusive access to the channel by opening the barriers on the "start I/O" entries.

The RP forbids such a protected object because it has more than one entry and because many tasks might be queued on the same entry. Moreover, splitting the tasks into "starter" and "continuer" tasks as explained above is no longer enough to satisfy our single-suspension-point rule, because the calls of the "start I/O" entries would also add suspension points to the "starter" tasks. We need two new mechanisms: one to serialize the I/O operations without suspending the "starter" tasks, and another to avoid many "continuer" tasks waiting on the same "wait for I/O to stop" entry.

The serialization problem suggests a queue to store waiting I/O requests. The enqueue operation can be non-suspending if the queue is given enough space to hold one request from each "starter" task (a static number in RP). The queue is consumed by a single "driver" task that actually starts the I/O operations.

The single-waiter problem suggests creating one protected synchronization object for each "continuer" task, with each I/O request referring to the associated synchronization object. The synchronization object has a "signal" operation and a "wait for signal" entry, and also buffers one I/O result for the "continuer" task.

This means that I/O requests and I/O results must be encoded or *reified* as data objects that contain the kind of I/O to be performed, buffer pointers, synchronization object pointers and so on. This resembles how operating systems reify I/O requests, building and enqueuing "I/O control blocks" to be handled by device drivers.

In the PASW OBRT example, we need a synchronization object that is accessible to the "starter" and "continuer" tasks:

```
OBRT_Synch : aliased IO_Channel.Synch_Object_T;
```

where we now assume that IO_Channel is a package which declares, among other things, the protected type Synch_Object_T and types for I/O requests and results. The "starter" task is:

```
loop
    Compute Next_Time;
    delay until Next_Time;
    Request := (
        Kind  => IO_Channel.Read_OBRT,
        Synch => OBRT_Synch'Access);
    IO_Channel.Enqueue (Request);
end loop;
```

The "driver" task is triggered when there are enqueued I/O requests and the channel is idle. It takes a request from the queue and starts the I/O. When the I/O is done, the interrupt handler signals the synchronization object referred to in the Synch component of the request, passing the I/O result as a parameter. The "driver" task can then start the next I/O. The "continuer" task waits on the synchronization object:

```
loop
    OBRT_Synch.Wait_For_Completion (Result);
    case Result.Kind is
        when IO_Channel.Read_OBRT =>
            Check and use Result.OBRT;
        when others =>
            Handle unexpected kind of result;
    end case;
end loop;
```

This design complicates the static analysis in two ways. Firstly, the time offset between the "starter" task and the "continuer" task depends on the variable number and kind of requests in the queue. Secondly, the interrupt handler calls the synchronization object via an access (the Synch component ) which makes it hard to find the call graph. We can avoid this "dynamic" call only if the IO_Channel package knows more about its client processes. For example, if we know that only one process will ever read the OBRT, the synchronization object for this kind of I/O can be encapsulated in IO_Channel and statically called from a case statement based on the Kind component of the I/O request. However, this would increase the architectural coupling. The PASW design currently uses the dynamic approach.

## 4.5    Pacing Data Flow

A task is generally part of a data-flow network, taking input from "upstream" tasks or other data sources and generating output for "downstream" tasks or other data sinks. The task should be activated only when the sources can provide input. We assume that data must not be lost. If the sinks are not immediately ready to receive the resulting output, the task is often designed to suspend itself until the sinks can accept the output. Our tasking profile does not allow this, because the task would have two suspension points: one waiting for input data, the other waiting for output capacity. Adding an output buffer queue only shifts the problem to suspending on a full queue.

The "downstream" suspension is avoided if the design ensures that there is *always* capacity to handle or buffer the output. However, this is not robust against overload.

Another possibility is to activate the task only when there is output capacity as well as input data. This is possible only for sporadic tasks and only when the upstream sources can buffer data. This method propagates the downstream congestion upwards, which may improve the resource utilization if the input is smaller than the output.

For example, the PASW can be sent telecommands to dump stored telemetry from the system log to ground. The dumping process contains a task that reads system log memory. This task is activated only when a dump TC is in progress (upstream) *and* there is buffer space (downstream) to handle the TM, all the way down to the I/O channel leading to the TM hardware. The I/O interrupt on this channel signals that some buffer space is released. This signal propagates upwards in the task chain and eventually activates the task to read more TM from the system log memory.

If more dump telecommands arrive when a dump is in progress, they are held in a queue. Since a dump TC is typically much smaller than the TM data in the dump, much less memory is needed for the TC queue than would be needed for all the TM, if all the data were read from system log memory at once.

This "downstream" data-flow pacing complicates the schedulability analysis because the interval between activations of a sporadic task is now determined both by "upstream" events and "downstream" events.

## 4.6    Queues with Release Control

Data queues are generally used to convey data from task to task. In full Ada, a queue would normally be coded as a protected object that contains an array holding the queued data and two operations, Put and Get, to enqueue and dequeue data. Both operations would usually be entries, with Put suspending the upstream producer task until the queue has room for more data, and Get suspending the downstream consumer task until the queue has some data to be taken out. This provides flow control (execution pacing) for both producer and consumer. The RP, however, allows only one entry, so either Put or Get must be just a protected procedure with no barrier. The queue can then pace its producers or its consumers, but not both.

In our profile only one task can be paced in this way, since only one task can call the entry. Queues in the PASW usually have multiple producers but one consumer, so we make Put a procedure and Get an entry.

To take downstream capacity into account, each queue has a *release quotum* which sets the number of elements that can be released. The Get barrier is logically "(queue not empty) and (release quotum > 0)" but the RP requires that this condition is held in one Boolean variable that is updated whenever any part of the condition changes. Each call of Get decrements the release quotum. The release quotum is initialized to reflect the initially available downstream capacity. When the downstream capacity is replenished (e.g. buffer space is released), the release quotum is increased by means of a protected procedure Release of the queue object.

### 4.7 Complete Process Example

We can now describe a complete PASW process, for example the execution of system log dump telecommands. When such a TC arrives, it is placed in a TC queue which initially has a release quotum of 1 (one). The first TC to arrive in the queue is thus immediately released and activates a "starter" task which starts to read the first block of system log data into a ring buffer in RAM. Even if there are more TCs in the queue, the "starter" task is not activated again because the release quotum is now zero.

When a block of system log data has been read, the I/O interrupt activates the "continuer" task which passes the data to a "packet scanner" task and starts reading the next data block. The "packet scanner" divides the data stream into TM packets and sends them to a specific queue in the set of TM queues. The "sender" task selects packets from these queues according to a priority algorithm and transmits them on the I/O channel to the TM hardware.

Each task in this chain is chiefly activated by the arrival of inputs from the upstream task, but the activation is inhibited until downstream capacity is available. Thus, the "sender" task is inhibited when the TM I/O channel is busy; the "packet scanner" task is inhibited when the TM queue for system log data is full; the "continuer" task is inhibited when the ring buffer is full and also when the system log has been fully read; and the "starter" task is inhibited (via the release quotum of the TC queue) until the "continuer" task has finished reading the system log. The RP one-entry limit means that the inhibition condition must be embedded in the barrier of the upstream entry and new protected procedures (such as Release) are needed by which the downstream tasks can control the inhibition.

## 5  Summary and Conclusions

We have described how the tasking design of the PASW has been influenced and constrained by the chosen tasking profile, which is the Ravenscar profile further limited to one suspension point per task and one call per protected entry. This profile was chosen to match the analysis tools supported by ESA and SSF.

We expected the profile to complicate the design and the parametrization of the analysis, but the impact was larger than we thought, mainly because the single-suspension-point rule limits procedural abstraction of suspending operations. This forces us to split tasks. The RP rules on protected objects increase the number of protected objects. Our limit of one call per entry creates dynamic accesses to protected objects which complicates the static analysis of the call-graph and task interactions.

The greater number of tasks demands more memory for task stacks. The number of task switches is not increased, but the CPU time for a switch may grow since the kernel data structures are larger. However, we do not expect this to be critical for the PASW. The increase in protected objects seems to have small time and space impact.

Much more severe is the impact on the algorithm design. Because the tasking profile limits the syntax and control flow within a task, the designer often cannot use control flow to express algorithms, but must instead use data structures and state variables. What was a loop within a task body, clearly shown by "for" and "loop" keywords, now becomes a cycle of states with the loop-counter hidden in a protected object and often accessed by several tasks. The algorithms are harder to understand and it is harder to find loop-bounds by automatic analysis. We also suspect that WCET bounds may become more pessimistic because the actual execution paths will depend more on global data than on local control flow and local data. For functional verification, the structural test coverage metrics become less useful and may need to be supported by data-state coverage metrics.

Splitting tasks also means splitting real-time requirements. A single function of the application, such as executing one telecommand, for which the user has a well-defined response-time requirement, is split into a process that invokes several tasks, possibly several times. The designer must manually define the real-time task attributes so that schedulability implies the required end-to-end response-time.

We have described some design patterns that can help the designer. However, the impact of the single-suspension-point rule is so severe that we would really like to get rid of it in favour of the full RP (perhaps with the single-entry-call rule). This would require either better analysis tools, able to analyse an RP design, or a method to translate an RP Ada task-set into a set of "analysis-level" model tasks that obey the single-suspension-point rule and can be analysed by our current tools.

Perhaps we were too ambitious and should have divided the PASW design into a hard-real-time part, using our strict profile, and a soft real-time part, using the RP. The problem here is that the same I/O channels are used by both parts, making it hard to analyse the parts separately.

# References

1. http://www.xgc.com
2. Burns, A., Dobbing, B., Vardanega, T.: Guide for the use of the Ada Ravenscar Profile in high integrity systems. University of York Technical Report YCS-2003-348, January 2003
3. Burns, A., Wellings, A.J.: HRT-HOOD: A design method for hard real-time Ada. Real-Time Systems, 6(1), 73-114, 1994
4. http://www.bound-t.com
5. Holsti, N., Långbacka, T., Saarinen, S.: Using a Worst-Case Execution Time Tool for Real-time Verification of the DEBIE Software. Proceedings of the Conference 'DASIA 2000 - Data Systems in Aerospace', Montreal, Canada, 22-26 May 2000, (ESA SP-457, September 2000) 307-312

# Booch's Ada vs. Liskov's Java: Two Approaches to Teaching Software Design

Ehud Lamm

The Open University of Israel, Max Rowe Educational Center, P.O.B 39328
Ramat-Aviv, Tel-Aviv 61392, Israel
ehudla@openu.ac.il

**Abstract.** We study two textbooks for teaching undergraduate software engineering, both focusing on software design and data abstraction in particular. We describe the differences in their didactic approaches. We analyze how the subject matter is influenced by the choice of programming language, one book using Ada and the other book using Java. Java is a relatively new candidate for teaching software engineering. How does it compare to Ada?

**Keywords:** education and training, evaluation and comparison of languages, object-oriented programming, software development methods

## 1 Background

The undergraduate software engineering course at The Open University of Israel, called *Software Engineering with Ada* is based on a Hebrew translation [2] of Booch's book of the same name [1].[1] This course has been offered since 1989. I have been involved in teaching this course since 1999. The course is elementary and focuses on classic programming techniques in Ada (e.g., designing ADTs, generic units, and tasking). Since the book deals with Ada83 the course study guide includes a detailed chapter about tagged records and inheritance in general. The course is practical and the students are expected to do a fair amount of coding in order to solve their problem sets.

We are in the process of thinking whether this course provides the best introduction to software engineering for our computer science students. Changes to the curriculum make it necessary to recheck how the course interacts with other courses (e.g., the introduction to CS course and the data structures course).

In this paper I will try to analyze the differences between the existing course and a possible replacement based on Liskov's *Program Development in Java* [10] which is used in some other schools (e.g., MIT).

---

[1] A third edition of *Software Engineering with Ada* coauthored by Doug Bryan was published in 1993.

J.-P. Rosen and A. Strohmeier (Eds.): Ada-Europe 2003, LNCS 2655, pp. 102–112, 2003.

## 2  Reasons for Change

Booch's book is example oriented (the book includes five extended examples), but lacks theoretical foundations. Teaching the course for several semesters led me to believe that the lack of theoretical grounding makes learning harder for students (e.g., the notion of an ADT *invariant* is not defined nor is the algebraic interpretation of ADTs). Essentially, we want students to come and appreciate the notion of software abstraction, but this concept is never explained in detail or analyzed sufficiently.

A more practical reason for replacing the current course is, alas, the use of Ada. First, our introduction to CS which was previously Pascal based now has a C++ version[2] and a Java course is planned. Students arriving at the software engineering course have a hard time adjusting to the Pascal-ish syntax and terminology of Ada (e.g., the syntactic distinction between procedures and functions).

Second, students want to learn skills that are useful in industry. Their impression is that Ada is hardly ever used, which makes them question the choice of language (causing endless debates). Since the software engineering course is not mandatory, many students choose not to take it, preferring other advanced programming courses. This is problematic, since several important concepts are only taught in the software engineering course. The paradoxical result is that many students who plan to work as programmers and software designers in industry, skip the course most relevant to their needs.

## 3  Course Goals

In this section I discuss the *main* objectives I see for the undergraduate software engineering course.

*The course should develop program design skills, teaching basic software engineering concepts needed for exploring and analyzing software design alternatives.*

Students are expected to have limited programming experience, and are likely not to have been part of an industrial scale software project. Most students take the introductory software engineering course after completing only two programming courses: Introduction to Computer Science, and the Data Structures and Algorithms course. Thus, the course should provide opportunities for practical experience, in the form of exercises and – if possible – a small project. Theoretical concepts (e.g., information hiding, robustness) should be introduced and explored in the context of the programming assignments. The practical exercises should provide opportunities for software testing and include small maintenance tasks.

As this is an academic course, students are expected to be able to reason about simple design issues using accepted terminology, and to express their reasoning succinctly and precisely.

---

[2] C++ is used in this course as a type safe C. No OOP is taught.

## 3.1    Software Abstractions

The course focuses on modularity. Students should be able to design a module structure from a given problem statement, and design adequate module interfaces.

Thus, the course elaborates on the notion of software abstractions, and in particular data abstractions. The students are expected to feel comfortable with procedural abstractions before taking the course, but the term itself, and the connection to software abstraction in general, should be introduced. Control abstractions (e.g., non deterministic choice) are for the most part beyond the scope of the course.[3]

The course is not an exhaustive survey of software engineering concepts and methodologies, nor does it cover the full scope of software design techniques. Instead, it focuses on the fundamental notions of abstraction and modularity.

# 4    Didactic Approach

Booch's and Liskov's books follow a similar didactic approach. They give small code examples (snippets), and some larger scale design examples. Some of the noticeable differences:

- **Explicit vs. Implicit Learning.** Liskov defines several concepts that help reasoning about software abstractions. The most important ones are: adequacy ([10], Sect. 5.8.3), invariant, and abstraction function.
- Throughout the book Liskov gives tips and rules for effective design (these are highlighted by using grey background). These can be seen as beneficial, but can lead students to a cookbook approach to programming. They should realize it is not enough to "follow the rules."
- Booch gives a useful list of software engineering goals and principles.
- Liskov dedicates a whole chapter to procedural abstraction ([10], Chap. 3)

In our experience students find the existing course quite difficult. Some examples of tasks students find difficult:

- Designing abstraction interfaces. Not surprisingly, this is the main obstacle for students. Booch talks about this extensively, in each of the design problems. Liskov provides fewer examples, but summarizes the issues nicely in Sect. 5.8, *Design Issues*, which discusses mutability, operation categories and adequacy.
- Proposing alternative designs. Students often fall in love with the first design they think of, and find it impossible to think of competing designs for solving the problem at hand.

---

[3] Control abstraction is often discussed in courses about programming language theory and design.

– Comparing different designs. Even when presented with two alternative designs many students find it hard to compare the designs, unless given a list of specific criteria (e.g., "How sparse does the matrix have to be, so as to justify the space overhead of using pointers?")
– Succinctly expressing their design rationale, especially doing so precisely.

## 5   Language Differences

The main question this paper tries to attack is the impact of the choice of Java versus Ada on the ease of teaching and on achieving the goals of an undergraduate software engineering course. We do this by comparing language features we find relevant.

Though used mainly as a vehicle for introducing language agnostic programming techniques, the programming language chosen is a major influence on the design of the course, since the order of presentation has to be consistent with relations between language constructs (e.g., in Ada, tagged types rely on packages).

It should be noted immediately that language comparisons are problematic, inflammatory, and often biased. Comparing languages by listing language features is in itself a questionable technique, seeing as there are often several possible language features that can be used to achieve any one design goal – each feature with its own advantages and disadvantages. We are not trying to compare the languages in general. We highlight language features that seem important from our experience teaching software engineering with Ada.

### 5.1   Problems with Using Java

Since we are thinking of moving from Ada to Java, we start by noting Ada features we will miss.

*The Type System.* Perhaps the most glaring difference is in the type system. Ada has a rich and expressive type system that structures the whole language. Ada makes it easy to understand the software engineering benefits of strong typing. The integration of the type system with generic units and inheritance is very enlightening (e.g., by showing the difference between static and dynamic polymorphism).

Java doesn't support records, which are so basic that Liskov has to define and explain them, in the chapter dealing with software abstraction ([10], Sect. 5.3.4). They are, of course, a simple language feature obvious to any student with a Pascal, C, or Ada background.

*Reference Semantics.* Java uses reference semantics. Reference semantics are especially bothersome in the context of a course dealing with data abstraction, since they easily lead to abstraction breaking. Liskov is, of course, well aware of this problem. Section 5.6.2 deals with "exposing the rep" and gives examples of this problem. Chapter 2, which is a quick Java refresher, also covers the pertinent Java semantics.

*Genericity.* Another crucial feature is Ada's support for generic units. Indeed, Liskov dedicates a whole chapter to the software design notion of *polymorphic abstractions* ([10], Chap. 8), whereas Booch's related chapter is focused on the Ada feature generic units ([1], Chap. 12). It is important to note that Booch deals with Ada83, and thus does not talk about using inheritance for polymorphism. This omission can be fixed with supplemental material, which indeed we provide for our students. Java's lack of genericity, however, is harder to overcome. If we move to Java, and use Liskov's book, it is likely that when genericity is added to Java [9], we will have to provide supplemental material covering genericity, until the book is revised or a new book is chosen.

Teaching genericity (i.e., *parameterized polymorphism*) before introducing inheritance seems to ease the understanding of both topics. The compile-time nature of Ada generic units makes them an ideal stepping stone to the more complicated mixture of dynamic and static properties that exists in the presence of inheritance (e.g., class wide programming and dispatching).

The importance of polymorphism, from a software engineering point of view, is apparent. It is an important design technique, that enhances reuse, clarity and flexibility. The question is whether polymorphic abstractions based on inheritance as provided by Java, are good enough for the purpose of our software engineering course, or is a templating mechanism as provided by Ada generic units essential.

Genericity is a tool for building parameterized units. Parameterization is an important abstraction method. Ada generic units, indeed the Ada syntax, emphasize this fact.[4]

Coupled with tagged types for dynamic polymorphism (i.e., dispatching), the Ada model which allows for nesting of generic and non-generic program units is very appealing, see figure 1 (a different scenario, which is handled with exclusive use of inheritance can be seen in the exercise shown in figure 2. Students are expected to be able to understand the differences between the two designs.). However, it seems to me that the treatment of polymorphic abstractions in Liskov's book is acceptable. Much as I like Ada's generic units, I think we can achieve the course goals without them.[5]

*Concurrency.* Ada provides higher level concurrency constructs than Java. Booch takes advantage of this and dedicates two chapters to tasking (a technical chapter and a detailed design problem). Liskov does not cover concurrency.

By teaching Ada's tasking constructs we are able to explore interesting kinds of software abstractions. For example, we show active objects (e.g., a self sorting array for fast lookup). Ada tasks are also helpful for showing simple parallel algorithms.

---

[4] Visible discriminants of private types, are another useful way to introduce students to parameterization.

[5] Java's interfaces will, of course, come in handy.

In this exercise students implement a *Set* ADT.

1. Write a generic package, exporting an abstract tagged type Set. The type should provide the standard Set interface. The type of the elements in the set will be provided as a generic parameter.
2. Write a child package, in which you inherit from the abstract Set type, and implement the set type as a sorted linked list. *Note: The < parameter is only required by the child package.*
3. Compare the uses of inheritance and genericity in this exercise.

Students are often asked to use an ADT they wrote previously (e.g., a TRIE) to implement the abstract type (layered design).

**Fig. 1.** Genericity and Inheritance

In this exercise students implement a polymorphic *Stack* ADT.

1. Write a package exporting an abstract tagged type Stack. The type should provide the standard Stack interface. The Stack should be polymorphic: it should be possible to store values of different types in the same Stack *Note: The exercise explains how to achieve this by defining a parent type for all items. Genericity is used to wrap any type, in a type derived from this parent type.*
2. Write two child packages, in which you inherit from the abstract Stack type. One package will implement the Stack using a linked list. The second package will implement the Stack using an array.
3. Inheritance is used for two reasons (and to achieve two different goals) in this exercise. What are these uses, and how does inheritance help achieve them?

**Fig. 2.** Interface inheritance, and heterogeneous collections

Combined with inheritance and other Ada language features, Ada tasks make it easy to show simple patterns for concurrent programming (e.g., thread pools, synchronization objects [3] etc.)

It is possible to do similar things in Java, but Ada task and protected types make implementing them much easier.

*Miscellaneous.* Two of the mini-projects I give involve building program analysis utilities. In one, the students have to traverse an Ada source code tree, and collect interesting metrics (e.g., number of procedures, number of with statements etc.). In the second project, the students have to display the package dependency graph. Some of the students were shown how to use ASIS (*Ada Semantic Information Specification*). This reduced the effort needed to complete the project, and learning the ASIS API was a useful learning experience. Though similar tools for Java do exist, ASIS is standard Ada, and the Gnat implementation is relatively easy to use.

## 5.2   Problems with Using Ada

Perhaps the most problematic thing about using Ada is the relative lack of freely available tools, as compared to C++ and Java. Ideally, students learning about software engineering should come in contact with various kinds of CASE tools. Because of the nature of the course, I am thinking specifically about IDEs, testing tools and frameworks, measurement utilities, diagraming and code generation tools, and pretty printers.

There are tools of these kinds for Ada, of course. It would be nicer if there was a larger choice, so we could have students compare tools etc. The main problem, however, is to package a *stable* set of quality tools.

Tools we currently use are: Gnat and AdaGide (our basic setup). Students occasionally use GRASP for pretty printing.

*Reusable Code.* Available Ada libraries are not always easy to find and use. I would like to encourage students to use reusable code as much as possible. However, of the available libraries some are pretty hard to use, and require knowing advanced Ada techniques, which students learn towards the end of the semester. Perhaps the best example is using the Booch Components (the Ada95 version), which require knowing about tagged types and nested generic instantiations. This is a shame, since using standard data structures is a classic reuse scenario, which comes naturally when learning about data abstraction. For example, a standard exercise we give is building a priority queue, which is based on an array of linked lists of elements. Naturally, standard linked lists can be reused, thus saving the students time, and teaching them about the process of reusing publicly available code. Indeed, I want students to compare several reuse scenarios: The priority queue can be seen as an array (indexed by priority) of queues, rather than as an array of lists. It can also be seen as a map from priorities to queues. Having a standard collection library, as part of the Ada standard library in Ada200X, would make giving such exercises a bit easier, and could also help improve Ada textbooks.

Students are encouraged to explore the Internet for reusable code. In my experience this often leads to better educational results than giving students code to reuse. The main advantage with this approach is that it teaches about some of the problems with achieving reuse in real life (e.g., finding the appropriate package is not easy, not all kinds of documentation help clients use a package, packages you find may interfere with each other etc.)

Liskov's book makes use of the extensive Java library as a source of examples (e.g., [10], Sect. 7.10).

I currently recommend to students wanting to build GUIs, to use JEWL [6] (some students have used GWindows). A simplified unit testing framework was recently made available to students.

I have plans for using AWS ("Ada Web Server") and XML/Ada[6] for extended exercises (mini-projects) which will involve building web based applications (e.g., a browser based user interface, and an RSS news aggregator[7]).

---

[6] Both available from *http://libre.act-europe.fr*

[7] See *http://radio.userland.com/whatIsANewsAggregator*

*Interfaces.* One of the important objectives of the course it to teach students to write interface oriented code. By that we mean polymorphic code that works for a variety of abstractions that supports a common interface. A simple example may be printing the contents of a container (e.g., tree) using an active iterator. The print procedure should be agnostic to the details of the container, and be compatible with any container that provides an iterator with the required set of operations.

Ada provides quite a few ways to specify the interface such a routine relies on. At times generic formal types are enough (e.g., when the routine works for any *array*). Another technique is to pass a private formal, and specify additional formal subprogram parameters. The interface can be encapsulated as a signature package (and passed as a formal package parameter). Another approach is to represent the interface as an abstract tagged type, in which case the routine may at times be written as a class wide routine (however, the lack of multiple inheritance makes this approach problematic).

All these techniques have their uses, of course.[8] But the fact that the basic and essential concept of "interface" can be represented programmatically in so many ways can be confusing to students. It is not possible to restrict our attention to only one of the relevant Ada constructs, since their uses are quite different. Specifically, package specifications must obviously be introduced, but this interface definition is not appropriate for parameterization purposes. Likewise, signature packages cannot be the only interface definition discussed, because packages in Ada are not first class and this prevents useful design techniques [7]. The usefulness of abstract tagged types as interfaces is marred by the lack of multiple inheritance.

Java, unlike Ada, provides a special language construct for interfaces.

*Constructors and Destructors.* Data abstractions often require special set up and cleaning code. This functionality can be implemented by defining constructors and destructors, which are automatically called by the language. In Ada this is done using `Ada.Finalization`. This requires knowledge of tagged types. Combining controlled types with genericity is awkward, since the generic units must then by instantiated at library level. The rationale for this technical aspect of the language is lost on students, and they find the compiler error messages hard to understand.[9]

When students first learn about building ADTs, they often want to overload assignment. Instead of doing so they are told to write a `Copy` routine, since at that point they haven't yet learned about tagged types. When they are taught about `Ada.Finalization` it turns out that changing their packages to use controlled types requires a fair amount of work.

---

[8] We, in fact, try to teach students to choose the most appropriate technique in each case.

[9] Recall that the course deals with software engineering principles, and we do not have enough time to explore subtle language specific issues.

There are situations where manual cleanup is impossible or hard to do correctly, and the use of finalization is especially important (e.g., multitasking programs).

Coupled with the fact that Ada doesn't mandate garbage collection[10] the issue of manual memory management must be addressed. In a sense, this is a good thing, since memory management is something programmers are expected to understand. However, the effort to do it correctly can be substantial.[11]

*Visibility.* It has been argued that the methods of controlling visibility of methods and fields to subclasses (i.e., types extended via inheritance) are less than ideal. Other languages, including Java, support designating properties as either public, private or protected. The `protected` designator granting visibility to subclasses.

In Ada, you achieve this sort of visibility either by using child packages (that have visibility over the parent's private part) or by including the derived type in the parent package.[12] Students may find these techniques confusing and awkward. Moreover, they are taught to include as few declarations as possible in the `private` part of package specifications (to avoid recompilations) and put helper routines that are implementation specific in the body of the package. When hierarchical units are introduced, they are told that in order to provide for easy extendibility some routines are better declared in the private part, so as to allow child unit visibility. Another classic example is the state variables of an abstract state machine, which are defined in the body, but must be moved to the `private` part, if the package is to be extended. These situations are confusing, because when you see a declaration in the private part, there is no way to tell if it is there by mistake, or whether it was placed there for a reason.

The appropriate uses of the explicit `protected` definitions in Java are discussed in [10] (sidebar 7.5, for example). Personally, I find the Ada model more flexible, however this issue was raised by instructors, and I did encounter students that were confused by it.

Perhaps hierarchical units should be introduced immediately following the chapter on packages (as done in [5]), instead of being introduced after tagged types, the way we do today. This may help overcome some of the problems discussed above. However, Ada's lack of *explicit* `protected` definitions is a language expressiveness and readability concern.

*Miscellaneous.* Ada95 was developed from Ada83 and is mostly backwards compatible. This causes redundancy. We teach implementing ADTs as encapsulated private types exported from packages, making no use of inheritance. Students

---

[10] Students are often confused by this, failing to grasp the difference between what the language specifies, and implementation particulars.

[11] Garbage collection is known to help with modularity, due to the complexity of manually managing memory between module boundaries (e.g., because of aliasing). Discussing this issue is beyond the scope of this paper.

[12] A more complicated scenario involves genericity.

are then shown how to implement ADTs that can be derived from, leaving them wondering when to use non-tagged ADTs.

Ada's inheritance is not class based. This has important advantages (e.g, no need for special treatment of friend classes, binary methods are more readable etc.) However, since the students are likely to move to a class based object oriented programming language (either C++ or Java), this can be seen as a liability. However, from a didactic point of view, this difference is a bonus, since it makes students review their implicit assumptions, especially those who have prior OOP experience.

Ada exceptions, that are used as part of an abstraction interface, are declared in the package specification. However, there is no way to specify in Ada which exceptions are raised by each routine. We found that requiring students to specify this information in comments helps them think about using exceptions as part of an interface. It would be better if this information could be specified in the code and checked by the compiler.[13] This would also improve integration between exceptions and signature packages.

### 5.3   Features Missing from Both Languages

Both languages are missing some features that would have enriched the course. First and foremost is *Design by Contract* (DbC) [8] which is especially important for a course focused on data abstraction. There are several DbC tools for Java, but contracts are not part of the language. Indeed, Liskov consistently provides preconditions and postconditions, in the form of source code comments (Booch doesn't do this).

Other language features that would have been helpful: anonymous and first class functions, and laziness. These features help support a style of programming closer to functional programming. For example, they help create high order functions, which are useful and interesting abstraction and reuse mechanisms. It is possible to program in this style in both Ada and Java, but in both languages this style of programming is quite awkward. Laziness provides an alternative and often useful solution to the problem of iterating over data structures (a common data abstraction issue, for which Liskov dedicates an entire chapter, [10] Chap. 6). Another classic use is for defining infinite data structures (e.g., streams), which can be used to implement powerful abstractions, and which are quite hard to implement otherwise [4].

## 6   Conclusions

The books discussed [1,10] are very different in outlook. Liskov's book is much more suited for academic use, as it provides tools for reasoning about data abstraction, and software abstraction in general. However, this difference is not the result of the choice of programming language.

---

[13] Java has checked and unchecked exceptions, which give the programmer finer control on the way exception consistency is checked ([10], Sect. 4.4.2).

The problems encountered with Ada are mainly the popularity of the language, and library and packaging issues. The problems with Java are with essential properties of the language (e.g., reference semantics). Though Java is a new player in the field, Ada still seems at least as good for teaching software engineering.

The decision whether to make the move to Java has not been made yet. If it will be made, it will not be the result of shortcomings of the Ada language.

**Acknowledgements.** I had valuable discussions with instructors teaching Open University course 20271, *Software Engineering with Ada*.

# References

1. Grady Booch. Software Engineering with Ada. The Benjamin Cummings Publishing Company, 2nd edition, 1987.
2. Grady Booch. Software Engineering with Ada (Hebrew). The Open University of Israel, 1989. Translated from the second edition (1987).
3. Alan Burns and Andy Wellings. Concurrency in Ada. Cambridge University Press, 2nd edition, 1998.
4. Arthur G. Duncan. Reusable Ada libraries supporting infinite data structures. In Proceedings of the annual ACM SIGAda international conference on Ada, pages 89–103. ACM Press, 1998.
5. John English. Ada 95: The Craft of Object-Oriented Programming. Prentice Hall, 1996.
6. John English. JEWL: A GUI library for educational use. In Dirk Craeynest and Alfred Strohmeier, editors, Reliable Software Technologies – Ada-Europe 2001, volume 2043 of Lecture Notes in Computer Science, pages 266–277. Springer-Verlag, May 2001.
7. Ehud Lamm. Component libraries and language features. In Dirk Craeynest and Alfred Strohmeier, editors, Reliable Software Technologies – Ada-Europe 2001, volume 2043 of Lecture Notes in Computer Science, pages 215–228. Springer-Verlag, May 2001.
8. Ehud Lamm. Adding design by contract to the Ada language. In Johann Blieberger and Alfred Strohmeier, editors, Reliable Software Technologies – Ada- Europe 2002, volume 2361 of Lecture Notes in Computer Science, pages 205–218. Springer-Verlag, June 2002.
9. Sun Microsystems. JSR 14 – Add Generic Types To The Java Programming Language. http://jcp.org/jsr/detail/14.jsp.
10. Barbara Liskov with John Guttag. Program Development in Java. Abstraction, Specification, and Object-Oriented Design. Addison-Wesley, 2001.

# A Comparison of the Asynchronous Transfer of Control Features in Ada and the Real-Time Specification for Java[TM]

Benjamin M. Brosgol[1] and Andy Wellings[2]

[1] Ada Core Technologies, 79 Tobey Road, Belmont MA 02478, USA
brosgol@gnat.com
[2] Dept. of Computer Science, University of York, Heslington, York YO10 5DD, UK
andy@cs.york.ac.uk

**Abstract.** Asynchronous Transfer of Control ("ATC") is a transfer of control within a thread,[1] triggered not by the thread itself but rather from some external source such as another thread or an interrupt handler. ATC is useful for several purposes; e.g. expressing common idioms such as timeouts and thread termination, and reducing the latency for responses to events. However, ATC presents significant issues semantically, methodologically, and implementationally. This paper describes the approaches to ATC taken by Ada [2] and the Real-Time Specification for Java [3,4], and compares them with respect to safety, programming style / expressive power, and implementability / latency / efficiency.

## 1 Introduction

Asynchronous Transfer of Control is a rather controversial feature. It is methodologically suspect, since writing correct code is difficult if control transfers can occur at unpredictable points. It is complicated to specify and to implement, and in the absence of optimizations ATC may incur a performance penalty even for programs that don't use it.

Despite these difficulties, there are several situations that are common in real-time applications where ATC offers a solution:

**Timing Out on a Computation.** A typical example is a function that performs an iterative calculation where an intermediate approximation after a given amount of time is required.

**Terminating a Thread.** An example is a fault-tolerant application where because of a hardware fault or other error a thread might not be able to complete its assigned work. In some situations the termination must be "immediate"; in other cases it should be deferred until after the doomed thread has had the opportunity to execute some "last wishes" code.

---

[1] We use the term "thread" generically to refer to a concurrent activity within a program. When discussing a particular language's mechanism we use that language's terminology (e.g., "task" in Ada).

J.-P. Rosen and A. Strohmeier (Eds.): Ada-Europe 2003, LNCS 2655, pp. 113–128, 2003.
© Springer-Verlag Berlin Heidelberg 2003

**Terminating One Iteration of a Loop.** The ability to abort one iteration but continue with the next is useful in an interactive command processor that reads and executes commands sequentially, where the user can abort the current command.

It is possible, and generally preferable stylistically, to program such applications via "polling" (the thread can check synchronously for "requests" to perform the given action, and take the appropriate action). However this may introduce unwanted latency and/or unpredictability of response time, a problem that is exacerbated if the thread can block.

The basic issue, then, is how to resolve the essential conflict between the desire to perform ATC "immediately" and the need to ensure that certain sections of code are executed to completion[2] and that relevant finalization is performed.

In brief, ATC in Ada is based on the concept of aborting either a task (via an explicit **abort** statement) or a syntactically-distinguished sequence of statements (as the result of the completion of a triggering action, such as a timeout, in an asynchronous select statement). The fundamental principle is that ATC should be safe: it is postponed while control is in an *abort-deferred* operation (e.g., a rendezvous), and when it takes place any needed finalizations are performed. ATC does not entail raising or handling exceptions.

In contrast, the Real-Time Specification for Java captures ATC via asynchronous exceptions, which are instances of the `AsynchronouslyInterrupted-Exception` (abbreviated "AIE") class. An ATC request is posted to a target thread by a method call rather than a specialized statement, making an `AIE` instance *pending* on the thread. In the interest of safety, several contexts are *ATC-deferred*: synchronized code, and methods and constructors lacking a `throws AIE` clause. An asynchronous exception is only thrown when the target thread is executing code that is not ATC-deferred, and special rules dictate how it is propagated / handled. Finalization can be arranged via appropriately placed explicit `finally` clauses. The ATC mechanism can be used for thread termination, timeout, and other purposes.

The remaining sections will elaborate on these points, with a focus on uniprocessor systems; multiprocessors present some issues that are beyond the scope of this paper.

An extended version of this paper appears in [1].

## 2    ATC in Ada

This section summarizes, illustrates, and critiques Ada's[3] ATC features.

### 2.1    Semantics

One of Ada's ATC constructs is the **abort** statement, which is intended to trigger the termination of one or more target tasks. However, termination will not

---

[2] Note that "executed to completion" does not imply non-preemptability.

[3] It is assumed that the implementation complies with the *Real-Time Systems Annex.*

necessarily be immediate, and in some situations (although admittedly anomalous in a real-time program) it will not occur at all. An aborted task becomes "abnormal" but continues execution as long as it is in an *abort-deferred operation* (e.g., a protected action). When control is outside such a region the aborted task becomes "completed". Its local controlled objects are finalized, and its dependent tasks are aborted. The task terminates after all its dependent tasks have terminated. (This description was necessarily a simplification; full semantics are in [2], Sects. 9.8 and D.6).

Ada's second ATC feature – the *asynchronous select* statement – offers finergrained ATC. This syntactic construct consists of two branches:

- A *triggering alternative*, comprising a sequence of statements headed by a *triggering statement* that can be either a delay statement (for a timeout) or an entry call;
- The *abortable part*, which is the sequence of statements subject to ATC.

Conceptually, the abortable part is like a task that is aborted if/when the triggering statement completes.[4] In such an event, control passes to the statement following the triggering statement. Otherwise (i.e., if the abortable part completes first) the triggering statement is cancelled and control resumes at the point following the asynchronous select statement.

The following example illustrates a typical use for an asynchronous select statement. The context is a **Sensor** task that updates **Position** values at intervals no shorter than 1 second. It times out after 10 seconds and then displays a termination message. **Position** is a protected object declared elsewhere, with an **Update** procedure that modifies its value.

```
task body Sensor is
   Time_Out        : constant Duration := 10.0;
   Sleep_Interval  : constant Duration := 1.0;
begin
   select
     delay Time_Out;
     Put_Line("Sensor terminating");
   then abort
     loop
       Position.Update;
       delay Sleep_Interval;
     end loop;
   end select;
end Sensor;
```

The asynchronous select statement deals with nested ATCs correctly. For example, if the delay for an outer triggering statement expires while an inner

---

[4] It is thus subject to the rules for abort-deferred operations, and finalization of controlled objects.

delay is pending, the inner delay will be cancelled and an ATC will be promulgated out of the inner abortable part (subject to the semantics for abort-deferred operations) and then out of the outer abortable part.

The Ada ATC facility does not rely on asynchronous exceptions. It thereby avoids several difficult semantic issues; these will be described below in Sect. 5.1, in conjunction with the RTSJ's approach to ATC.

## 2.2  Safety

The Ada rules for abort-deferred operations give precedence to safety over immediacy: e.g., execution of `Update` in the `Sensor` task will be completed even if the timeout occurs while the call is in progress. Abort-deferred regions include other constructs (e.g. finalization) that logically must be executed to completion. However, the possibility for ATC must be taken into account by the programmer in order to ensure that the program performs correctly. For example, the idiom of locking a semaphore, performing a "block assignment" of a large data structure, and then unlocking the semaphore, is susceptible to data structure corruption and/or deadlock if an ATC occurs while the assignment is in progress.

Ada's two ATC features take different approaches to the issue of whether the permission for ATC is implicit or explicit. Any task is susceptible to being aborted; thus the programmer needs to program specially so that regions that logically should be executed to completion are coded as abort-deferred constructs. In any event the `abort` statement is intended for specialized circumstances: to initiate the termination of either the entire partition or a collection of tasks that are no longer needed (e.g., in a "mode change").

The asynchronous select statement takes the opposite approach. The abortable part of such a statement is syntactically distinguished, making clear the scope of the effect of the triggering condition. However, since subprograms called from the abortable part are subject to being aborted, the author of the asynchronous select construct needs to understand and anticipate such effects.

Some Ada implementations provide facilities for specifying local code regions that are abort deferred (e.g., pragma `Abort_Defer` in GNAT).

## 2.3  Style and Expressiveness

The Ada ATC mechanisms are concise and syntactically distinguished, making their effects clear and readable. For example, the timeout of the `Sensor` thread is captured succinctly in the asynchronous select statement.

While a task is performing an abort-deferred operation, an attempt to execute an asynchronous select statement is a bounded error. Thus a programmer implementing an abort-deferred operation needs to know the implementation of any subprograms that the operation calls.

Ada does not have a construct that immediately/unconditionally terminates a task. This omission is a good thing. Such a feature would have obvious reliability problems; e.g., if it were to take place during a protected operation, shared data might be left in an inconsistent state.

Several gaps in functionality should be noted:

**Triggering "Accept" Statement.** Ada allows an entry call but not an `accept` statement as a triggering statement. This restriction can lead to some stylistic clumsiness and the need for additional intermediate tasks.

**Cleaner Finalization Syntax.** Capturing finalization by declaring a controlled type and overriding `Finalize` can be somewhat awkward. An explicit control structure (e.g. something like Java's `finally` clause) would be cleaner.

**Awakening a Blocked Task.** There is no way for one task to awaken a second, blocked task, except by aborting it. Such a capability would sometimes be useful.

## 2.4   Implementability, Latency, and Efficiency

There are two basic approaches to implementing the asynchronous select statement, referred to as the "one thread" and "two thread" models [6,7,8]. In the one-thread model, the task does not block after executing the triggering statement but instead proceeds to execute the abortable part. The task is thus in a somewhat schizophrenic state of being queued on an entry or a timeout while still running. If the triggering statement occurs, the task is interrupted; it performs the necessary finalizations and then resumes in the triggering alternative. These effects can be implemented via the equivalent of a signal and `setjmp/longjmp`.

With the two-thread model, the task executing the asynchronous select is blocked at the triggering statement (as suggested by the syntax), and a subsidiary task is created to execute the abortable part. If the triggering condition occurs, the effect is equivalent to aborting the subsidiary task. Conceptually, this model seems cleaner than the one-thread approach. However, adding implicit threads of control is not necessarily desirable; e.g., data declared in the outer task and referenced in the abortable part would need to be specified as `Volatile` or `Atomic` to prevent unwanted caching. Moreover, aborting the subsidiary thread is complicated, due to the required semantics for execution of finalizations [7,8].

As a result of these considerations, current Ada implementations generally use the one-thread model.

In addition to the basic question of how to implement the ATC semantics there are issues of run-time responsiveness (i.e., low latency) and efficiency. As noted above, the effect of ATC is sometimes deferred, in the interest of safety. If the abortable part declares any controlled objects or local tasks, or executes a rendezvous, then this will introduce a corresponding latency. The amount incurred is thus a function of programming style.

The efficiency question is somewhat different. Unlike its "pay as you go" effect on latency, ATC imposes a "distributed overhead"; i.e., there is a cost even if ATC isn't used. As an example, the epilog code for rendezvous and protected operations needs to check is there is a pending ATC request (either an abort or the completion of a triggering statement). Somewhat anticipating this issue, the Ada language allows the user to specify restrictions on feature usage, thus allowing/directing the implementation to use a more specialized and more efficient version of the run-time library, omitting support for unused features.

## 3   ATC in Java

This section briefly summarizes and critiques the ATC capabilities provided in Java [9] itself;[5] Java's asynchrony facilities inspired the RTSJ in either what to do or what not to do.

Java's support for asynchronous communication is embodied in three methods from the `Thread` class: `interrupt`, `stop`, and `destroy`. It also has a limited form of timeout via overloaded versions of `Object.wait`.

When `t.interrupt()` is invoked on a target thread[6] `t`, the effect depends on whether `t` is blocked (at a call for `wait`, `sleep`, or `join`). If so, an ATC takes place: an `InterruptedException` is thrown, awakening `t`. Otherwise, `t`'s "interrupted" state is set; it is reset either when `t` next calls the `interrupted` method or when it reaches a call on `wait`, `sleep`, or `join`. In the latter cases an `InterruptedException` is thrown.

Despite the ATC aspects of `interrupt`, it is basically used in polling approaches: each time through a loop, a thread can invoke `interrupted` to see if `interrupt` has been called on it, and take appropriate action if so.

When `t.stop()` is invoked on a target thread `t`, a `ThreadDeath` exception is thrown in `t` wherever it was executing, and normal exception propagation semantics apply. This method was designed to terminate `t` while allowing it to do cleanup (via `finally` clauses as the exception propagates). However, there are several major problems:

- If `t.stop()` is called while `t` is executing synchronized code, the synchronized object will be left in an inconsistent state
- A "catch-all" handler (e.g. for `Exception` or `Throwable`) in a `try` statement along the propagation path will catch the `ThreadDeath` exception, preventing `t` from being terminated.

As a result of such problems, the `Thread.stop()` method has been deprecated.

When `t.destroy()` is invoked, `t` is terminated immediately, with no cleanup. However, if `t` is executing synchronized code, the lock on the synchronized object will never be released. For this reason, even though `destroy()` has not been officially deprecated, its susceptibility to deadlock makes it a dangerous feature. In any event, `destroy()` has not been implemented in any JVM released by Sun.

Java's timeout support is limited to several overloaded versions of class `Object`'s `wait` method. However, there is no way to know (after awakening) which condition occurred: the timeout, or an object notification. Indeed, there

---

[5] This description is based on [10].

[6] In this section "thread" means an instance of `java.lang.Thread`.

are race conditions in which both may have happened; an object notification may occur after the timeout but before the thread is scheduled.

## 4   An Overview of the Real-Time Specification for Java

ATC is one of several facilities provided by the RTSJ. To establish a perspective, this section summarizes the problems that the RTSJ sought to address, and the main aspects of its solution.

The Real-Time Specification for Java is a class library (the package `javax.-realtime`) that supplements the Java platform to satisfy real-time requirements. It was designed to particularly address the following shortcomings in regular Java:

**Incompletely Specified Thread Model.** Java places only loose requirements on the scheduler.[7] There is no guarantee that priority is used to dictate which thread is chosen on release of a lock or on notification of an object. Priority inversions may occur; moreover, the priority range is too narrow.

**Garbage Collector Interference.** Program predictability is compromised by the latency induced by the Garbage Collector.

**Lack of Low-Level Facilities.** Java (although for good reasons) prevents the program from doing low-level operations such as accessing physical addresses on the machine.

**Asynchrony Shortcomings.** As mentioned above, Java's features for asynchronous thread termination are flawed, and it lacks a general mechanism for timeouts and other asynchronous communication.

The RTSJ provides a flexible scheduling framework based on the `Schedulable` interface and the `Thread` subclass `RealtimeThread` that implements this interface. The latter class overrides various methods with versions that add real-time functionality, and supplies new methods for operations such as periodic scheduling. The `Schedulable` interface is introduced because certain schedulable entities (in particular, handlers for asynchronous events) might not be implemented as threads.

The RTSJ mandates a default POSIX-compliant preemptive priority-based scheduler that supports at least 28 priority levels, and that enforces Priority Inheritance as the way to manage priority inversions. The implementation can provide other schedulers (e.g., Earliest Deadline First) and priority inversion control policies (e.g., Priority Ceiling Emulation).

To deal with Garbage Collection issues, the RTSJ provides various *memory areas* that are not subject to Garbage Collection: "immortal memory", which

---

[7] This lack of precision may seem strange in light of Java's well-publicized claim to portability ("Write Once, Run Anywhere"). However, in the threads area there is considerable variation in the support provided by the operating systems underlying the JVM implementations. If the semantics for priorities, etc., were tighter, that would make Java difficult or inefficient to implement on certain platforms.

persists for the duration of the application; and "scoped memory", which is a generalization of the run-time stack. Restrictions on assignment prevent dangling references. The RTSJ also provides a `NoHeapRealtimeThread` class; instances never reference the heap, may preempt the Garbage Collector at any time (even when the heap is in an inconsistent state), and thus do not incur GC latency.

The RTSJ provides several classes that allow low-level programming. "Peek and poke" facilities for integral and floating-point data are available for "raw memory", and "physical memory" may be defined with particular characteristics (such as flash memory) and used for general object allocation.

Java's asynchrony issues are addressed through two main features. First, the RTSJ allows the definition of asynchronous events and asynchronous event handlers – these are basically a high-level mechanism for handling hardware interrupts or software "signals". Secondly, the RTSJ extends the effect of `Thread.-interrupt` to apply not only to blocked threads, but also to real-time threads[8] and asynchronous event handlers whether blocked or not. How this is achieved, and how it meets various software engineering criteria, will be the subject of the next section.

# 5    ATC in the Real-Time Specification for Java

This section describes, illustrates and critiques the RTSJ's ATC facilities.

## 5.1    Semantics

ATC in the RTSJ is defined by the effects of an asynchronous exception that is thrown in a thread[9] t as the result of invoking `t.interrupt`. However, asynchronous exceptions raise a number of issues that need to be resolved:

**Inconsistent State.** If the exception is thrown while the thread is synchronized on an object – or more generally, while the thread is in a code section that needs to be executed to completion – then the object (or some global state) will be left inconsistent when the exception is propagated.

**Unintended Non-termination.** If the purpose of the ATC request is to terminate t, but the resulting exception is thrown while t is in a `try` block that has an associated catch clause for, say, `Exception` or `Throwable`, then the exception will be caught; t will not be terminated.

**Unintended Termination.** If the purpose of the ATC request is, say, to make t timeout on a computation, but the exception is thrown either before t has entered a `try` statement with an associated handler, or after it has exited from such a construct, then the exception will not be handled. It will propagate out, and eventually cause t to terminate.

---

[8] A *real-time thread* is an instance of the `RealtimeThread` class.

[9] For ease of exposition, we refer to the target of `interrupt` as a "thread", but in fact for ATC effects it must be a real-time thread or asynchronous event handler. Regular Java threads – instances of `java.lang.Thread` – do not have ATC semantics.

**Nested ATCs / Competing Exceptions.** A thread may receive an ATC request while another ATC is in progress. This raises the issue of propagating multiple exceptions or choosing which one should be discarded.

Indeed, these kinds of problems motivated the removal of the asynchronous 'Failure exception from an early pre-standard version of Ada.

The RTSJ's approach to ATC addresses all of these issues. It is based on the class AsynchronouslyInterruptedException, abbreviated "AIE", a subclass of the checked exception class InterruptedException. An ATC request always involves, either explicitly or implicitly, a target thread t and an AIE instance aie. For example, the method call t.interrupt() posts an ATC request to the explicit target thread t, but the AIE instance (the system-wide "generic" AIE) is implicit.

Key to the semantics are the complementary concepts of *asynchronously interruptible* (or *AI*) and *ATC-deferred* sections. The only code that is asynchronously interruptible is that contained textually within a method or constructor that includes AIE on its **throws** clause, but that is not within synchronized code or in inner classes, methods, or constructors. Synchronized statements and methods, and also methods and constructors that lack a **throws** AIE clause, are ATC-deferred.

Posting an ATC request for AIE instance aie on thread t has the following effect:

1. aie is made *pending* on t.[10]
2. If t is executing within ATC-deferred code, t continues execution until it either invokes an AI method, or returns to an AI context.
3. If t is executing within AI code, aie is thrown (but stays pending).

Note that if control never reaches an AI method, then aie will stay pending "forever"; it will not be thrown.

If/when control does reach an AI method, then aie is thrown. However, the rules for handling AIE are different from other exceptions; this exception class is never handled by catch clauses from AI code. Instead, control transfers immediately – without executing **finally** clauses in AI code as the exception is propagated – to the catch clause for AIE (or any of its ancestor classes) of the nearest dynamically enclosing **try** statement that is an an ATC-deferred section. Unless the handling code resets the "pending" status, the AIE stays pending.

These rules address two of the previously-noted issues with asynchronous exceptions (and thus avoid the problems with Thread.stop):

- Since AI code needs to be explicitly marked with a **throws** AIE clause, and synchronized code is abort-deferred, the RTSJ prevents inconsistent state (or at least forces the programmer to make explicit the possibility of inconsistent state).
- Since handling the AIE does not automatically reset the "pending" status, the RTSJ rules prevent unintended non-termination.

---

[10] A special case, when there is already an AIE pending on t, is treated below.

Since the RTSJ provides no mechanism for immediately terminating a thread, it avoids the difficulties inherent in `Thread.destroy`.

Here is an example of typical RTSJ style for thread termination:

```
class Victim extends RealtimeThread{
  private void interruptibleRun()
    throws AsynchronouslyInterruptedException{
    ... // Code that is asynchronously interruptible
  }

  public void run(){
    try{
      this.interruptibleRun();
    }
    catch (AsynchronouslyInterruptedException aie){
    System.out.println( "terminating" );
    }
  }
}
```

To create, start, and eventually terminate a `Victim` thread:

```
Victim v = new Victim();
v.start();
...
v.interrupt();
```

In order to address the problem of unintended termination – i.e., throwing an `AIE` outside the target thread's `try` statement that supplies a handler – the `AIE` class declares the `doInterruptible` and `fire` instance methods. The invocation `aie.doInterruptible(obj)` in a thread `t` takes an instance `obj` of a class that implements the `Interruptible` interface. This interface declares (and thus `obj` implements) the `run` and `interruptAction` methods. The `run` method (which should have a `throws AIE` clause) is invoked synchronously from `doInterruptible`. If `aie.fire()` is invoked – presumably from another thread that holds the `aie` reference – then an ATC request for `aie` is posted to `t`. The implementation logic in `doInterruptible` supplies a handler that invokes `obj.interruptAction` and resets aie's *pending* state. If `aie.fire()` is invoked when control is not within `aie.doInterruptible` then the ATC request is discarded – thus `t` will only receive the exception when it is executing in a scope that can explicitly handle it (through user-supplied code).

A timeout is achieved by combining an asynchronous event handler, a timer, and an `AIE`. Since the details can be tedious, the RTSJ supplies the `Timed` subclass of `AIE`; a constructor takes a `HighResolutionTime` value. The application can then perform a timeout by invoking `timed.doInterruptible(obj)` where `timed` is an instance of `Timed`, and `obj` is an `Interruptible`. Here is an example, with the same effect as the Ada version in Sect. 2.1. It assumes that a `Position` object (passed in a constructor) is updated via the synchronized method update. Instead of declaring an explicit class that implements the Interruptible interface, it uses an anonymous inner class.

```
class Sensor extends RealtimeThread{
  final Position pos;
  final long sleepInterval  =  1000;
  final long timeout = 10000;

  Reporter(Position pos){ this.pos = pos; }

  public void run(){
    new Timed(new RelativeTime( timeout, 0 )).
      doInterruptible(
        new Interruptible(){
          public void run(AsynchronouslyInterruptedException e)
            throws AsynchronouslyInterruptedException{
            while (true){
              pos.update();  // synchronized method
              try {
                sleep(sleepInterval);
              }
              catch(InterruptedException ie) {}
            }
          }
          public void interruptAction(
            AsynchronouslyInterruptedException e){
            System.out.println("Sensor instance terminating");
          }
        });
  }
}
```

If the timeout occurs while the Interruptible's run() method is in progress, execution of this method is abandoned (but deferred if execution is in synchronized code) and interruptAction is invoked, here simply displaying a termination message.

Nested ATCs raise the issue of multiple ATC requests in progress simultaneously. The RTSJ addresses this problem by permitting a thread to contain at most one pending AIE, and by defining an ordering relation between AIE instances. The rules give a "precedence" to an AIE based on the dynamic level of its "owning" scope, where ownership is established through an invocation of doInterruptible. Shallower scopes have precedence over deeper ones, and the generic AIE has the highest precedence. Thus a pending AIE is replaced by a new one only if the latter is "aimed" at a shallower scope.

## 5.2  Safety

Like Ada, the RTSJ opts for safety over immediacy and thus defers ATC in synchronized code (the Java analog to Ada's protected operations and rendezvous). However, ATC is not deferred in `finally` clauses,[11] thus leading to potential problems where essential finalization is either not performed at all or else only done partially.

Since the RTSJ controls asynchronous interruptibility on a method-by-method basis (i.e., via the presence or absence of a `throws AIE` clause), legacy code that was not written to be asynchronously interruptible will continue to execute safely even if called from an asynchronously-interruptible method.

## 5.3  Style and Expressiveness

A major issue with the RTSJ is the rather complex style that is needed to obtain ATC. Some examples:

- Aborting a thread requires splitting off an `AIE`-throwing method that is invoked from `run` (`interruptibleRun` in the `Victim` example above).
- Achieving timeout is somewhat obscure, typically involving advanced features such as anonymous inner classes or their equivalent.
- Programming errors that are easy to make – e.g., omitting the `throws AIE` clause from the `run` method of the anonymous `Interruptible` object – will thwart the ATC intent. This error would not be caught be the compiler.
- If the `run` method for an `Interruptible` invokes `sleep`, then the method has to handle `InterruptedException` even though the RTSJ semantics dictate that such a handler will never be executed.

These stylistic problems are due somewhat to the constraint on the RTSJ to not introduce new syntax. ATC is a control mechanism, and modeling control features with method calls tends to sacrifice readability and program clarity.

Another issue with the RTSJ is its mix of high-level and low-level features. An advantage is generality, but the disadvantage is that the programmer needs to adhere to a fairly constrained set of idioms in order to avoid writing hard-to-understand code.

The rules for `AIE` propagation diverge from regular Java semantics; this does not help program readability. For example, if an asynchronously interruptible method has a `try` statement with a `finally` clause, then the `finally` clause is not executed if an ATC is triggered while control is in the `try` block.

## 5.4  Implementability, Latency, and Efficiency

The conceptual basis for implementing ATC is a slot in each "thread control block" to hold an `AIE`, a flag indicating whether the `AIE` is pending, and a flag

---

[11] This is because the RTSJ was designed to affect only the implementation of the Java Virtual Machine, and not the compiler. The syntactic distinction for the `finally` clause is not preserved in the class file, and there is no easy way for the implementation to recognize that such bytecodes should be ATC-deferred.

in each stackframe indicating if the method is currently asynchronously interruptible. Posting an AIE instance aie to a real-time thread t involves setting aie as t's AIE (subject to the precedence rules) and setting the pending flag to true. Entering / leaving synchronized code affects the AI flag. The bytecodes for method call/return and exception propagation use these data values to implement ATC semantics.

As with Ada, the latency to asynchronous interruption depends on style.

Also as with Ada, ATC incurs a cost even if not used; the implementation of various bytecodes needs to check the ATC data structures.

# 6  Comparison

Table 1 summarizes how Ada and the RTSJ compare with respect to specific ATC criteria; for completeness it also includes the basic Java asynchrony mechanisms. The following subsections will elaborate on the Ada and RTSJ entries.

**Table 1.** Comparison of ATC mechanisms

|  | Ada | | Java | | | RTSJ |
|---|---|---|---|---|---|---|
|  | *abort* | *asynch select* | *interrupt* | *stop* | *destroy* | *AIE* |
| *Semantic basis* | Task abort | | Synchronous exception | Asynch exception | Immediate terminate | Asynch exception |
| *Safety* | Good | Good | Good | Poor | Poor | Good |
| *Defer in synchronized code* | Yes | Yes | No | No | No | Yes |
| *Defer in finalization* | Yes | Yes | No | No | No | No |
| *Defer unless explicit* | No | No | No | No | No | Yes |
| *Style* | Good | Good | Good | Poor | Poor | Fair |
| *Expressiveness* | Good | Good | Fair | Fair | Poor | Good |
| *Implementability* | Good | Fair | Good | Fair | Poor | Fair |
| *Latency* | Fair | Fair | Poor | Good | Good | Fair |
| *Efficiency* | Fair | Fair | Good | Fair | Poor | Fair |

## 6.1  Semantics

Ada and the RTSJ take very different approaches to ATC. Ada defines the semantics for asynchronous task termination (the abort statement) and then applies these semantics in another context (the asynchronous select) to model ATC triggered by a timeout or the servicing of an entry call. Ada does not define ATC in terms of exceptions. Indeed, aborting a task t does not cause an exception to be thrown in t; even if a Finalize procedure throws an exception, this exception is not propagated ([2], ¶7.6.1(20)).

In contrast, the RTSJ's ATC mechanism is based on asynchronous exceptions, a somewhat natural design decision given the semantics of Java's interrupt() facility. Thus the RTSJ has a general ATC approach, and realizes real-time thread termination as a special case of ATC. However, a side effect is a rather complicated set of rules, e.g., the precedence of exceptions.

## 6.2   Safety

Both Ada and the RTSJ recognize the need to define regions of code where ATC is inhibited, in particular in code that is executed under a "lock". Ada is safer in specifying additional operations (e.g. finalization of controlled objects) as abort-deferred; in the RTSJ finally clauses are asynchronously interruptible. The RTSJ, however, offers finer granularity of control than Ada; asynchronous interruptibility can be specified on a method-by-method basis, and the default is "noninterruptible". It thus avoids the problem of aborting code that might not have been written to be abortable.

## 6.3   Style and Expressiveness

Ada and the RTSJ are roughly comparable in their expressive power but they differ significantly with respect to their programming styles. Ada provides distinguished syntax – the abort and asynchronous select statements – whose effects are clear. The RTSJ realizes ATC via method calls, which sacrifices readability, as is evidenced by comparing the two versions of the Sensor timeout example. There are also a number of non-intuitive points of style that programmers will need to remember.

On the other hand, Java's interrupt() mechanism allows the programmer to awaken a blocked thread by throwing an exception; Ada lacks a comparable feature. Moreover, the RTSJ is somewhat more regular than Ada with respect to feature composition. ATC-deferred code can call an AI method, and an ATC can thus be triggered in the called method. In Ada it is a bounded error if a subprogram invoked by an abort-deferred operation attempts an asynchronous select statement.

## 6.4   Implementability, Latency, and Efficiency

ATC in both Ada and the RTSJ requires non-trivial implementation support; these features are complicated semantically (in the details if not the main concepts) and are among the most difficult to implement and to test.

ATC latency is roughly equivalent in Ada and the RTSJ, and is a function of programming style. Heavy use of abort/ATC-deferred constructs will induce high latency, and inversely. However, since subprograms by default are abortable in Ada, the latency from ATC in Ada is likely to be less than in the RTSJ.

Efficiency will be a challenge for both Ada and the RTSJ; this seems intrinsic in ATC rather than a flaw in either design. Non-optimizing implementations will likely impose an overhead even if ATC is not used. Possible approaches include sophisticated control and data flow analysis, hardware support, and the definition of restricted profiles.

# 7   Conclusions

ATC is a difficult issue in language design, and both Ada and the RTSJ have made serious attempts to provide workable solutions. They share a common philosophy in opting for safety as the most important objective, and thus in defining ATC to be deferred in certain regions that must be executed to completion. They offer roughly comparable expressive power, but they differ significantly in how the mechanism is realized and in the resulting programming style. Ada bases ATC on the concept of abortable code regions, integrates ATC with the inter-task communication facility, and provides specific syntax for ATC in general and for task termination in particular. The RTSJ bases ATC on an asynchronous exception thrown as the result of invoking interrupt() on a real-time thread; termination is a special case. The RTSJ does not introduce new syntax for ATC, so the effect must be achieved through new classes and method calls.

Since the RTSJ is so new, there is not much experience revealing how it compares with Ada in practice. As implementations mature, and developers gain familiarity with the concepts, it will be interesting to see whether ATC fulfills its designers' expectations and its users' requirements.

**Acknowledgements.** Detailed comments from an anonymous referee helped us improve the organization and exposition.

# References

1. B.M. Brosgol and A. Wellings; *A Comparison of the Asynchronous Transfer of Control Features in Ada and the Real-Time Specification for Java$^{TM}$*; Univ. of York Technical Report YCS-350; February 2003.
2. S.T. Taft, R.A. Duff, R.L. Brukardt, and E. Ploedereder; *Consolidated Ada Reference Manual, Language and Standard Libraries, International Standard ISO/IEC 8652/1995(E) with Technical Corrigendum 1*; Springer LNCS 2219; 2000
3. G. Bollella, J. Gosling, B. Brosgol, P. Dibble, S. Furr, D. Hardin, and M. Turnbull; *The Real-Time Specification for Java*, Addison-Wesley, 2000
4. The Real-Time for Java Expert Group; *The Real-Time Specification for Java, V1.0*; Sun Microsystems JSR-001; http://www.rtj.org; November 2001.
5. ISO/IEC 9945-1: 1996 (ANSI/IEEE Standard 1003.1, 1996 Edition); *POSIX Part 1: System Application Program Interface (API) [C Language]*
6. E.W. Giering and T.P. Baker; "The GNU Runtime Library (GNARL): Design and Implementation", *WAdaS '94 Proceedings*, ACM SIGAda; 1994.
7. E.W. Giering and T.P. Baker; "Ada 9X Asynchronous Transfer of Control: Applications and Implementation", *Proceedings of the SIGPLAN Workshop on Language, Compiler, and Tool Support for Real-Time Systems*, ACM SIGPLAN; 1994.
8. J. Miranda; *A Detailed Description of the GNU Ada Run Time (Version 1.0)*; http://www.iuma.ulpgc.es/users/jmiranda/gnat-rts/main.htm; 2002.
9. J. Gosling, B. Joy, G. Steele, G. Bracha; *The Java Language Specification (2nd ed.)*; Addison Wesley, 2000

10. B. Brosgol, R.J. Hassan II, and S. Robbins; "Asynchronous Transfer of Control in the Real-Time Specification for Java", *Proc. ISORC 2002 – The 5th IEEE International Symposium on Object-oriented Real-time distributed Computing*, April 29–May 1, 2002, Washington, DC, USA

# Exposing Memory Corruption and Finding Leaks: Advanced Mechanisms in Ada

Emmanuel Briot[1], Franco Gasperoni[1], Robert Dewar[2], Dirk Craeynest[3], and Philippe Waroquiers[3]

[1] ACT Europe, 8 rue de Milan, 75009 Paris, France
{briot,gasperon}@act-europe.fr
[2] Ada Core Technologies
104 Fifth Avenue, New York, NY 10011, USA
dewar@gnat.com
[3] Eurocontrol/CFMU, Development Division
Rue de la Fusée, 96, B-1130 Brussels, Belgium
{dirk.craeynest,philippe.waroquiers}@eurocontrol.int

**Abstract.** This article discusses the tools that Ada offers to deal with dynamic memory problems. The article shows how the storage pools mechanism of Ada 95 can be extended to enpower developers when tracking memory leaks and memory corruption in their code. This Ada extension rests on the notion of "checked pools", i.e. storage pools with an additional **Dereference** operation. The paper describes how a particular instance of the checked pool, called the "debug pool", is implemented in the GNAT technology. Performance measurements for the use of debug pools are provided in the context of the Air Traffic Flow Management application at Eurocontrol.

## 1 Introduction

### 1.1 About Garbage Collection

Any system confronted with the possibility of memory leaks and memory corruption has to consider why garbage collection (GC) cannot be used as a solution to its potential dynamic memory problems. While it is true that, when practical, GC removes incorrect deallocations and dangling pointer problems, it is not a systematic panacea. For one thing, GC cannot deal completely with the problem of memory leaks. Consider, for instance, the following Java code:

```
global = new Huge (100000000);
```

If `global` is a static class field that is never set to `null` or updated with another reference, the memory allocated in `new Huge (...)` cannot be freed until the class in which `global` is declared is finalized, even though the data allocated in `new Huge (...)` may never be accessed.

Another potential and more serious problem with GC exists for systems requiring a guaranteed (predictable) response time. We will come back to the issue of GC in the conclusion.

J.-P. Rosen and A. Strohmeier (Eds.): Ada-Europe 2003, LNCS 2655, pp. 129–141, 2003.

## 1.2   Dynamic Memory Allocation and Ada Safety Nets

Ada provides a number of safety nets to help programmers catch some common dynamic memory handling mistakes. All pointers are, for instance, set to `null` upon creation or deallocation and an exception is raised if a program tries to dereference a `null` pointer. Furthermore, the accessibility rules of access types of Ada 95 are designed to prevent dangling references when the scope of the pointee is inside that of the access type. These rules help to ensure that the most obvious cases of memory corruption are avoided.

When it comes to memory deallocation, Ada makes it possible to allocate on the stack (and hence automatically deallocate) all objects created through an access type local to a subprogram. In GNAT, for instance, the storage allocated for type `String_Access` in the following code excerpt, will be allocated on P's stack and automatically freed upon P's return.

```
procedure P is
   type String_Access is access all String;
   for String_Access'Storage_Size use 100_000;
   A : String_Access := new String (1 .. 1_000);
begin
   ...
end P;
```

## 1.3   User-Defined Storage Managment in Ada 95

Ada 95 gives the programmer complete freedom for the allocation/deallocation algorithms to use for a given access type through the mechanism of storage pools [1]. A storage pool is an abstract limited controlled type with 3 additional abstract operations: `Allocate`, `Deallocate`, and `Storage_Size`. In a concrete storage pool type `SP_Type`, the programmer must define the bahavior of `Allocate`, `Deallocate`, `Storage_Size`, and possibly `Initialize` and `Finalize`. Then when the programmer writes

```
Pool : SP_Type;
type A_Type is access ...;
for  A_Type'Storage_Pool use Pool;
```

it requests the Ada compiler to use `SP_Type`'s `Allocate` and `Deallocate` every time memory is allocated or deallocated for `A_Type`. See Sect. 13.11 in [1] for details.

Ada implementations are free to provide specialized storage pools in addition to the standard (default) one. GNAT, for instance, provides 2 specialized pools:

- `System.Pool_Size.Stack_Bounded_Pool`, a stack-bounded pool where dynamic memory allocation is done on the stack and memory is globally reclaimed by normal stack management. GNAT uses this pool for access types with a `Storage_Size` representation clause as shown in the example of the previous section.

- System.Pool_Local.Unbounded_Reclaim_Pool, a scope-bounded pool where dynamic memory allocation is done using the system malloc() routine but which is automatically freed when the scope where the storage pool object is declared is exited. A programmer can use this pool for access types that are locally declared and for which he is not in a position to provide a maximum Storage_Size, or whose maximum Storage_Size would exceed the overall stask size.

When writing user-defined storage pools the main catch is thc issue of alignment, i.e. how to ensure that the portion of memory returned by Allocate is aligned on a given byte boundary [2]. Note that the requested byte boundary could be greater than the maximum memory alignement for the underlying processor. If the programmer is trying to allocate data to be placed in a data cache, for instance, he may wish to specify the data cache line size as the alignment.

## 2 Extending Ada's Storage Pools

Despite the help and safety checks that Ada 95 provides, there are three type of programming problems that Ada does not address in full generality:

- Memory leaks (i.e. forgetting to deallocate dynamically allocated storage);
- Incorrect deallocation (i.e. deallocating unallocated memory);
- Dangling references (i.e. accessing deallocated or unallocated memory).

Because storage pools provide no means to check dereferences, GNAT offers a special type of storage pool, called a "checked pool", with an additional abstract primitive operation called Dereference. Dereference has the same parameter profile as the subprograms Allocate and Deallocate and is passed the same information. Dereference is invoked before dereferencing an access type using a checked pool. Its intended role is to do some checking on the reference, e.g. check that the reference is valid.

   Checked pools are abstract types. They only act as a framework which other tagged types must extend. A concrete implementation is provided in the GNAT library package GNAT.Debug_Pools. This package was developed in collaboration with Eurocontrol.

## 3 Debug Pools

The goal of a debug pool is to detect incorrect uses of memory, specifically: incorrect deallocations, access to invalid memory, and memory leaks. To use a debug pool developers need to instrument their code for each access type they want to monitor, as shown below in the lines marked with the special comment -- Add. The debug pool reports errors in one of two ways: either by immediately raising an exception, or by logging a message that can be printed on standard output, which is what we have decided to do in the following example. Thye example contains a number of typical errors that the debug pool will point out.

```
File p.ads
----------
 1. with GNAT.Debug_Pools; use GNAT.Debug_Pools;          -- Add
 2. with Ada.Unchecked_Deallocation;
 3. with Ada.Unchecked_Conversion; use Ada;
 4.
 5. procedure P is
 6.    D_Pool : GNAT.Debug_Pools.Debug_Pool;               -- Add
 7.
 8.    type IA is access Integer;
 9.    for  IA'Storage_Pool use D_Pool;                    -- Add
10.
11.    procedure Free is new Unchecked_Deallocation (Integer, IA);
12.    function Convert is new Unchecked_Conversion (Integer, IA);
13.
14.    Bogus : IA := Convert(16#0040_97AA#);
15.    A, B  : IA;
16.    K     : Integer;
17.
18.    procedure Nasty is
19.    begin
20.       A := new Integer;
21.       B := new Integer;
22.       B := A;           -- Error: Memory leak
23.       Free (A);
24.       K := B.all;       -- Error: Accessing deallocated memory
25.       K := Bogus.all;   -- Error: Accessing unallocated memory
26.       Free (B);         -- Error: Freeing deallocated memory
27.       Free (Bogus);     -- Error: Freeing unallocated memory
28.    end Nasty;
29.
30. begin
31.    Configure (D_Pool, Raise_Exceptions => False);      -- Add
32.    Nasty;
33.    Print_Info_Stdout (D_Pool, Display_Leaks => True); -- Add
34. end P;
```

For each faulty memory use the debug pool will print several lines of information
as shown below[1]:

```
Accessing deallocated storage, at p.adb:24 p.adb:32
   First deallocation at p.adb:23 p.adb:32
Accessing not allocated storage, at p.adb:25 p.adb:32
Freeing already deallocated storage, at p.adb:26 p.adb:32
```

---

[1] The actual output shows full backtraces in hexadecimal format that we have post-
processed with the tool addr2line to display the information in symbolic format.

```
Memory already deallocated at p.adb:23 p.adb:32
Freeing not allocated storage, at p.adb:27 p.adb:32

Total allocated bytes:  8
Total logically deallocated bytes:  4
Total physically deallocated bytes:  0
Current Water Mark:  4
High Water Mark:  8

List of not deallocated blocks:
Size:  4 at: p.adb:21 p.adb:32
```

The debug pool reports an error in the following four cases:

1. Accessing deallocated storage
2. Accessing not allocated storage
3. Freeing already deallocated storage
4. Freeing not allocated storage

As the reader can see the debug pool displays the traceback for each faulty memory access, memory free, and potential memory leaks. The depth of the traceback is programmer-configurable. Note how the information reported when accessing deallocated storage or when freeing already deallocated storage is much richer than what one can get with a debugger since the debug pool indicates the backtrace of the program location where the original deallocation occurred.

In addition to the above, the debug pool prints out the traceback for all memory allocations that have not been deallocated. These are potential memory leaks. The debug pool also displays the following memory usage information :

1. High water mark: The maximum amount of memory that the application has used at the point where `Print_Info_Stdout` is called.
2. Current water mark: The current amount of memory that the application is using at the point where `Print_Info_Stdout` is called.
3. The total number of bytes allocated and deallocated by the application at the point where `Print_Info_Stdout` is called.
4. Optionally the tracebacks of all the locations in the application where allocation and deallocation takes place (this information is not displayed in the previous output). This can be used to detect places where a lot of allocations are taking place.

It is worth noting that debug pools can be used in several important situations:

− One does not have to wait for the program to terminate to collect memory corruption information, since the debug pool can log problems in a file and developers can look at this file periodically to ensure that no problems have been detected so far by the debug pool mechanism.
− Various hooks are provided so that the debug pool information is available in the debugger. A developper can, for instance, interactively ask the status

of a given memory reference: is the reference currently allocated, where has it been allocated, is the reference logically deallocated, where has it been deallocated, etc.

# 4  Debug Pool Implementation

The debug pools package was designed to be as efficient as possible, but has an impact on the code performance. This depends on the number of allocations, deallocations and, somewhat less, dereferences that the application performs.

## 4.1  Debug Pool Memory Release Strategy

Physical allocations and deallocations are done through the usual system calls. However, in order to provide proper checks, the debug pool will not immediately release a memory block when asked to. The debug pool marks the memory block as "logically deallocated" but keeps the released memory around (the amount kept around is configurable) so that it can distinguish between memory that has not been allocated and memory that has been allocated but freed. This allows detection of dangling references to freed memory, which would not be possible if memory blocks were immediately released as this memory could be reused by a subsequent call to the system `malloc()`.

Retaining memory could be a problem for long-lived applications or applications that do a lot of allocations and deallocations. To address this the following parameters were added to the debug pool:

1. **Maximum Logically Freed Memory:** This parameter sets the limit of the amount of memory that can be logically deallocated, but not released to the system. When this limit is reached, the debug pool will start freeing memory.
2. **Minimum to Free:** This parameter indicates how much memory the debug pool should try to release to the system at once. For performance reasons it is better to free several blocks of memory at the same time.

The debug pool can use one of two algorithms to select the memory blocks to hand back to system memory:

1. **First Deallocated – First Released:** The first block that was deallocated by the application is the first to be released to the system.
2. **Advanced Block Scanning:** This more expensive algorithm parses all the blocks currently allocated, and finds all values that look like pointers. If these values match a currently deallocated block, that block will not be physically released. This ensures that dangling pointers are properly detected when the access type is dereferenced. This algorithm is only a good approximation: it is not guaranteed to detect all dangling pointers since it doesn't check task stacks or CPU registers.

## 4.2   Data Structures

The debug pool will respect all alignments specified in the user code by aligning all objects using the maximum machine alignment. This limits the performance impact of using the debug pool and, as we will show below, allows to quickly compute the validity of a memory reference.

**Global Structures.** The debug pool contains a packed boolean array. Each entry in this array matches a location in memory to indicate whether the corresponding address is under control of the debug pool (1 bit per address). Because all the addresses returned by the debug pool are aligned on the Maximum_Alignment of the underlying machine, the array index of each memory address Addr can be quickly computed as follows:

    array index = (Addr - Heap_Addr) / Maximum_Alignment

where Heap_Addr is the address of the beginning of the application's heap.

The initial size of the global array is small. During program execution the array grows, doubling its size every time more room is needed.

**Local Structures.** For each allocated memory block, the debug pool stores the following data in a header, located just before the memory block returned by the debug pool. The overall size of this header is a total of 16 bytes on 32-bit machines and includes:

1. The size of the allocated memory block. This is needed for the advanced block scanning algorithm described in the previous section. This value is negated when the block of memory has been logically freed by the application but has not yet been physically released.
2. A pointer to the next allocated or logically deallocated memory block.
3. A pointer to the allocation traceback, i.e. the traceback of the program location where this block was allocated.
4. A pointer to the first deallocation traceback. For memory blocks that are still allocated this pointer is used to point back to the previously allocated block for algorithmic convenience.

To save memory, the tracebacks are not stored in the header itself, but in a separate hash table. That way, only one instance of the traceback is stored no matter how many allocation are done at that program location.

All the allocated blocks are stored in a double-linked list, so that the advanced block scanning algorithm can find all of them and look for possible dangling pointers. This list is also used to report potential memory leaks.

When a block is deallocated by the application code, it is removed from the allocated blocks linked list and moved to the deallocated blocks list. This is the list from which, if needed, memory blocks will be returned to system memory.

The debug pool must be usable in a multi-tasking application, and has therefore been made thread-safe. Any time a new memory block is allocated or an existing block deallocated, the GNAT runtime is locked for concurrent accesses.

## 5    Debug Pool and General Access Types

A debug pools is a powerful mechanism to help debugging memory problems
in an application. There is, currently, a limitation with general access types. As
shown in the following example access to local variables can not be properly
handled by debug pools:

```
with GNAT.Debug_Pools; use GNAT.Debug_Pools;
procedure Q is
   D_Pool : GNAT.Debug_Pools.Debug_Pool;
   type IA is access all Integer;
   for  IA'Storage_Pool use D_Pool;

   Ptr : IA;
   K   : aliased Integer;
begin
   Configure (D_Pool);
   Ptr := K'Access;
   Ptr.all := 4;
   --  Exception GNAT.Debug_Pools.Accessing_Not_Allocated_Storage
         raised
end Q;
```

Because the memory pointed by Ptr wasn't allocated on the heap, the debug
pool will consider this as an invalid dereference, and will report an error.

## 6    Partition-Wide Storage Pool

The intention of the storage pool design is that an access type without a storage
pool clause use a default storage pool with appropriate characteristics. An obvious idea is to provide a facility for changing this default, and indeed given the
usefulness of debug pools in finding memory corruption problems, wanting to use
a partition-wide debug pool by default would be sensible. There are, however, a
number of difficulties in providing this feature.

The first difficulty is that this requires the entire run-time and third party
libraries to be recompiled for consistency. This is because an object allocated
with a given allocator must be deallocated with the matching deallocator. This
can only be guaranteed if all the application libraries are compiled in the same
configuration, e.g. using a debug pool as the default. In the case of third party
libraries this may not be possible since the sources of such libraries may not be
available.

Another issue is that certain new operations may malfunction when the default pool is changed. A most obvious and notable example is that the body
of the debug pool itself contains allocators, so if the wish is to change the default storage pool to be this debug pool there will be an infinite recursion. This

can be fixed by making the pool to be used for each of these types within the implementation of the debug pool explicit.

However, again the issue of third party libraries arises in an even fiercer form. It may be quite impractical to analyze a third party library to find those cases (e.g. performance requirements) where the use of the debug pool would disrupt the correct operation of the third party library, even if sources were available.

The design of the GNAT runtime has introduced the `System.Memory` abstraction partly to allow a completely different approach to replacing the default storage pool, which is to essentially replace the interface to the system `malloc()` and `free()` by a set of user-supplied routines.

## 7  Debug Pools in a Real Application

The development of the debug pool was sponsored by Eurocontrol. The CFMU (Central Flow Management Unit) is in charge of the European flight plan processing and air traffic flow management.

The total down-time for the CFMU has to be kept to a strict minimum, and therefore it is important to eliminate as many memory leaks in the application as possible.

In addition, access to invalid blocks of memory has proved to be difficult to find in more than 1.5 million lines of code. For example, several years ago such a bug took approximately 3 person-weeks of work to isolate; another one early last year required roughly 3 person-days.

To detect such problems as early as possible, a specialized build environment is now set up by CFMU using a common debug pool for all access types.

### 7.1  Use of Debug Pools

To give an idea of the CFMU development environment: at the time of writing, the set of Ada sources for TACT, one of the CFMU applications, consists of over 4,354 files: 1,993 specifications and 2,362 bodies and subunits. The total size of these sources is roughly 1.25 million lines of code. In these units, 675 access types are defined in 427 different files.

As the use of the GNAT debug pools has a performance impact, we obviously want to make it optional when building the TACT application. An emacs script was developed to automatically insert the appropriate code snippets to activate the debug pool.

### 7.2  Special Cases

For a small subset of all access types in the TACT code, the use of GNAT debug pools is not appropriate. This is because the access type is used for:

1. access to parts of untyped memory chunks (5 access types);
2. conversion to untyped memory (1 type);

3. access to memory allocated by C code in the OS or by system calls (13 types);
4. access to shared memory (4 types);
5. access to objects allocated in block via an array (1 type);
6. test code of a reference counting package (1 type), which was created to enable the detection of dangling pointers before the debug pool mechanism was available.

In retrospect, only 25 of the 675 access types in our application (less than 4%) cannot be associated to the debug pool.

## 7.3   Impact on Executable Size

One of the build options in the TACT development environment, is to create one executable for the whole system instead of creating separate executables for each logical processing. This configuration contains all code of the complete system and gives us a good measure of the impact of debug pools on the size of executables. The impact on the size of executables and on the image of processes is only in the order of 5 percent and hence negligible.

**Table 1.** Size of Executables (sizes are in bytes, bss = uninitialized data)

|  | file size 'ls' | section sizes - 'size' | | | |
|---|---|---|---|---|---|
|  |  | text | data | bss | total |
| no debug pool | $195, 904, 512$ | $76, 192, 889$ | $12, 199, 776$ | $70, 560$ | $88, 463, 225$ |
| debug pool | $201, 839, 848$ | $80, 988, 283$ | $12, 343, 312$ | $70, 560$ | $93, 402, 155$ |
| increase | +3.0% | +6.3% | +1.2% | 0.00% | +5.6% |

## 7.4   Run-Time Performance Impact

To get an indication of the run-time performance impact of debug pools, a realistic test is used. The test consists of running the TACT application and setting up its normal environment including meteo forecast data for the complete pan-European airspace. Furthermore, 853 flight plans are injected in the system, and 55 regulations are created all over Europe to force several of these flights being delayed.

Then, the arrival of a large number of radar reports is simulated over time, who, just as in real operations, provide real-time corrections to the predicted positions of these flights. These radar reports were generated to induce different kinds of shifts, such as in position, flight level or time of overflying a point. Each of these reports implies a new calculation of the remaining flight profile of that specific flight, which could mean a change of the load distribution in the different airspaces and eventually a reallocation of flight slots for several flights.

This is a rather heavy test which exercises a reasonably large part of the TACT system. The Unix "time" system call is used to measure the performance: "real" is the elapsed clock time in seconds, "user" is the CPU time spent executing user code and "system" is the CPU time spent in the OS itself (I/O operations, memory allocations, etc.). Typically, user + system is a good measure of the time needed for a job, regardless of the system load (within reason).

With the default CFMU debug pool configuration, the test runs roughly 4 times slower, hence the impact of extensively using debug pools on the performance is quite large.

**Table 2.** Execution times averaged over multiple runs

|              | real    | user    | system  | user+system |
|--------------|---------|---------|---------|-------------|
| no debug pool | 1689.49 | 1213.90 | 17.54   | 1231.44     |
| debug pool   | 5670.13 | 3766.97 | 1400.99 | 5167.96     |
| increase     | *3.36   | *3.10   | *79.87  | *4.20       |

**Depth of Stack Traces.** An important element in this slow-down is the computation of backtraces in the debug pool implementation. This is controllable with the `Stack_Trace_Depth` parameter, described as:

```
-- Stack_Trace_Depth. This parameter controls the maximum depth
-- of stack traces that are output to indicate locations of
-- actions for error conditions such as bad allocations. If set
-- to zero, the debug pool will not try to compute backtraces.
-- This is more efficient but gives less information on problem
-- locations.
```

The CFMU default value for this parameter is 10. When set to 0 or to the GNAT default of 20, respectively, the timing results of the test are given in the following table.

**Table 3.** Impact of Stack Trace Depth

|                        | real    | user    | system   | user+system |
|------------------------|---------|---------|----------|-------------|
| no debug pool          | 1689.49 | 1213.90 | 17.54    | 1231.44     |
| debug pool (stack=0)   | 1898.30 | 1635.51 | 17.20    | 1652.71     |
| increase               | *1.12   | *1.35   | *0.98    | *1.34       |
| debug pool (stack=20)  | 8575.56 | 5442.96 | 2553.74  | 7996.70     |
| increase               | *5.08   | *4.48   | *145.60  | *6.49       |

These results clearly show the majority of the slow-down is due to the computation of backtraces. When backtraces are disabled in the debug pool implementation, execution time only goes up one third. On the other hand, when the maximum depth of backtraces is set to the GNAT default of 20, execution time increases with a factor of more than six!

So a compromise needs to be found. Depending on the performance of an application without debug pools and on the available resources, it can be interesting to have regular tests with debug pools enabled but without backtrace computation. If these indicate a heap problem, the test can then be rerun with a large value for the Stack_Trace_Depth parameter.

**Scanning Memory before Releasing.** Another useful configuration parameter is Advanced_Scanning, described as:

```
-- Advanced_Scanning: If true, the pool will check the contents
-- of all allocated blocks before physically releasing
-- memory. Any possible reference to a logically free block will
-- prevent its deallocation.
```

CFMU sets this parameter to the non-default value of True. To get an idea of the overhead this entails, the same tests are run but now with the GNAT-default value of False. The timing results are given in the following table.

**Table 4.** Impact of No Advanced Scanning

|  | real | user | system | user+system |
|---|---|---|---|---|
| debug pool (stack=0) | 1825.70 | 1507.33 | 17.32 | 1524.65 |
| vs. with scanning | −72.60 | −128.18 | +0.12 | −128.06 |
|  | 96.2% | 92.2% | 100.7% | 92.3% |
| debug pool (stack=10) | 5507.90 | 3621.19 | 1397.66 | 5018.85 |
| vs. with scanning | −118.18 | −142.73 | −3.60 | −146.34 |
|  | 97.9% | 96.2% | 99.7% | 97.7% |
| debug pool (stack=20) | 8385.11 | 5301.28 | 2530.61 | 7831.89 |
| vs. with scanning | −190.45 | −141.68 | −23.13 | −164.81 |
|  | 97.8% | 97.4% | 99.1% | 97.9% |

This shows that the performance cost of "Advanced Scanning" is quite small: it only requires between 2% of the user plus system time, if the Stack_Trace_Depth parameter is set to the default GNAT value of 20, and less than 8% if the computation of backtraces is disabled completely. For these test runs, this accounts for roughly 2-3 minutes of CPU time on a total between 25 and 130 minutes.

## 7.5   Results Obtained

One of the important results obtained through the use of debug pools, is that its extensive reporting has indicated some bizarre heap usage in our application, which caused a serious performance drain.

All processes on the TACT server and all MMIs on various workstations need access to a large amount of changing environment data (definitions of aerodromes, points, routes, etc.). Each of these processes maintains a local cache of the data it needs, and these caches are kept synchronised by inter-process communication via encoded buffer transfers.

Closer examination showed that due to an incorrectly set boolean variable, the encoding and decoding of these buffers was not done in the intended compact binary format but in the bulky textual format (intended for easy human interpretation, though approx. two orders of magnitude larger). This not only implied a lot of unneeded heap usage, detected through the debug pool reporting, but also very inefficient inter-process communication.

Another important result of running our large suite of regression tests with the debug pool enabled, is that our code now is shown to be free of heap corruptions.

And as our system builds always include the systematic execution of the full regression test suite, regular builds with the debug pool enabled offer a protection against the introduction of new heap corruptions in our code.

There are some limitations. Allocations in non-Ada code fall outside the scope of the GNAT debug pools. Nor are allocations on the stack through general access types taken into account: a possible future enhancement?

# 8    Conclusion

This paper has shown how the notion of checked pools, an extension of the storage pools concept of Ada 95, can be put to profit to implement debug pools to track down memory corruptions and possible memory leaks in a user application. Are debug pools the ultimate solution to dynamic memory problems in Ada code?

Probably not: like for all things if all you have is a hammer all your problems look like a nail and it is important to offer developers a choice of tools and approaches when tackling memory managment issues.

Ada was designed to allow garbage collection. Currently no native Ada implementation offers it. Wouldn't it be nice to combine the power of debug pools with the flexibility of a real-time garbage collecting implementation?

# References

1. Taft, S.T., Duff, R.A., Brukardt, R.L. and Plödereder, E.; *Consolidated Ada Reference Manual. Language and Standard Libraries*, ISO/IEC 8652:1995(E) with COR.1:2000, Lecture Notes in Computer Science, vol. 2219, Springer-Verlag, 2001.
2. Barnes, J.; *Storage Pool Alignment*, Ada User Journal, pp. 182-187, vol. 19, number 3, October 2001.

# Busy Wait Analysis

Johann Blieberger[1], Bernd Burgstaller[1], and Bernhard Scholz[2]

[1] Institute for Computer-Aided Automation, TU Vienna
Treitlstr. 1–3, A-1040 Vienna, Austria
{blieb,bburg}@auto.tuwien.ac.at
[2] Institute of Computer Languages, TU Vienna
Argentinierstr. 8/4, A-1040 Vienna, Austria
scholz@complang.tuwien.ac.at

**Abstract.** A busy wait loop is a loop which repeatedly checks whether an event occurs. Busy wait loops for process synchronization and communication are considered bad practice because (1) system failures may occur due to race conditions and (2) system resources are wasted by busy wait loops. In general finding a busy wait loop is an undecidable problem. To get a handle on the problem, we introduce a decidable predicate for loops that will spot most important classes of busy waiting although false alarms may occur. The loop predicate for detecting busy wait loops is based on control flow graph properties (such as loops) and program analysis techniques.

## 1   Introduction

Although for efficiency reasons busy waiting is employed in operating system kernels for process synchronization ("spin locks"), it is considered bad practice to use it for task synchronization and task communication in application programs. Busy waiting results in a waste of system resources. Programs, that actively wait for an event, may cause a severe overhead in a multi-tasking environment and can cause system failure due to race conditions. Therefore, programmers should use higher communication facilities such as semaphores, monitors, rendezvous, etc. [Hoa85,Ada95].

However, it is hard to detect busy waiting in existent code and therefore it is of great importance to have a static analysis tool that targets the detection of busy waiting. Such a tool significantly improves the quality of software in order to prevent programs that use busy waiting.

Before we discuss busy waiting, we have to note that this term is extremely vague. In fact, there is no definition of busy waiting, everybody agrees upon. For example in [And91] it is defined as

> " ... a form of synchronization in which a process repeatedly checks a condition until it becomes true ... ".

We start by defining what we mean by busy waiting: A program that within a loop constantly reads a value from a certain variable, where the loop exit

J.-P. Rosen and A. Strohmeier (Eds.): Ada-Europe 2003, LNCS 2655, pp. 142–152, 2003.

condition is dependent on this value, imposes busy waiting and we say that the loop is a *busy wait loop*. We assume that only another thread/task can terminate the loop by altering the value of the variable. A variable being responsible for busy waiting is called a *wait variable*.

Since busy wait loops may loop forever, the general problem of spotting them is equivalent to the halting problem and thus undecidable. Hence we cannot find all busy wait loops automatically. On the other hand, our analysis will raise false alarms in certain cases (e.g. with *blocking assignments* described below). However, we believe that our analysis will find a large class of busy wait loops and even false alarms may stimulate the programmer to improve the program code.

The paper is structured as follows. In Sect. 2 we give definitions used throughout the paper. In Sect. 3 we motivate our analysis. In Sect. 4 we describe the algorithm for detecting busy wait loops. Finally we draw our conclusions in Sect. 6.

## 2    Background

A control flow graph $G < N, E, e, x >$ is a directed graph [ASU86] with node set $N$ and edge set $E \subseteq N \times N$. Nodes $n \in N$ represent basic blocks consisting of a linear sequence of statements. Edges $(u, v) \in E$ represent the non-deterministic branching structure of $G$, and $e$ and $x$ denote the unique *start* and *end node* of $G$, respectively. Moreover, $succ(u) = \{v \mid (u, v) \in E\}$ and $pred(u) = \{v \mid (v, u) \in E\}$ represent the *immediate successor* and *predecessor* of node $u$. A *finite path* of $G$ is a sequence $\pi =< u_1, u_2, \ldots, u_k >$ of nodes such that $u_{i+1} \in succ(u_i)$ for all $1 \leq i < k$. Symbol $\varepsilon$ denotes the empty path.

A path $\pi =< u_1, \ldots, u_k >$ is said to be a member of a node set $X$ ($\pi \in X$), if all nodes in the path are members of $X$.

Let node $u$ *dominate* [Muc00] node $v$, written $u \operatorname{dom} v$, if every possible path from the start node $e$ to node $v$ includes $u$. The domination relation $u \operatorname{dom} v$ is reflexive ($u \operatorname{dom} u$), transitive ($u \operatorname{dom} v \wedge v \operatorname{dom} w \Rightarrow u \operatorname{dom} w$), and anti-symmetric ($u \operatorname{dom} v \wedge v \operatorname{dom} u \Rightarrow u = v$).

For every node $u \in N \setminus \{e\}$ there exists an immediate dominator $v$, written as $v = \operatorname{idom}(u)$ such that there exists no dominator $w \neq v$ of $u$ which is dominated by $v$. The immediate dominators construct a tree also known as the dominator tree. The dominator tree is a compressed representation of the domination relation.

A *back edge* $(m, n) \in E$ in a control flow graph $G$ is defined to be an edge whose target dominates its source. The set of back edges is defined to be $B = \{(m, n) \in E \mid n \operatorname{dom} m\}$.

## 3    Motivation

Our intuition of a busy wait loop is based on the notion of loops and the read/write semantics of program variables inside the loop. According to [ASU86]

```
1    Turn : Integer := 1;
2    Flag0 : Boolean := False;
3    Flag1 : Boolean := False;

4    procedure P0 is
5    begin
6       --  Claim Critical Section:
7       Flag0 := True;                              --  Node 1
8       while Flag1 = True loop                     --  Node 2
9          if Turn = 1 then                         --  Node 3
10            Flag0 := False;                       --  Node 4
11            while Turn = 1 loop                   --  Node 5
12               null;                              --  Node 6
13            end loop;
14            Flag0 := True;                        --  Node 7
15         end if;
16      end loop;
17      --  Critical Section:
18      null;                                       --  Node 8
19      --  Leave Critical Section:
20      Turn := 1;                                  --  Node 8
21      Flag0 := False;                             --  Node 8
22   end P0;
```

**Fig. 1.** Running Example: Dekker's Algorithm

a *definition* of a variable $x$ is a statement that assigns $x$ a value. Contrary, a variable *declaration* is a syntactic constructs which associates information (e.g. type information) with a given name. A variable declaration usually implies also an initial definition for the variable itself.

With our analysis we are interested in program variables which determine whether the loop is terminated or iterated again. We assume that we find these program variables in the exit-condition of loops. If such a variable is only read, without employing higher communication facilities or being *defined* inside the loop, this variable might be responsible for inducing busy waiting and we call this variable a *wait variable*.

We illustrate these terms by the mutual exclusion algorithm given in Fig. 1. Dijkstra [Dij68] attributes this algorithm to the Dutch mathematician T. Dekker, and it is in fact the first known solution to the mutual exclusion problem for two processes that does not require strict alternation. It is worth noting that this algorithm only assumes mutual exclusion at the memory access level which means that simultaneous memory access is serialized by a memory arbiter. Beyond this, no further support from the hardware (e.g. atomic test and set instructions), operating system, or programming language is required. Since, for this reason, the algorithm solely relies on global variables (cf. lines 1 ... 3) that are written and read for synchronization purposes, we have found it to be an instructive example of busy waiting behavior. Note that we have omitted the code for the second process (P1), as it is dual to P0.

In order to demonstrate the overhead in CPU processing time induced by busy waiting we have implemented Dijkstra's $N$-way generalization [Dij65] of Dekker's 2-way mutual exclusion algorithm for the RTAI [$M^+00$] real-time Linux executive run on a uni-processor platform. Our investigation focused on the average execution time of a busy waiting real-time task that has to enter the critical section a constant number of times in the presence of $N-1$ busy waiting competitors with the same assignment. In Fig. 2 these results are compared to an implementation utilizing a semaphore to ensure mutual exclusion between tasks. Due to the blocking nature of the semaphore the measured execution times show only linear growth in terms of an increasing number of participating tasks. This behavior is clearly exceeded by the busy waiting task ensemble that spends most of its execution time polling to gain access to the critical section.

It is clear that our example contains busy waiting and our objective is to design an algorithm that detects this busy waiting behavior just by inspecting the loops of the program and the read/write semantics of program variables inside the loop.

*Non-dangerous statements* are statements which prevent a variable from being a wait variable. We call non-dangerous assignment statements *blocking assignments*. If a variable is defined by some of these statements, we assume that busy waiting is improbable to occur. Statements which are not non-dangerous are called *dangerous*. We discriminate between tasking statements of Ada and other statements.

1. All calls to a (guarded[1]) entry of a task or protected object are considered non-dangerous.
2. Timed entry calls are non-dangerous iff the expiration time does not equal zero.
3. Asynchronous transfer of control (ATC) is non-dangerous if the triggering statement is an entry call (to a guarded entry).
4. A timeout realized by ATC is considered non-dangerous.
5. Conditional entry calls are generally considered dangerous.
6. In general we assume that file operations are non-dangerous; the same applies to Terminal I/O. We do not consider cases such as a file actually being a named pipe which is fed by a program providing output in an infinite loop.
7. Read/Write attributes may or may not block depending on the actual implementation. For this reason we consider read/write attributes dangerous.
8. Assignments via function or procedure calls are dangerous even if inside the subprogram there is a blocking assignment (we do not perform interprocedural analysis).

---

[1] A call to a guarded entry is a non-dangerous statement with high probability except if the guard equals *true*; a call to a task entry without a guard has high probability to be dangerous, except if the corresponding accept statement is located in a part different from the main "select" loop in the task body (which makes sense for task initialization and finalization).

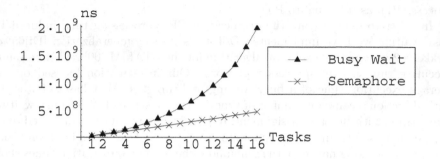

**Fig. 2.** Task Execution Times: Busy Waiting vs. High-Level Synchronization

## 4   Algorithm

For detecting busy wait loops we analyze loops of a program. We use the control flow graph as an underlying data structure for the detection. Based on control flow properties and semantic properties of statements, we decide whether a loop is a busy wait loop or not.

In general it is not easy to find loops in control flow graphs [Ram99]. However, if those graphs are *reducible* (cf. [ASU86]), loops are simple to define [Muc00]. In the following we introduce *natural loops*, which are defined by their back-edges[2].

A *natural loop* of a back-edge $(m, n)$ is the sub-graph consisting of the set of nodes containing $n$ and all the nodes which can reach $m$ without passing $n$. Let $L_{(m,n)}$ denote the set of nodes which induce the sub-graph that forms the natural loop of back-edge $(m, n)$. Then,

$$L_{(m,n)} = \{u \mid \exists \pi = < u, \dots, m >: n \notin \pi\} \cup \{n\}. \tag{1}$$

Node $n$ is said to be the *loop header* of loop $L_{(m,n)}$ because the loop is entered through $n$. Therefore, the loop header dominates all nodes in the loop.

The algorithm for computing $L_{(m,n)}$ can be found in Fig. 4. It is a simple work-list algorithm. It computes the immediate and intermediate successors of node $m$. The algorithm stops after all successors of $m$ excluding the successors of $n$ have been found. The immediate and intermediate successors excluding the successor of $n$ represent the set of nodes for the loop.

Recall our example of Fig. 1. The control flow graph of our running example is given in Fig. 3a. In addition the dominator tree is depicted in Fig. 3b. To compute

---

[2] Note that this definition is only valid for reducible control flow graphs. However, Ada programs result in reducible control flow graphs only, and this is no restriction for the analysis.

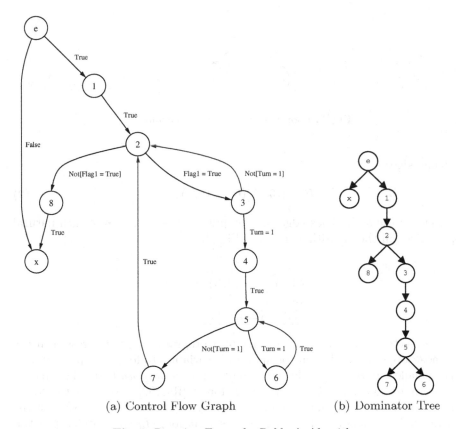

(a) Control Flow Graph                    (b) Dominator Tree

**Fig. 3.** Running Example: Dekker's Algorithm

```
1:  W := {m};
2:  L_(m,n) := {n};
3:  repeat
4:     select u ∈ W;
5:     L_(m,n) := L_(m,n) ∪ {u};
6:     W := (W ∪ succ(u)) \ L_(m,n));
7:  until W = ∅
```

**Fig. 4.** Algorithm for Computing $L_{(m,n)}$

the dominator tree several algorithms have been proposed in literature [Muc00, LT79,AHLT99]. The complexity of the algorithms varies from cubic to linear[3].

For finding the loops of our example we determine the back edges occurring in the control flow graph of Fig. 3a. The set $B$ of back edges consists of the

---

[3] It seems to be the case that if a lower complexity is desired more effort has to be put into implementing the algorithm.

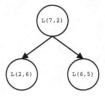

**Fig. 5.** Loop Forest of Running Example

following edges:

$$B = \{(3,2),(6,5),(7,2)\}. \tag{2}$$

Based on the set of back edges we compute the set of nodes of each natural loop as given in the algorithm shown in Fig. 4.

| $(m,n)$ | $L_{(m,n)}$ |
|---------|-------------|
| $(3,2)$ | $\{2,3\}$ |
| $(6,5)$ | $\{5,6\}$ |
| $(7,2)$ | $\{2,3,4,5,6,7\}$ |

For detecting busy waiting behavior of a program we are interested in the innermost loops containing one or more wait variables to focus the programmer on the smallest portion of code which may induce a *busy wait loop*. For finding the innermost loops we construct a loop forest [Ram99]. Nodes in the forest represent loops and an edge denotes a nesting relation. For our example the loop forest is depicted in Fig. 5. It shows that loop $L_{(7,2)}$ contains the two inner loops, i.e., $L_{(6,5)}$ and $L_{(3,2)}$. Note that the loop forest represents the set-relationship $L_1 \subset L_2$ between two loops $L_1$ and $L_2$. In the example the following holds: $L_{(6,5)} \subset L_{(7,2)}$ and $L_{(3,2)} \subset L_{(7,2)}$.

For locating the busy wait loop we will analyze the loops in reverse topological order of the loop forest, i.e. from the inner-most loops to the top-level loops of a program. The reverse topological order guarantees that the busy wait behavior of the program can be localized quite precisely.

Now, we want to deduce the set of statements, which influence the termination of a loop. These are statements inside the loop, which have at least one immediate successor that is outside the loop. For example an `exit when` ... statement might terminate the loop.

The statements that influence the termination of the loop are given by the following set of edges:

$$T_{(m,n)} = \{(u,v) \in E \mid u \in L_{(m,n)} \land v \notin L_{(m,n)}\} \tag{3}$$

The definition establishes those edges whose sources but not its destinations are part of the loop $L_{(m,n)}$. Note that a statement in loop $L_{(m,n)}$ must have at least two successors for its out-going edges to contribute to the set $T_{(m,n)}$. Therefore, the statement must be a branching node and there must be a branch predicate that decides whether to stay in the loop or to exit the loop. The branch predicate consists of program variables and we call these variables candidates for *wait variables*. These wait variables might cause a busy wait loop and must be checked.

The set of candidates, which might be wait variables, are given as follows,

$$V_{(m,n)} = \{\mathbf{var} \in bp(u) \mid (u,v) \in T_{(m,n)}\} \tag{4}$$

where $bp(u)$ denotes the branch predicate of branching node $u$ and **var** is a candidate for a wait variable and hence needs to be added to the set $V_{(m,n)}$. For our example given in Fig. 1 the T-sets and V-sets are as follows:

| $(m,n)$ | $T_{(m,n)}$ | $V_{(m,n)}$ |
|---------|-------------|-------------|
| $(3,2)$ | $\{(3,4),(2,8)\}$ | $\{\text{Turn}, \text{Flag1}\}$ |
| $(6,5)$ | $\{(5,7)\}$ | $\{\text{Turn}\}$ |
| $(7,2)$ | $\{(2,8)\}$ | $\{\text{Flag1}\}$ |

In the example we have three branching nodes which might cause a busy wait loop, i.e. nodes 2, 3, and 5. For example the branch predicate of branching node 2 is given as `Flag1 = true`. Variable `Flag1` occurs in the branch predicate and, therefore, it is candidate for a busy wait variable, i.e. `Flag1` $\in V_{(3,2)}$ and `Flag1` $\in V_{(7,2)}$.

Now, we have to check if a candidate for a busy wait variable might be a wait variable and might cause the loop to loop forever without any interaction from another thread/task. This might happen if there is a path in the (natural) loop that does not contain a definition for a canditate.

For a basic block $u$ we introduce a local predicate $defined_{\mathbf{var}}(u)$. The predicate holds for a blocking assignment for variable **var** in basic block $u$. For example statement `Flag0:=False;` in node 4 of our running example contains a blocking assignment for variable `Flag0`. Therefore, predicate $defined_{\texttt{Flag0}}(4)$ holds.

We extend the definition of $defined_{\mathbf{var}}$ for paths. If in a path $\pi$ there exists at least one blocking assignment for variable **var**, the predicate $defined_{\mathbf{var}}(\pi)$ holds. The extension is obtained as follows:

$$defined_{\mathbf{var}}(<u_1, u_2, \ldots, u_k>) = \bigvee_{1 \le i \le k} defined_{\mathbf{var}}(u_i) \tag{5}$$

**Definition 1.** *If for a variable* **var** *the* **busy-var***-predicate*

$$\mathbf{busy\text{-}var}(L_{(m,n)}, \mathbf{var}) =_{\text{defs}} \exists \pi = <n, \ldots, n> \in L_{(m,n)} : \neg defined_{\mathbf{var}}(\pi) \tag{6}$$

*holds in loop* $L_{(m,n)}$, *the loop is supposed to be a busy wait loop.*

For each program variable **var** $\in V_{(m,n)}$ we construct the induced subgraph of the nodes in the loop where $defined_{\mathbf{var}}(u)$ is false. If $m$ is reachable from $n$ in this induced sub-graph, it is simple to show that the predicate **busy-var**$(L_{(m,n)}, \mathbf{var})$

holds. The check boils down to a reachability check in the induced subgraph of the node set $L_{(m,n)} \setminus \{u \mid \mathit{defined}_{\mathbf{var}}(u)\}$. In graph theory there are very efficient algorithms for doing so [Meh84,Sed88].

If we consider our example, the loop $L_{(3,2)}$ is such a loop. For both variables Turn and Flag1 in $V_{(3,2)}$ there is no blocking assignment in nodes 2 and 3. Therefore, node 3 is reachable from node 2 and the predicate **busy-var**$(L_{(m,n)}, \text{Turn})$ and **busy-var**$(L_{(m,n)}, \text{Flag1})$ is true. Similar accounts for the other loops and their variables of our running example, which implies that all loops of our running example are busy wait loops.

Finally, we put together our detection algorithm in Fig. 6. In the pre-phase we build the dominator tree and determine the back-edges. Then, we compute the loop forest which tells us the loop order of the analysis. In reverse topological order of the loop forest, we check loop by loop. For each loop we compute the set of nodes which terminate the loop $(T_{(m,n)})$ and the variables inside the loop exit condition $(V_{(m,n)})$. For every variable in $V_{(m,n)}$ we compute the busy wait predicate and if it is true, we output a warning for this particular loop.

> 1:  *compute dominator tree*
> 2:  *determine back edges*
> 3:  *compute loop forest*
> 4:  **for** $(m, n)$ in reverse topological order of loop forest **do**
> 5:      *compute* $T_{(m,n)}$
> 6:      *compute* $V_{(m,n)}$
> 7:      **for var** $\in V_{(m,n)}$ **do**
> 8:          *determine* **busy-var**$(L_{(m,n)}, \mathbf{var})$
> 9:          **if busy-var**$(L_{(m,n)}, \mathbf{var})$ **then**
> 10:             *output warning for* **var** *and loop* $L_{(m,n)}$
> 11:         **end if**
> 12:     **end for**
> 13: **end for**

**Fig. 6.** Algorithm for Detecting Busy Waiting

## 5   Refinement

By a slight modification we can improve (sharpen) our analysis, i.e., we can detect more busy wait loops.

Consider the code fragment given in Fig. 7. In this case our algorithm finds an assignment to the variable $i$ within the loop body and concludes that there is no busy waiting, which is plainly wrong.

This behavior can be improved by considering all variables appearing on the right hand side of assignments to candidate variables to be candidates as well.

```
i,j: integer := 0;

loop
    i := j;
    exit when i=1;
end loop;
```

**Fig. 7.** Example: Indirect Busy Wait Variable

Formally we have to redefine $V_{(m,n)}$ in the following way:

$$V_{(m,n)}^0 = \{\mathbf{var} \in bp(u) \mid (u,v) \in T_{(m,n)}\}$$
$$V_{(m,n)}^{k+1} = \{\mathbf{var} \in \text{rhs of assignments in } L_{(m,n)} \text{ to } \mathbf{var} \in V_{(m,n)}^k\}$$
$$V_{(m,n)} = \bigcup_{k \geq 0} V_{(m,n)}^k$$

This refinement of our analysis sharpens its results in that more busy wait loops are detected, but on the other hand more false alarms can be raised. For example replace the assignment i := j; with i := j+i-j+1; in Fig. 7. In this case busy waiting will be reported by our refined algorithm although this is not true.

## 6  Conclusion and Future Work

We have presented an algorithm for detecting busy waiting that can be used either for program comprehension or for assuring code quality criteria of programs. Specifically, if processes or threads are part of a high level language (as with Ada or Java), the programmer should be aware of synchronization mechanisms. A tool that detects busy waiting is of great importance for saving system resources and making a program more reliable.

Symbolic methods such as those introduced in [CHT79] will certainly improve analysis in that less false alarms will be raised and more busy wait loops can be found. We will consider symbolic busy wait analysis in a forthcoming paper.

**Acknowledgments.** One of the authors (JB) wants to thank Manfred Gronau for pointing him to the subject of busy wait loops and the fact of missing computer-aided analysis.

## References

[Ada95]    ISO/IEC 8652. *Ada Reference Manual*, 1995.
[AHLT99]  S. Alstrup, D. Harel, P. W. Lauridsen, and M. Thorup. Dominators in linear time. *SIAM Journal on Computing*, 28(6):2117–2132, 1999.

[And91]   G. R. Andrews. *Concurrent Programming, Principles & Practice.* Benjamin/Cummings, Redwood City, California, 1991.

[ASU86]   A. V. Aho, R. Sethi, and J. D. Ullman. *Compilers.* Addison-Wesley, Reading, Massachusetts, 1986.

[CHT79]   T. E. Cheatham, G. H. Holloway, and J. A. Townley. Symbolic Evaluation and the Analysis of Programs. *IEEE Trans. on Software Engineering*, 5(4):403–417, July 1979.

[Dij65]   E. W. Dijkstra. Solution of a problem in concurrent programming control. *Communications of the ACM*, 8(9):569, 1965.

[Dij68]   E. W. Dijkstra. Co-operating sequential processes. In F. Genuys, editor, *Programming Languages: NATO Advanced Study Institute*, pages 43–112. Academic Press, 1968.

[Hoa85]   C. A. R. Hoare. *Communicating Sequential Processes.* Prentice Hall, Englewood Cliffs, NJ, 1985.

[LT79]   T. Lengauer and R. E. Tarjan. A fast algorithm for finding dominators in a flow graph. *ACM Transactions on Programming Languages and Systems*, 1(1):121–141, July 1979.

[M+00]   P. Mantegazza et al. *DIAPM RTAI Programming Guide 1.0.* Lineo, Inc., Lindon, Utah 84042, US, 2000. http://www.rtai.org.

[Meh84]   K. Mehlhorn. *Graph Algorithms and NP-Completeness*, volume 2 of *Data Structures and Algorithms*. Springer-Verlag, Berlin, 1984.

[Muc00]   S. S. Muchnick. *Advanced Compiler Design and Implementation.* Morgan Kaufmann, San Francisco, 2000.

[Ram99]   G. Ramalingam. Identifying loops in almost linear time. *ACM Transactions on Programming Languages and Systems*, 21(2):175–188, March 1999.

[Sed88]   R. Sedgewick. *Algorithms.* Addison-Wesley, Reading, MA, 2nd edition, 1988.

# Eliminating Redundant Range Checks in GNAT Using Symbolic Evaluation

Johann Blieberger and Bernd Burgstaller

Department of Computer-Aided Automation
Technical University Vienna
A-1040 Vienna, Austria
{blieb,bburg}@auto.tuwien.ac.at

**Abstract.** Implementation of a strongly typed language such as Ada95 requires range checks in the presence of array index expressions and assignment statements. Range checks that cannot be eliminated by the compiler must be executed at run-time, inducing execution time and code size overhead. In this work we propose a new approach for eliminating range checks that is based on symbolic evaluation. Type information provided by the underlying programming language is heavily exploited.

## 1 Introduction

Strongly typed programming languages impose non-trivial requirements on an implementation in order to ensure the properties of the type system as specified by the language semantics. For discrete types these properties are often expressed as range constraints that must be met by values of a given type. For example, [Ada95] requires for signed integer types that Constraint_Error is raised by the execution of an operation yielding a result that is outside the base range of the respective type. In the case of constrained subtypes the constraints imposed by the requested ranges have to be enforced. In the case of array access, [Ada95] requires that each index value belongs to the corresponding *index range* of the array. Integrity of a given program with respect to type system properties is usually enforced through checks performed for each program statement that can potentially violate them. Every check that cannot be proved redundant at compile-time has to be postponed until run-time involving two additional comparison and conditional branch instructions that need to be executed in order to ensure that a given value is within a required range. The resulting impact both in execution time and code size is high enough to justify the existence of compile-time options and language constructs that allow (partial) suppression of checks but open up a potential backdoor to erroneous program execution in turn.

On the contrary, an approach that increases the number of checks that can be proved redundant at compile-time aids in avoiding this backdoor by reducing run-time overhead and thus supporting "acceptance" of those checks for which either no proof of redundancy can be found or no such proof exists at all.

J.-P. Rosen and A. Strohmeier (Eds.): Ada-Europe 2003, LNCS 2655, pp. 153–167, 2003.

It is the goal of our range check elimination method to identify redundant range checks that cannot be spotted by existing state-of-the-art compiler technology.

The outline of this paper is as follows. In Section 2 we present the notions of the symbolic data flow analysis framework used. In Section 3 we demonstrate the effectiveness of our framework through an example. Section 4 is devoted to the implementation of our dataflow framework within GNAT. Section 5 presents the experimental results gathered so far. In Section 6 we survey related work. Finally we draw our conclusion in Section 7.

## 2   Symbolic Evaluation

### 2.1   Motivation

Consider the code fragment given in Fig. 1 which swaps the values stored in $u$ and $v$ in place. It contains standard Ada95 code except for the numbered curly braces "$\{\dots\}^{R<\text{number}>}$" around those expressions for which GNAT will issue a range check (cf., for instance, line 5).

Fig. 1 shows the intervals that GNAT derives by the standard interval arithmetic ([Hag95]) of Fig. 2[‡]. Three range checks, one for each assignment, are necessary because the computed intervals of the expressions on the right hand side exceed the range of subtype $sint$.

| | | Interval Arithmetic | | Symbolic Values | |
|---|---|---|---|---|---|
| | | u | v | u | v |
| 1 | **subtype** sint **is integer range** -10 .. +10; | | | | |
| 2 | ... | | | | |
| 3 | **procedure** Swap (u, v : **in out** sint) **is** | | | | |
| 4 | **begin** | $[-10,+10]$ | $[-10,+10]$ | $\underline{u}$ | $\underline{v}$ |
| 5 |     u := $\{$u+v$\}^{R1}$; | $[-20,+20]$ | $[-10,+10]$ | $\underline{u}+\underline{v}$ | $\underline{v}$ |
| 6 |     v := $\{$u−v$\}^{R2}$; | $[-10,+10]$ | $[-20,+20]$ | $\underline{u}+\underline{v}$ | $\underline{u}$ |
| 7 |     u := $\{$u−v$\}^{R3}$; | $[-20,+20]$ | $[-10,+10]$ | $\underline{v}$ | $\underline{u}$ |
| 8 | **end** Swap; | | | | |

**Fig. 1.** Swapping the contents of two variables

$$[a,b] + [c,d] = [a+c, b+d]$$
$$[a,b] - [c,d] = [a-d, b-c]$$
$$[a,b] * [c,d] = [\min(ac, ad, bc, bd), \max(ac, ad, bc, bd)]$$
$$[a,b]/[c,d] = [a,b] * [1/d, 1/c], 0 \notin [c,d]$$

**Fig. 2.** Interval Arithmetic

---

[‡] In fact GNAT applies simplified rules for $[a,b] * [c,d]$ and $[a,b]/[c,d]$.

In contrast, the symbolic values in Fig. 1 show that range checks $R2$ and $R3$ are redundant. In line 6 variable $v$ is assigned the value of the actual parameter of $u$, namely $\underline{u}$, and in line 7 $u$ is assigned $\underline{v}$. If one of $\underline{u}$ or $\underline{v}$ would not be in the range of subtype *sint*, the range checks due to the Ada95 parameter passing mechanism would raise exception *Constraint_Error* and control flow would not reach the body of procedure *Swap*. It is this property, denoted as $\underline{u}, \underline{v} \in [-10, +10]$, together with the fact that the values for $u$ and $v$ are computed symbolically, that lets us actually derive that range checks $R2$ and $R3$ are redundant.

As can be seen by this introductory example symbolic analysis derives tighter bounds for the ranges of expressions and thus reduces the number of necessary range checks. The underlying theory of this approach is sketched in the remaining part of this section.

## 2.2  Preliminaries

*Symbolic evaluation* is an advanced static program analysis in which symbolic expressions are used to denote the values of program variables and computations (cf. e.g. [CHT79]). A path condition describes the impact of the program's control flow onto the values of variables and the condition under which control flow reaches a given program point. In the past, symbolic evaluation has already been applied to the reaching definitions problem [BB98], to general worst-case execution time analysis [Bli02], to cache hit prediction [BFS00], to alias analysis [BBS99], to optimization problems of High-Performance Fortran [FS97], and to pointer analysis for detecting memory leaks [SBF00].

The underlying program representation for symbolic evaluation is the *control flow graph (CFG)*, a directed labeled graph. Its nodes are basic blocks containing the program statements, whereas its edges represent transfers of control between basic blocks. Each edge of the CFG is assigned a condition which must evaluate to true for the program's control flow to follow this edge. *Entry (e)* and *Exit (x)* are distinguished nodes used to denote the start and terminal node.

In the center of our symbolic analysis is the *program context*, which includes states $S_i$ and *path conditions* $p_i$. A program context completely describes the variable bindings at a specific program point together with the associated path conditions and is defined as

$$\bigcup_{i=1}^{k}[S_i, p_i],\tag{1}$$

where $k$ denotes the number of different program states. State $S$ is represented by a set of pairs $\{(v_1, e_1), \ldots, (v_m, e_m)\}$ where $v_s$ is a program variable, and $e_s$ is a symbolic expression describing the value of $v_s$ for $1 \le s \le m$. For each variable $v_s$ there exists exactly one pair $(v_s, e_s)$ in $S$. A path condition $p_i$ specifies a condition that is valid for a given state $S_i$ at a certain program point.

*Example 1.* The context

$$[\{(x, \underline{n}^2 - 1), (y, \underline{n} + 3)\}, x = y - 2]\tag{2}$$

consists of two variables $x$ and $y$ with symbolic values $\underline{n}^2 - 1$ and $\underline{n} + 3$, respectively. Variable $\underline{n}$ denotes some user input. The path condition ensures that $x = y - 2$ at the program point where the context is valid.

## 2.3   Type System Information and Range Checks

Informally a type $\overline{\mathcal{T}}$ is characterized by a set $\mathcal{T}$ of values and a set of primitive operations. To denote the type of an entity $j$ we write $\overline{\mathcal{T}}(j)$, to denote the set of values for $\overline{\mathcal{T}}(j)$ we write $\mathcal{T}(j)$. The fact that a given value $e \in \mathbb{Z}$ is contained in the set $\mathcal{T}(j)$ of a (constrained) integer type[§] is written as

$$e \in \mathcal{T}(j) \Leftrightarrow e \in \left[ \overline{\mathcal{T}}(j)\text{'First}, \overline{\mathcal{T}}(j)\text{'Last} \right] \tag{3}$$

where $\overline{\mathcal{T}}(j)$'First and $\overline{\mathcal{T}}(j)$'Last denote the language-defined attributes for scalar types (cf. [Ada95]). A range check is a test of Equation (3) and is denoted as

$$e \in^? \mathcal{T}(j). \tag{4}$$

In general the test is not performed on a value $e$ but on an expression $E$. Predicate *val* evaluates range checks symbolically within program contexts, which means that the check $E \in^? \mathcal{T}(j)$ is evaluated for each pair $[\mathcal{S}_i, p_i]$:

$$\mathrm{val}(E \in^? \mathcal{T}(j), [\mathcal{S}_1, p_1] \cup \cdots \cup [\mathcal{S}_k, p_k]) \mapsto$$

$$[\mathcal{S}_1, p_1 \wedge \mathrm{val}(E \in^? \mathcal{T}(j), [\mathcal{S}_1, p_1])] \cup \cdots \cup [\mathcal{S}_k, p_k \wedge \mathrm{val}(E \in^? \mathcal{T}(j), [\mathcal{S}_k, p_k])].$$

This evaluation involves the symbolic evaluation of $E$ for $[\mathcal{S}_i, p_i]$, denoted as $\mathrm{val}(E, [\mathcal{S}_i, p_i])$. If the result of this evaluation can be shown to be contained in the set $\mathcal{T}(j)$, then the evaluation of check $E \in^? \mathcal{T}(j)$ yields *true*, which means that the check is *redundant* for $[\mathcal{S}_i, p_i]$:

$$\mathrm{val}(E \in^? \mathcal{T}(j), [\mathcal{S}_i, p_i]) = \begin{cases} true \Leftrightarrow \mathrm{val}(E, [\mathcal{S}_i, p_i]) \in \mathcal{T}(j) \\ false \text{ else.} \end{cases} \tag{5}$$

Otherwise the range check is *required* or cannot be proved to be redundant. Deciding on the truth value of the above equation represents the center-piece of our range check elimination method. It depends on the data flow framework presented in Section 2.4, its exact treatment is thus deferred until Section 2.5.

Based on Equation (5) we define the necessity of a range check at a given program context via predicate ?rc. It evaluates to *false* only iff the range check is *redundant* for every pair $[\mathcal{S}_i, p_i]$ of the program context:

$$?\mathrm{rc}\left( E \in^? \mathcal{T}(j), \bigcup_{i=1}^{k} [\mathcal{S}_i, p_i] \right) = \begin{cases} false \Leftrightarrow \bigvee_{1 \leq i \leq k} (\mathrm{val}(E \in^? \mathcal{T}(j), [\mathcal{S}_i, p_i]) = true) \\ true \text{ else.} \end{cases}$$
$$\tag{6}$$

For the compiler backend we map predicate ?rc to the node-flag *Do_Range_Check* of GNAT's *abstract syntax tree* (cf. also Section 4).

---

[§]   In Ada terms [Ada95] a combination of a type, a constraint on the values of the type, and certain specific attributes is called *subtype*.

## 2.4   A Data-Flow Framework for Symbolic Evaluation

We define the following set of equations for the symbolic evaluation framework:

$$\text{SymEval}(B_{\text{entry}}) = [\mathcal{S}_0, p_0]$$

where $\mathcal{S}_0$ denotes the initial state containing all variables which are assigned their initial values, and $p_0$ is true,

$$\text{SymEval}(B) = \bigcup_{B' \in \text{Preds}(B)} \text{PrpgtCond}(B', B, \text{SymEval}(B')) \mid \text{LocalEval}(B)$$

$$(7)$$

where $\text{LocalEval}(B) = \{(v_{i_1}, e_{i_1}), \ldots, (v_{i_m}, e_{i_m})\}$ denotes the symbolic evaluation local to basic block $B$. The variables that get a new value assigned in the basic block are denoted by $v_{i_1}, \ldots, v_{i_m}$. The new symbolic values are given by $e_{i_1}, \ldots, e_{i_m}$. The *propagated conditions* are defined by

$$\text{PrpgtCond}(B', B, \text{PC}) = \text{Cond}(B', B) \odot \text{PC},$$

where $\text{Cond}(B', B)$ denotes the condition assigned to the CFG-edge $(B', B)$. Denoting by PC a program context, the operation $\odot$ is defined as follows:

$$\text{Cond}(B', B) \odot \text{PC} = \text{Cond}(B', B) \odot [\mathcal{S}_1, p_1] \cup \cdots \cup [\mathcal{S}_k, p_k]$$

$$= [\mathcal{S}_1, \text{Cond}(B', B) \wedge p_1] \cup \cdots \cup [\mathcal{S}_k, \text{Cond}(B', B) \wedge p_k]$$

The following definition gives rewrite rules for the | operator, which integrate local changes of a basic block into the program state and path conditions.

**Definition 1.** *The semantics of the | operator is as follows:*

1. *We replace $\{\ldots, (v, e_1), \ldots\} \mid \{\ldots, (v, e_2), \ldots\}$ by $\{\ldots, (v, e_2), \ldots\}$.*
2. *Furthermore* $\{\ldots, (v_1, e_1), \ldots\} \mid \{\ldots, (v_2, e_2(v_1)), \ldots\}$, *where* $e(v)$ *is an expression involving variable* $v$, *is replaced by* $\{\ldots, (v_1, e_1), \ldots, (v_2, e_2(v_1)), \ldots\}$.
   *For the above situations it is important to apply the rules in the correct order, which is to elaborate the elements of the right set from left to right.*
3. *If a situation like $[\{\ldots, (v, e), \ldots\}, C(\ldots, v, \ldots)]$ is encountered during symbolic evaluation, we replace it with $[\{\ldots, (v, e), \ldots\}, C(\ldots, e, \ldots)]$.*

This data-flow framework has been introduced in [Bli02].

**Solving the Data-Flow Problem** We solve the equation system defining the data-flow problem using an elimination algorithm for data-flow analysis [Sre95]. It operates on the *DJ* graph (DJG), which essentially combines the control flow graph and its dominator tree into one structure. Node $n$ *dominates* node $m$, if every path of CFG edges from *Entry* to $m$ must go through $n$. Node $n$ is the *immediate dominator* of $m$ if $n \neq m$, $n$ dominates $m$, and $n$ does not dominate any other dominator of $m$. The *dominator tree* is a graph containing every node of the CFG, and for every node $m$ an edge from its immediate dominator $n$ to $m$.

The elimination algorithm given in [Sre95] consists of two phases. The first phase performs DJ graph reduction and variable substitution of the equation system until the DJ graph is reduced to its dominator tree. Cycles (e.g. due to loops) are treated by the *loop-breaking* rule [RP86], [Bli02]. After the first phase the equation at every node is expressed only in terms of its parent node in the dominator tree. After determining the solution for the equation of Node *Entry*, the second phase of the algorithm is concerned with propagation of this information in a top-down fashion on the dominator tree to compute the solution for the other nodes. Every node corresponds to a basic block of the program under investigation, and its solution is expressed in terms of a program context as stated by Equation (1). By its definition such a context describes the possible variable bindings valid at this program point, and in this way it provides the information required for the range check elimination decision of Section 2.5.

**Hierarchical Data-Flow Frameworks** Since range checks are part of certain programming language constructs such as array access or assignment statements, a program analysis method that incorporates these checks has to be aware of control flow occurring on intra-statement level. The abstraction level of intra-statement control flow is in the same order of magnitude lower than inter-statement control flow as assembly language is compared to high-level language code. It is not desirable to spend the complete analysis of a program on intra-statement level since one gets easily overwhelmed by the amount of detail and tends to loose the view of the "big picture". For this reason we introduce a two-level hierarchy in our data-flow framework where the two levels correspond to analysis incorporating intra- and inter-statement control flow. We avoid intra-statement control flow as much as possible, which means that it is only considered for statements for which the compiler inserts a range check.

As a notational convenience we collapse the intra-statement control flow subgraph into one single *compound* node of the inter-statement CFG. In this way Fig. 3 depicts the code associated with Node 4 of Fig. 4 as a collapsed compound node (left) and as a compound node expanded to its subgraph (right). Note that we use circular shapes for ordinary nodes and boxes for compound nodes to distinguish between the two.

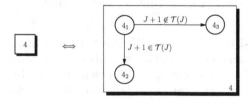

**Fig. 3.** Levels of Abstraction: Inter-Statement ⇔ Intra-Statement

## 2.5   The Range Check Elimination Decision

Consider Equation (8) which is an example of a valid symbolic equation according to the symbolic data-flow framework introduced in Section 2.4. Construction of an input program yielding this equation is straight-forward and suppressed for space considerations. The types of the used variables are $\overline{T}(c) = Boolean$, and $\overline{T}(n, x, y, z) = Positive$.

$$[\{(c, \perp), (n, \underline{n}), (x, \underline{x}), (y, \underline{y}), (z, \underline{z})\}, x^n + y^n = z^n] \mid \{(c, 3 - n \in^? T(x))\}. \quad (8)$$

The proposed range check can only be removed, if, according to Equation (5),

$$\mathrm{val}\big(3 - n, [\{(c, \perp), (n, \underline{n}), (x, \underline{x}), (y, \underline{y}), (z, \underline{z})\}, x^n + y^n = z^n]\big) \in T(x).$$

This formula is valid if it it can be shown that $n \leq 2$, which requires a proof of

$$(\forall n)(\forall x)(\forall y)(\forall z)[x^n + y^n = z^n \Rightarrow n \leq 2],$$

also known as Fermat's last theorem. While for this specific theorem our data-flow framework could be aware of the recently discovered proof, in general there exists no algorithm capable of determining the validity of a formula such as Equation (5) stated in elementary arithmetic built up from $+$, $*$, $=$, constants, variables for nonnegative integers, quantifiers over nonnegative integers, and the sentential connectives $\neg$, $\wedge$, $\vee$, $\Rightarrow$, $\Leftrightarrow$ subject to the requirement that every variable in such a formula be acted on by some quantifier. This follows from a conclusion from Gödel's incompleteness theorem, [Rog87, p. 38] contains the corresponding proof.

However, for a subclass of elementary arithmetic called *Presburger arithmetic*, validity is decidable [Sho79]. Presburger formulas are those formulas that can be constructed by combining first degree polynomial (*affine*) constraints on integer variables with the connectives $\neg$, $\wedge$, $\vee$, and the quantifiers $\forall$ and $\exists$. Constraints are affine due to the fact that Presburger arithmetic permits addition and the usual arithmetical relations $(<, \leq, >, \geq, =)$, but no arbitrary multiplication of variables[¶].

The Omega test [Pug92] is a widely used algorithm for testing the satisfiability of arbitrary Presburger formulas. We can use it as a range check elimination decision procedure if we are able to translate Equation (5) into such a formula. We split this translation into two steps, each yielding a conjunction $\Gamma$ of constraints. Initially we set $\Gamma_1 = \Gamma_2 = true$.

**Step 1:** We derive constraints from the path-condition $p_i$ of state $S_i$ as follows. The path-condition essentially is a conjunction of predicates $P_l$ that are *true* for state $S_i$:

$$p_i = P_1 \wedge \cdots \wedge P_N.$$

Each predicate $P_l$ corresponds to a CFG condition $C$ that is an expression involving program variables $v_c \in V_c$, where $V_c$ denotes the set $\{v_1, \dots, v_m\}$ of possible

---

[¶] Although it is convenient to use multiplication by *constants* as an abbreviation for repeated addition.

program variables (cf. also Equation (1)). This can be written as $P_l = C(V_c)$. Once we evaluate $C(V_c)$ for state $S_i$ (cf. Definition 1), we get $P_l = C(E_c)$ as a condition over symbolic expressions. Solving $C(E_c)$ yields the solution $L_l(V_c)$ for which $P_l = true$. Thus each predicate $P_l$ yields a constraint for $\Gamma_1$:

$$\Gamma_1 ::= \Gamma_1 \wedge \bigwedge_{1 \leq l \leq N} L_l(V_c). \tag{9}$$

*Example 2.* Starting from the context given in Equation (2), we have predicate $P_1 = C(V_c) : x = y - 2$, and $C(E_c) : \underline{n}^2 - 1 = \underline{n} + 1$ which yields the solutions $\underline{n}_1 = 2$, and $\underline{n}_2 = -1$ resulting in the constraint $(\underline{n} = 2 \vee \underline{n} = -1)$.

*Example 3.* Another example involves a predicate that arises from iteration-schemes of *for* and *while* loops. Context $[\{(x, 1), (y, 1), (z, \underline{z})\}, x$ in $y..z]$, yields the constraint $C(E_c) : 1 \leq x \leq \underline{z}$.

**Step 2:** While the previous step dealt with the information captured in the path condition $p_i$ of Equation (5), Step 2 addresses the translation of the proposed range check $\text{val}(E \in^? \mathcal{T}(j))$ into a conjunction $\Gamma_2$ of constraints. Like the conditions $C$ of Step 1, expression $E$ is a symbolic expression involving program variables $v_c \in V_c$. Again we evaluate $E(V_c)$ for state $S_i$ to get $E(E_c)$. We can now set up a constraint that requires $E(E_c)$ to be outside the range of type $\overline{\mathcal{T}}(j)$:

$$\Gamma_2 ::= \Gamma_2 \wedge j = E(E_c) \wedge \left(j < \overline{\mathcal{T}}(j)\text{'First} \vee j > \overline{\mathcal{T}}(j)\text{'Last}\right). \tag{10}$$

We then check by means of the Omega test whether there exists a solution satisfying the conjunction $\Gamma_1 \wedge \Gamma_2$. Non-existence of such a solution means that for $[S_i, p_i]$ the expression $E(E_c)$ will be within the range of type $\overline{\mathcal{T}}(j)$. Completing Equation (5), we finally get

$$\text{val}(E \in^? \mathcal{T}(j), [S_i, p_i]) = \begin{cases} true \Leftrightarrow \text{val}(E, [S_i, p_i]) \in \mathcal{T}(j) \Leftrightarrow \Gamma_1 \wedge \Gamma_2 = false \\ false \text{ else.} \end{cases}$$
$$\tag{11}$$

**Non-affine Expressions and Conservative Approximations** A method capable of transforming certain classes of general polynomial constraints into a conjunction of affine constraints has been presented by Maslov [MP94].

For non-transformable nonaffine expressions that are part of (the solution of) a predicate $P_l$ (cf. Step 1 above), we can omit the constraint imposed by $P_l$ and hence generate a conservative approximation for $\Gamma_1$ for the following reason: given the set $\{v_1, \ldots, v_m\}$ of possible program variables, the mapping of the symbolic expression $E(E_c)$ to its value can be regarded as a function $f : \mathcal{T}(v_1) \times, \ldots, \times \mathcal{T}(v_m) \to \mathbb{Z}$. Each predicate $P_l$ potentially constrains the $m$-dimensional domain of $f$, which, depending on $E(E_c)$, potentially constrains the range of $f$. The smaller the range of $f$, the more likely it is that we can derive that the proposed range check of $E(E_c)$ is redundant. Omission of $P_l$ results in a conservative approximation of the range of $f$ in the sense that we might generate a *false positive* claiming that the proposed range check is needed whereas it is actually redundant (cf. also Equation (6)). False negatives are not possible since $P_l$ cannot *widen* the range of $f$.

# 3  Example

We demonstrate our range check elimination technique by an example for which it can be shown manually that no range check is needed. Therefore every language-defined range check is redundant and should be identified as such. Fig. 4 shows our example taken from the *Heapsort* algorithm as presented in [Sed88]. Fig. 5 shows the control flow graph of our example. It contains three compound nodes (3, 4, and 5) which correspond to the intra-statement analysis necessary to treat range checks $R1$, $R2$, and $R3$ (cf. Fig. 4). Compound Node 3 is expanded whereas compound Nodes 4 and 5 are collapsed due to space considerations. CFG edges are labelled with their associated conditions, edges without labels denote *"true"* conditions. The edge $e \rightarrow x$ is artificial in the sense that it is required by our elimination algorithm but does not correspond to "real" control flow. Symbol ⚡ denotes CFG edges that correspond to control flow of a *Constraint_Error* exception taken due to failure of a range check. Since procedure *Siftdown* contains no exception handler and since we assume there is no calling

```
1   Max: constant Positive := ??;              -- Number of elements to be sorted
2   subtype Index is Positive range 1 .. Max;
3   type Sort_Array is array(Index) of Integer;
4   Arr : Sort_Array;

5   procedure Siftdown (N,K:Index) is
6       J, H : Index;                                    -- Node 1
7       V : Integer;                                     -- Node 1
8   begin
9       V := Arr(K);                                     -- Node 1
10      H := K;                                          -- Node 1
11      while H in 1..N/2 loop                           -- Node 2
12          J := {2*H}^{R1};                             -- Node 3
13          if J<N then                                  -- Node 3
14              if Arr(J)<Arr({J+1}^{R2}) then           -- Node 4
15                  J := {J+1}^{R3};                      -- Node 5
16              end if;
17          end if;
18          if V >= Arr(J) then                          -- Node 6
19              Arr(H) := V;                             -- Node 7
20              exit;                                    -- Node 7
21          end if;
22          Arr(H) := Arr(J);                            -- Node 8
23          Arr(J) := V;                                 -- Node 8
24          H := J;                                      -- Node 8
25      end loop;
26      return;                                          -- Node 9
27  end Siftdown;
```

**Fig. 4.** Example: Procedure Siftdown

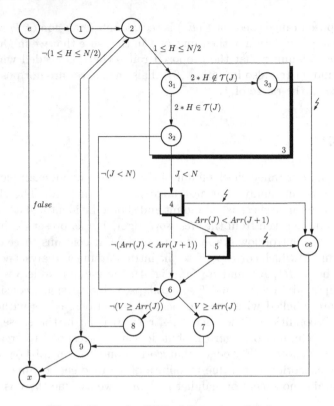

**Fig. 5.** Example: CFG of Procedure Siftdown

procedure with our example, these edges are simply "collected" by Node $ce$ that is connected to the procedure's *exit* node.

Table 1 shows the set of SymEval equations for our example procedure. The symbol $\perp$ is used to denote undefined values. $p_1$ denotes the predicate $\underline{N} \in \mathcal{T}(N) \wedge \underline{K} \in \mathcal{T}(K)$ that is due to the parameter association mechanism of [Ada95]. Table 2 shows the sequence of elimination steps performed during the elimination phase in order to reduce the system of equations so that each equation depends only on its immediate dominator. For the purpose of our example it suffices to consider the result of the application of the loop-breaking rule at Step 23, where each induction variable is replaced by an (indirect) recursion:

$$2 \varnothing: X_2 = X_1 \mid \{(Arr, Arr(\perp, \nu)), (H, H(\perp, \nu)), (J, J(\perp, \nu)), (C_{1\ldots7}, C_{1\ldots7}(\perp, \nu))\}.$$

For the purpose of our example it is furthermore sufficient to collapse all array assignments into one recursion $Arr(\perp, \nu)$.

In the propagation phase we propagate data flow information in a top-down manner on the dominator tree after the solution of the root node has been determined. Table 3 enumerates those steps for our example procedure.

For the sake of brevity we will focus on the examination of range check $R1$ located in Equation $X_{3_1}$. For this reason we are concerned with propagation

**Table 1.** Set of SymEval Equations for Example *Siftdown*

$$X_e = [\{(Arr, \underline{Arr}), (N, \underline{N}), (K, \underline{K}), (J, \perp), (V, \perp), (H, \perp), (C_{1...7}, \perp)\}, p_1]$$
$$X_1 = X_e \mid \{(V, Arr(K)), (H, K)\}$$
$$X_2 = (X_1 \cup X_8) \mid \{(C_1, (H \text{ in } 1 .. N/2))\}$$
$$X_{3_1} = C_1 \odot X_2 \mid \{(C_2, (2 * H \in^? \mathcal{T}(J))\}$$
$$X_{3_2} = C_2 \odot X_{3_1} \mid \{(J, 2 * H), (C_3, (J < N))\}$$
$$X_{3_3} = \neg C_2 \odot X_{3_1} \mid \{(J, \perp)\}$$
$$X_{4_1} = C_3 \odot X_{3_2} \mid \{(C_4, (J + 1 \in^? \mathcal{T}(Index))\}$$
$$X_{4_2} = C_4 \odot X_{4_1} \mid \{(C_5, (Arr(J) < (Arr(J + 1))))\}$$
$$X_{4_3} = \neg C_4 \odot X_{4_1}$$
$$X_{5_1} = C_5 \odot X_{4_2} \mid \{(C_6, (J + 1 \in^? \mathcal{T}(J))\}$$
$$X_{5_2} = C_6 \odot X_{5_1} \mid \{(J, J + 1)\}$$
$$X_{5_3} = \neg C_6 \odot X_{5_1} \mid \{(J, \perp)\}$$
$$X_6 = (\neg C_3 \odot X_{3_2} \cup \neg C5 \odot X_{4_2} \cup X_5) \mid \{(C_7, (V \geq Arr(J))\}$$
$$X_7 = C_7 \odot X_6 \mid \{(Arr(H), V)\}$$
$$X_8 = \neg C_7 \odot X_6 \mid \{(Arr(\{H\}), Arr(J)), (Arr(J), V), (H, J)\}$$
$$X_9 = \neg C_1 \odot X_2 \cup X_7$$
$$X_{ce} = X_{3_3} \cup X_{4_3} \cup X_{5_3}$$
$$X_x = X_9 \cup X_{ce}$$

**Table 2.** Elimination: from DJ-Graph to Dominator Tree

| | | | |
|---|---|---|---|
| 1.) $5_3 \to ce$ | 6.) $7 \to 9$ | 11.) $4_1 \to 6$ | 16.) $3_2 \to 2$ | 21.) $3_1 \to x$ |
| 2.) $5_2 \to 12$ | 7.) $4_3 \to ce$ | 12.) $6 \to 2$ | 17.) $3_2 \to 9$ | 22.) $9 \to x$ |
| 3.) $5_1 \to ce$ | 8.) $4_2 \to 6$ | 13.) $6 \to 9$ | 18.) $ce \to x$ | 23.) $2\ \emptyset$ |
| 4.) $5_1 \to 6$ | 9.) $4_2 \to ce$ | 14.) $3_3 \to ce$ | 19.) $3_1 \to 2$ | 24.) $2 \to x$ |
| 5.) $8 \to 2$ | 10.) $4_1 \to ce$ | 15.) $3_2 \to ce$ | 20.) $3_1 \to 9$ | 25.) $1 \to x$ |

**Table 3.** Propagation of the DFA-Solution

| | | | | | |
|---|---|---|---|---|---|
| 1.) $e \to x$ | 4.) $2 \to 3_1$ | 7.) $3_2 \to 4_1$ | 10.) $4_2 \to 5_1$ | 13.) $3_2 \to 6$ | 16.) $3_1 \to ce$ |
| 2.) $e \to 1$ | 5.) $3_1 \to 3_3$ | 8.) $4_1 \to 4_3$ | 11.) $5_1 \to 5_3$ | 14.) $6 \to 7$ | 17.) $2 \to 9$ |
| 3.) $1 \to 2$ | 6.) $3_1 \to 3_2$ | 9.) $4_1 \to 4_2$ | 12.) $5_1 \to 5_2$ | 15.) $6 \to 8$ | |

steps 2, 3, and 4 of Table 3:
$$e \to 1, 1 \to 2, 2 \to 3_1$$

$$X_{3_1} = [\{ \quad (Arr, Arr(\underline{Arr}, \nu)), (N, \underline{N}), (K, \underline{K}), (H, H(\underline{K}, \nu)), (J, J(\perp, \nu)),$$
$$(C_{1...7}, C_{1...7}(\perp, \nu))\}, p_1 \wedge H \text{ in } 1 .. N/2] \mid \{(C_2, (2 * H \in^? \mathcal{T}(J))\}.$$

In order to compute the solution for Node $X_{3_1}$, we evaluate predicate *val* according to Equation (5):

$$\text{val}\big(2 * H \in^? \mathcal{T}(J), [\{(Arr, Arr(\underline{Arr}, \nu)), (N, \underline{N}), (K, \underline{K}), (H, H(\underline{K}, \nu)),$$
$$(J, J(\bot, \nu)), (C_{1...7}, C_{1...7}(\bot, \nu))\}, p_1 \wedge H \text{ in } 1..N/2]\big). \quad (12)$$

Range check $R1$ is redundant if Equation (11) evaluates to *true*. From Equation (9) we get

$$\Gamma_1 ::= true \wedge 1 \leq \underline{N} \leq Max \wedge 1 \leq \underline{K} \leq Max \wedge 1 \leq H \leq \underline{N}/2,$$

where "/" denotes integer division and where the second conjunct is due to predicate $p_1$ and the third conjunct is due to predicate "$H$ in $1..N/2$". From Equation (10) we get

$$\Gamma_2 ::= true \wedge J = 2 * H \wedge \big(J < 1 \vee J > Max\big).$$

As expected the Omega test confirms that $\Gamma_1 \wedge \Gamma_2 = false$ which, according to Equation (11), means that the range check $2 * H \in^? \mathcal{T}(J)$ is redundant. Similarly we can derive at propagation steps 7 and 10 that range checks $R2$ and $R3$ are redundant.

## 4   Implementation

Fig. 6 is an extension of the structure of the GNAT compiler as explained in [SB94]. Our data-flow framework is situated between the frontend and the backend of the GNAT compiler. Its input is an abstract syntax tree (*AST*) that has been syntactically and semantically analyzed and expanded (cf. [SB94]). Every tree expression for which node flag *Do_Range_Check* is set requires the backend to generate a corresponding run-time check. The frontend makes use of this signaling mechanism for each range check that cannot be proved redundant at compile-time. When we build the CFG from the AST we generate an intra-statement control flow subgraph for every statement containing such an expression.

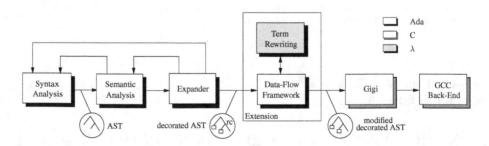

**Fig. 6.** Integration of the Data-Flow Framework Within GNAT

Once the CFG is extended by immediate domination edges, it is passed to the elimination algorithm which determines the sequence of equation insertion and loop-breaking steps in order to transform the DJG to its dominator tree (cf. e.g. Table 2). Equation insertion and loop-breaking itself is handled by the *term rewriting* component which, due to the nature of its requirements, depends on pattern matching and expression transformation capabilities available mainly in functional programming languages. An early prototype of this component has already been implemented as a *Mathematica* package [Mae90].

In order to compute the data-flow solution for each CFG node the elimination algorithm instructs the term rewriting component to perform the necessary propagation steps (cf. e.g. Table 3). If we can derive at a given CFG node that according to Equation (6) a given range check is redundant, we reset the AST flag *Do_Range_Check* which prevents the backend from generating the corresponding run-time check.

# 5  Experimental Results

In order to demonstrate the effectiveness of our approach we have considered several examples for which it can be manually proved that no range checks are necessary. The following programs have been examined so far: "Siftdown" (cf. Fig. 4) and the corresponding driver program "Heapsort" (cf. [Sed88]), "Mergesort", and "Quicksort". Table 4 compares the number of range checks required by the GNAT frontend to the number of checks that remain after symbolic analysis of the example programs. The number of assembler statements and the stripped object code size of the resulting executables are also given. For procedure "Siftdown" we have also measured the execution time overhead for range checks. We have found that the removal of superfluous range checks yields on average a performance increase of 11.7 percent, whereas the decrease in object code size is more than 24 percent. The performance figures given in Fig. 7 have been obtained on an ix86 Linux PC, based on GNAT 3.2 20020814. The left-hand picture shows absolute execution times, while the right-hand picture shows the relative performance gain and the decrease in object size.

**Table 4.** Experimental Results

| Source | Post-GNAT | | | Post-Symbolic Evaluation | | |
|---|---|---|---|---|---|---|
| | Range-Checks | Asm-Stmts | Obj.-Size | Range-Checks | Asm-Stmts | Obj.-Size |
| Siftdown | 3 | 89 | 800 | 0 | 55 | 604 |
| Heapsort | 2 | 73 | 736 | 0 | 51 | 596 |
| Quicksort | 4 | 122 | 908 | 2 | 100 | 824 |
| Mergesort | 6 | 183 | 1128 | 0 | 119 | 812 |

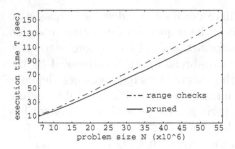 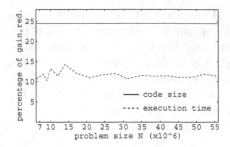

**Fig. 7.** Execution Times and Performance Gain/Obj. Code Size Reduction

## 6    Related Work

Determining how and when array items are accessed is of vital importance for *parallelizing compilers*. For this reason this problem has been studied extensively (see [Hag95] for a survey). Even symbolic approaches have been employed (again compare [Hag95]). On the other hand eliminating range checks has not been studied in this domain. Nevertheless it has been shown that expressions used for array indexes are linear expressions in most cases (cf. [SLY90]).

In [RR00] some kind of symbolic analysis is employed to determine upper and lower bounds of memory regions accessed by procedures and other pieces of programs. Linear programming is used to solve the problem. This approach can also be applied to our problem but it has some severe deficiencies; for example our procedure *Swap* (Fig. 1) cannot be handled correctly. In addition, if the sign of a variable is not constant (i.e., the variable does not assume only positive or negative values), the approach presented in [RR00] cannot be applied.

## 7    Conclusion and Future Work

We have presented a new method of range check elimination based on symbolic evaluation that incorporates type information provided by the underlying programming language. Our experiments showed that affine constraints can be handled without problems by the Omega test. As a next step we will conduct investigations on large Ada code bases in order to back the assumption that most constraints are affine in nature.

In addition, we will study validity checks which requires a delicate interprocedural analysis, and overflow checks which are very costly and for this reason turned off by GNAT in its default settings.

## References

[Ada95]   ISO/IEC 8652. *Ada Reference Manual*, 1995.
[BB98]    J. Blieberger and B. Burgstaller. Symbolic Reaching Definitions Analysis of Ada Programs. In *Proc. of the Ada-Europe International Conference on Reliable Software Technologies*, pages 238–250, Uppsala, Sweden, June 1998.

[BBS99]  J. Blieberger, B. Burgstaller, and B. Scholz. Interprocedural Symbolic Evaluation of Ada Programs with Aliases. In *Proc. of the Ada-Europe International Conference on Reliable Software Technologies*, pages 136–145, Santander, Spain, June 1999.

[BFS00]  J. Blieberger, T. Fahringer, and B. Scholz. Symbolic Cache Analysis for Real-Time Systems. *Real-Time Systems, Special Issue on Worst-Case Execution Time Analysis*, 18(2/3):181–215, 2000.

[Bli02]  J. Blieberger. Data-Flow Frameworks for Worst-Case Execution Time Analysis. *Real-Time Systems*, 22(3):183–227, May 2002.

[CHT79]  T.E. Cheatham, G.H. Holloway, and J.A. Townley. Symbolic Evaluation and the Analysis of Programs. *IEEE Trans. on Software Engineering*, 5(4):403–417, July 1979.

[FS97]  T. Fahringer and B. Scholz. Symbolic Evaluation for Parallelizing Compilers. In *Proc. of the ACM International Conference on Supercomputing*, July 1997.

[Hag95]  M.R. Haghighat. *Symbolic Analysis for Parallelizing Compilers*. Kluwer Academic, 1995.

[Mae90]  R. Maeder. *Programming in Mathematica*. Addison-Wesley, Reading, MA, USA, 1990.

[MP94]  V. Maslov and W. Pugh. Simplifying Polynomial Constraints Over Integers to Make Dependence Analysis More Precise. In *Proc. of the International Conference on Parallel and Vector Processing*, pages 737–748, Linz, Austria, 1994.

[Pug92]  W. Pugh. The Omega Test: A Fast and Practical Integer Programming Algorithm for Dependence Analysis. *Communications of the ACM*, 35(8):102–114, August 1992.

[Rog87]  H. Rogers Jr. *Theory of Recursive Functions and Effective Computability*. MIT Press, Cambridge, MA, 1987.

[RP86]  B. G. Ryder and M.C. Paull. Elimination Algorithms for Data Flow Analysis. *ACM Computing Surveys (CSUR)*, 18(3):277–316, 1986.

[RR00]  R. Rugina and M. Rinard. Symbolic Bounds Analysis of Pointers, Array Indices, and Accessed Memory Regions. In *Proc. of PLDI*, pages 182–195, 2000.

[SB94]  E. Schonberg and B. Banner. The GNAT Project: A GNU-Ada 9X Compiler. In *Proc. of the Conference on TRI-Ada '94*, pages 48–57. ACM Press, 1994.

[SBF00]  B. Scholz, J. Blieberger, and T. Fahringer. Symbolic Pointer Analysis for Detecting Memory Leaks. In *ACM SIGPLAN Workshop on "Partial Evaluation and Semantics-Based Program Manipulation"*, Boston, January 2000.

[Sed88]  R. Sedgewick. *Algorithms*. Addison-Wesley, Reading, MA, USA, 2$^{nd}$ edition, 1988.

[Sho79]  R. E. Shostak. A Practical Decision Procedure for Arithmetic with Function Symbols. *Journal of the ACM*, 26(2):351–360, April 1979.

[SLY90]  Z. Shen, Z. Li, and P.-C. Yew. An Empirical Study of Fortran Programs for Parallelizing Compilers. *IEEE Transactions on Parallel and Distributed Systems*, 1(3):356–364, July 1990.

[Sre95]  V.C. Sreedhar. *Efficient Program Analysis Using DJ Graphs*. PhD thesis, School of Computer Science, McGill University, Montréal, Québec, Canada, 1995.

# Quasar: A New Tool for Concurrent Ada Programs Analysis

Sami Evangelista, Claude Kaiser, Jean-François Pradat-Peyre, and
Pierre Rousseau

CEDRIC – CNAM Paris
292, rue St Martin, 75003 Paris
{evangeli,kaiser,peyre,lrousseau}@cnam.fr

**Abstract.** Concurrency introduces a high degree of combinatory which
may be the source of subtle mistakes. We present a new tool, Quasar,
which is based on ASIS and which uses fully the concept of patterns. The
analysis of a concurrent Ada program by our tool proceeds in four steps:
automatic extraction of the concurrent part of the program; translation
of the simplified program into a formal model using predefined patterns
that are combined by substitution and merging constructors; analysis of
the model both by structural techniques and model-checking techniques;
reporting deadlock or starvation results. We demonstrate the usefulness
of Quasar by analyzing several variations of a non trivial concurrent
program.

## 1 Introduction

Concurrency introduces at the same time design facilities and reliability problems. Indeed, the interleaving of tasks execution leads to a high degree of combinatory and may be the source of subtle mistakes that are difficult to detect by
simple simulations or human reasoning. Furthermore, a little modification in a
part of the code can produce a major transformation of the application behavior. Classically encountered problems are deadlock, starvation or more generally
race conditions on shared resources.

An automatic and easy to use tool is crucial for efficiently validating concurrent software and gaining in confidence with the code. Indeed such a tool will
act as a very clever front-end compiler which aim is not to check syntax nor
classical language semantic but to verify the concurrent semantic of the application with respect to safety or liveness properties. We have built such a tool,
named Quasar. It is based on the implementation of the ASIS interface for the
GNAT Ada 95 compiler and uses fully the concept of patterns. The analysis of
a concurrent Ada program by Quasar proceeds in four steps:

1. Automatic extraction of the relevant part of the program with respect to
   a given property; the present implementation is focused on the analyzis of
   global properties like deadlock or starvation. This step is denoted as program
   slicing [HS99], [DH99], [CDH+00] and reduces greatly the size of formal
   models produced in the second step;

J.-P. Rosen and A. Strohmeier (Eds.): Ada-Europe 2003, LNCS 2655, pp. 168–181, 2003.
© Springer-Verlag Berlin Heidelberg 2003

2. Translation of the simplified program into a formal model. We have chosen colored Petri nets [Jen91], [GPPch] as target formalism because their analysis may combine several techniques that are supported by experienced tools. For this translation we use the concept of pattern and Quasar produces nets with two basic constructors: substitution (that substitutes an abstract Petri net by a concrete one) and composition (that merges two different nets into a unique one).

3. Analysis of the model by combining structural techniques (like Petri nets reductions [Ber85], [HPP03], [PPP00]) and finite state verification methods (like temporal logic formula verification); in the current version of Quasar we use Maria [Mak02] as model checker.

4. Construction of a report; when the target property is not verified, the report indicates the sequence of actions that invalidates the property. For the moment, the produced report makes reference to the "sliced" program.

The originality of Quasar lies in the genericity of the analysis (a large part of Quasar is independent of the source language) and in the extensive use of recent structural Petri net analysis techniques (like extended agglomerations [HPP03]).

In the next section we present some patterns and show how they are used for producing colored Petri nets from Ada code source. Then we explain the use of the ASIS interface to perform the first two phases of analysis. We demonstrate in Sect. 4 the usefulness of our tool Quasar by analyzing a non trivial concurrent program proposed by L. Pautet in an advanced computing science course at ENST Paris. The paper ends briefly reviewing similar works and pointing out future actions.

## 2   Translation Patterns

The translation of a concurrent program to a high-level Petri net is similar to the compilation of a program in assembler. In both cases, the aim is to map the semantic of a program from a particular formalism towards another formalism. High-level Petri nets (with inhibitor arcs) and Ada have the same expressiveness. It is therefore possible to translate any Ada program to a Petri net; the translation does not raise any theoretical problem. However, we must keep in mind that concurrency may introduce combinatorial explosion. As our aim is to perform exhaustive analysis, the generated nets mustn't create useless combinatory.

The translation proceeds in two steps: first, each element of the program (statements, expressions, and declarations) is translated in a predefined Petri net pattern. Then all these patterns are combined by merging or by substitution to produce the final Petri net. For instance, Fig. 1 depicts three patterns used to respectively translate a call to an entry E, the accept statement of a server to an entry E and the translation of a loop statement without an iteration scheme. One can remark that place END_LOOP of this model exists but is not systematically relied to a transition: this will be the case only if the internal loop statement contains an exit statement.

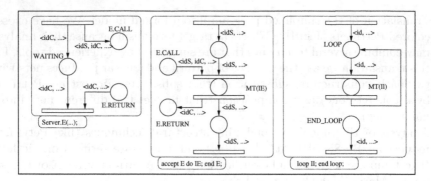

**Fig. 1.** Three patterns

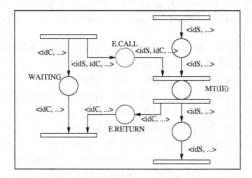

**Fig. 2.** Merging result

The merging operation consists in producing a Petri net by combining two Petri nets; this merging is done by the assimilation of places having the same name in both nets (we impose also that they share the same color domain, and that naming is an injective application: two different objects cannot have the same name). So places having the same name appear only once in the resulting net and all arcs of the merged nets take into account this assimilation. For instance, if in the models given Fig. 1, E denotes the same entry in the two nets of the left, then the merging of these two nets produces the model described in Fig. 2. In all these models, internal statements of compound statements are modeled by abstract transitions (for instance the abstract transition MT(IE) or MT(I1) in Fig. 1).

The substitution operation replaces an abstract transition by the corresponding concrete model. This operation consists in sticking input and output places of an abstract transition to the "enter" or to the "quit" transition of the concrete model (obviously, some constraints apply when performing this operation, especially on the existence of the "enter" and of the "quit" transition). For example, the substitution of the abstract transition MT(IE) of Fig. 2 by the model on the right of Fig. 1 in the model of Fig. 2, produces the net depicted on the left of Fig. 3. This model corresponds to a call of an entry that executes a loop state-

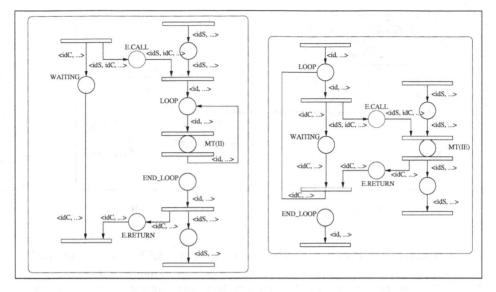

**Fig. 3.** Substitution result

ment. It is also possible to substitute the abstract transition MT(I1) (of Fig. 1) by the net of Fig. 2. In this case we obtain the net on the right of Fig. 3. This model corresponds to repetition of the call to the entry E.

As said previously, we must obtain a colored Petri net corresponding precisely to the analyzed concurrent program; but remember that produced models are of interest only if they can be analyzed. In particular, we must limit as much as possible the combinatory explosion. A key point for this limitation appeared in identifiers attribution when translating a task body. Indeed, a task must have a unique internal identifier; for instance this allows to distinguish responses of a server to its clients.

Although giving names at run time to task body instances can be modeled by colored Petri nets, it leads to $N!$ configurations when the number of tasks in the program is $N$. Thus we have excluded momentarily this Ada construction in our alpha version of Quasar.

## 3  Using ASIS

In this section, we describe how we get the information from Ada source code by using the Ada Semantic Interface Specification (ASIS for short) [ISO99].

### 3.1  A Brief ASIS Description

To implement Quasar, we need to be able to read and analyze a full Ada source code. To do this, we use the Ada Semantic Interface Specification (ASIS). It is an interface between an Ada environment and any tool requiring information from

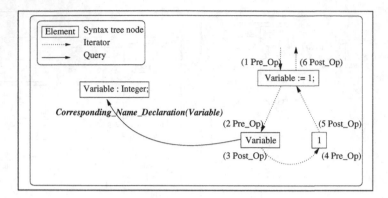

**Fig. 4.** Navigating in the syntax tree with ASIS

it. To represent the different elements of a program, ASIS defines three main abstractions: the context, which represents the set of compilation units on which the ASIS tools can be applied to (i.e., an Ada environment), the compilation units, which can be considered as the Ada compilation units defined in the Ada 95 RM (Sect. 10), and the element which represent any node of the syntax tree. An element can be a task body declaration, an integer expression, or any other component of a program. Element categorization is performed by ASIS functions like *Asis.Elements.Declaration_Kind(E)* which returns the type of declaration of *E* (for instance, a task declaration).

ASIS provides two methods for navigating in the syntax tree. The first one is an iterator which allows to traverse the syntax tree by using the depth-first method. During the traversal of each element, it permits to apply a pre and a post procedure. This method can be useful to get a list of elements (for example, the list of task declarations).

The second method consists in a set of queries. These queries give informations about a node of the tree (i.e. an ASIS element). But more than simple queries, these functions give the possibility to navigate through the syntax tree by the semantic correspondence between elements. For example, there is a query delivering the element in which is declared a variable used in an assignment. Fig. 4 shows the different navigation possibilities through a source code provided by ASIS.

## 3.2  A Specific Library Based on ASIS

Generating Petri nets from Ada programs requires the lists of all declared items of each kind (ordinary types, simple tasks, simple protected objects, tasks, protected objects, functions, procedures, entries, variables and constants). Moreover, each collected element requests some particular information for translating it into its corresponding Petri net part. For instance, the mapping of a procedure declaration needs its name, its parameters and their type and also the Petri net corresponding to its statements.

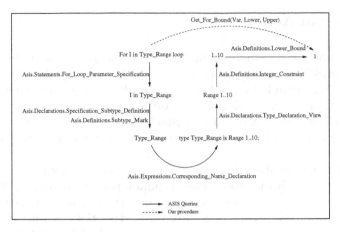

**Fig. 5.** Example of ASIS simplification

As a general library, ASIS is not so easy to use and it is sometimes difficult to find the right sequence of queries. Moreover, our tool uses often the same sequences of queries. Therefore each used sequence has been packed in a unique function. Our tool needs also some specific queries which are not directly provided by ASIS. So we have written our own specific and simplified library which provides a well-defined programming interface for Quasar programmers and which calls the ASIS powerful tools.

For instance, as shown on Fig. 5, our library provides a unique function which gives the lower and the upper bound of a "for loop" statement.

The information extracted from the source code is useful beyond the generation of the corresponding Petri net. It is also used to withdraw from this source code all the elements which are not related to the required property, i.e. in the present implementation, deadlock or starvation. For instance, in the procedure Log of the package Output (Appendix A), as the Put instruction is considered to be non blocking, lines 31, 32, 33 and 35 can be suppressed. So, the internal statements of the case block (line 30-34) become empty: they can be suppressed and then also the loop statement (lines 29-36). After these suppressions, the value of the global variable Info_Table becomes unused (we obtain this information with a specific Asis query). So, all the assignments performed on this variable (line 27) and also its declaration (line 8) can be withdrawn. The declaration of the type Info and the parameters I, S, T of the procedure Log become then unused and can be suppressed. These suppressions lead to the simplified package Output_Simple given in Appendix B.

In the same way, delay statements are of no interest when analyzing qualitative properties: we can suppress them without modifying the analyzed program with respect to the studied property. For instance, in Appendix A, lines 28 and 31 of the procedure diner can be suppressed; these suppressions lead to the suppression of the variable Max_Tick in the same procedure (line 10) and as a direct result, the package Random becomes unused.

# 4   Quasar in Action

We propose now to show that Quasar can be very useful for validating a concurrent program. We use a concurrent solution to the dining philosophers problem proposed as an example by L. Pautet in an advanced programming course [1]. We are interested in this solution because it is not ours, because it makes a rather complicated and clever use of requeue statements in a protected object and because it allows to experiment some little variations around the initial solution.

The dining philosophers problem is a paradigm for resource allocation without pre-emption with race conditions: a philosopher x competes with its two neighbors for obtaining its two sticks (the one numbered x and the one numbered x+1). Depending on the policy chosen for the sticks allocation, it may be possible to reach a deadlock (for instance all philosophers have taken one fork) or to suffer from the starvation problem (for instance two philosopher can starve a common neighbor). It's very common to reinsert a problem when attempting to suppress another one. For instance, suppressing starvation may reinsert deadlock while suppressing deadlock may introduce starvation. Although this paradigm is easy to express and can be implemented with comparatively short programs, it leads to very subtle and high combinatory solutions.

The solution that is presented can be classified as a solution with resources reservation and is composed of five modules: the package "types" that declares basic types, the package "random" that is used for randomizing the time elapsed by each philosopher in action (thinking, hungry, eating), the package "output" that provides fine outputs, the package "chop" that defines the chopsticks management and at last, the procedure "diner" that implements the behavior of the philosophers and which is the main procedure (it declares 5 philosophers). All these packages are given in Appendix A. In order to analyze different versions, we modify the boolean expressions assigned to the boolean variable Updating which is used as a guard (called barriers in Ada95) in the protected object Sticks.

The protected object semantic is the corner stone of the studied solution. In particular, each philosopher takes sticks one by one: first the left, by a call to the entry Get_Left, then the right one, managed by the entry Get_Right. It goes from the left to the right by a requeue statement. When the requested stick is not free, the philosopher requeues to an internal entry (Get_Left_Q or Get_Right_Q) depending on the needed stick (left or right). After executing the statements of these internal entries, philosophers are requeued again onto the corresponding external entry.

All these entries are guarded by a boolean variable named Updating (the guard is either "Updating" or its negation, "not Updating"). This variable is updated in three sub-programs of the protected object Sticks: in the entry Get_Left_Q, in the entry Get_Right_Q and in the procedure Free. The corresponding code is denoted in Appendix A by respectively EXP1, EXP2, and EXP3. The modification of this variable (and then of barriers values) must allow to

---

[1] http://www.infres.enst.fr/~pautet/sip/paca/

unblock waiting tasks or to block new arrival tasks. As an example of using our tool, let us find possible assignments to EXP1, EXP2, and EXP3 which define a correct solution (no deadlock, no starvation). Following the initial solution, let us use the queue size of the different entries. Four symmetric expressions can be expressed:

```
A1 :: ( Get_Left 'Count > 0)    or   ( Get_Right 'Count > 0)
A2 :: ( Get_Left 'Count > 0)    and  ( Get_Right 'Count > 0)
A3 :: ( Get_Left_Q 'Count > 0)  or   ( Get_Right_Q 'Count > 0)
A4 :: ( Get_Left_Q 'Count > 0)  and  ( Get_Right_Q 'Count > 0)
```

If we want to find the correct solution(s) without using a tool like Quasar we have to reason and/or to perform some simulations to gain in confidence with the kept solutions. If we want to be exhaustive, we have to perform $4^3 = 64$ different simulations. Having experimented automatically these simulations, with different action delays, we have observed 3 different behaviors: some solutions completely deadlock, some other ones deadlock partially (some tasks finish when others block) and some finish normally. After several hours of simulations and thinking we have the insight that at most three solutions may be correct: the solutions which assign A3 to EXP1 and to EXP3 and which assign either A2, A3 or A4 to EXP2.

However, using Quasar, only the solution which assigns A3 to EXP1, EXP2 and EXP3 is proved to be deadlock free (it's the original solution). The two others suffer from a deadlock that can be explained using execution traces provided by Quasar. For instance, a part of the trace corresponding to the deadlock of the solution EXP1 ← A3, EXP2 ← A4, EXP3 ← A3, is given in Appendix C. This trace shows that deadlock occurs when the queuing policy of the entry Get_Right_Q is not FIFO: philosopher 0 goes eating and obtains its two sticks; philosopher 4 takes its left fork and is queued in Get_Right_Q queue; philosopher 3 takes its left fork and is queued in Get_Right_Q queue; then philosopher 0 frees its sticks; the modification of the variable Updating and the egg-shell model implies that either 3 or 4 will be serviced. If queuing policy is not FIFO, then 3 may be chosen; A4 is evaluated to False, then Updating is assigned with False, and 3 is requeued on Get_Right; as its barrier is true, 3 is serviced but it cannot obtain its right stick and is then requeued to Get_Right_Q; as Updating is now equal to false, the barrier of Get_Right_Q is closed and 4 cannot be serviced. When philosopher 2, then 1, then 0 call Get_Left there is a deadlock. This deadlock is not detected by simulation because this simulation gives a biased view of possible executions (e.g. a FIFO entry queuing policy).

The correctness of the original solution relies on the egg-shell model of protected objects but is not depending on the queuing policy implemented by the run time and Quasar proves that it is deadlock free. Some others reliable solutions are discussed in [BKPP97] and [KPP97] in which a simple solution, using entry families, is discussed.

# 5   Related Works

Analyzing concurrent programs with formal methods is a recurrent challenge. Ada83 code analysis with Petri net techniques and particular with structural techniques like invariants or net reductions has been done by Murata and its team [MSS89], [TSM90], [NM94]. The use of ordinary Petri nets seems to us a major drawback of these works since it's very difficult with this formalism to express complex programming patterns. In the same perspective, [BW99], [GSX99] or [BWB+00] are intentions papers for Ada95 while in Vienna, J. Blieberger and its team use symbolic data flow analysis framework for detecting deadlocks in Ada programs with tasks [BBS00]. With a quite different approach, some very interesting work is being done at Bell Laboratories by G.J Holzmann's team [HS99], [Hol00b] [Hol00a], [GS02]. In particular this team developed the tool FeaVer [HS00], based on the Spin model-checker, that allows the verification of distributed applications written in ANSI-C source code. In parallel, P. Godefroid (also at Bell Laboratories) has developped a tool named VeriSoft [God97], [GHJ98].

At Kansas State University the Bandera project [CDH+00] aims to verify multi-threaded Java source code with model cheking techniques, using an adaptation of the Spin tool.

From our experience, all these tools suffer from a intermediate language (Promela or internal description language) less mature and studied than Petri nets. We claim that using colored Petri nets allows us to limit the combinatory explosion by combinig more easily and more efficiently structural techniques (that work directly on the model) and efficient techniques based on the underlying state graph exploration.

We have built a prototype tool [BPP99] based on colored Petri nets. However it was too monolithic and we were not able to choose amongst several possible construction patterns when producing formal models from source code. Furthermore, as we had defined our own grammar, and our own parser for an Ada subset, we had very important difficulties to augment the part of the language that we can analyze. The present Quasar is based on this experience and it uses also new results of Petri nets structural analysis.

# 6   Conclusion

In this paper we have presented the underlying ideas of Quasar, a new tool for concurrent Ada programs analysis, and an example of its use. We are experimenting it extensively in order to stabilize the implementation of our algorithms, and to get feedback of its users. The next steps have several purposes:

- extending the graphical interface and easing its use by practitioners;
- extending the scope of analyzable Ada constructions, such as tagged types or dynamic tasking creation;
- extending the analysis capacity using stochastic or temporal Petri nets.

All material used in this paper, as well as an alpha version of Quasar and related documentations, are available on http://quasar.cnam.fr.

# References

[BBS00]     J. Blieberger, B. Burgstaller, and B. Scholz. Symbolic Data Flow Analysis for Detecting Deadlocks in Ada Tasking Programs. In *Proc. of the Ada-Europe International Conference on Reliable Software Technologies*, Potsdam, Germany, 2000.

[Ber85]     G. Berthelot. Checking properties of nets using transformations. In G. Rozenberg, editor, *Advances in Petri nets*, volume No. 222 of *LNCS*. Springer-Verlag, 1985.

[BKPP97]    K. Barkaoui, C. Kaiser, and J.F. Pradat-Peyre. Petri nets based proofs of Ada95 solution for preference control. In *Proc. of the Asia Pacific Software Engineering Conference (APSEC) and International Computer Science Conference (ICSC)*, Hong-Kong, 1997.

[BPP99]     E. Bruneton and J.F. Pradat-Peyre. Automatic verification of concurrent ada programs. In Michael Gonzalez Harbour and Juan A. de la Puente, editors, *Reliable Software Technologies-Ada-Europe'99*, number 1622 in LNCS, pages 146–157. Springer-Verlag, 1999.

[BW99]      A. Burns and A. J. Wellings. How to verify concurrent Ada programs: the application of model checking. *ACM SIGADA Ada Letters*, 19(2):78–83, 1999.

[BWB⁺00]    A. Burns, A. J. Wellings, F. Burns, A. M. Koelmans, M. Koutny, A. Romanovsky, and A. Yakovlev. Towards modelling and verification of concurrent ada programs using petri nets. In Pezzé, M. and Shatz, M., editors, *DAIMI PB: Workshop Proceedings Software Engineering and Petri Nets*, pages 115–134, 2000.

[CDH⁺00]    James C. Corbett, Matthew B. Dwyer, John Hatcliff, Shawn Laubach, Corina S. Pasareanu, Robby, and Hongjun Zheng. Bandera: extracting finite-state models from java source code. In *International Conference on Software Engineering*, pages 439–448, 2000.

[DH99]      Matthew B. Dwyer and John Hatcliff. Slicing software for model construction. In *Partial Evaluation and Semantic-Based Program Manipulation*, pages 105–118, 1999.

[GHJ98]     Patrice Godefroid, Robert S. Hanmer, and Lalita Jategaonkar Jagadeesan. Model checking without a model: An analysis of the heart-beat monitor of a telephone switch using verisoft. In *International Symposium on Software Testing and Analysis*, pages 124–133, 1998.

[God97]     Patrice Godefroid. Verisoft: A tool for the automatic analysis of concurrent reactive software. In *Computer Aided Verification*, pages 476–479, 1997.

[GPPch]     C. Girault and J.F. Pradat-Peyre. Les réseaux de Petri de haut-niveau. In M. Diaz, editor, *Les réseaux de Petri: Modèles Fondamentaux*, number ISBN: 2-7462-0250-6, chapter 7, pages 223–254. Hermes, 2001 (French).

[GS02]      G.J.Holzmann and Margaret H. Smith. An automated verification method for distributed systems software based on model extraction. *IEEE Trans. on Software Engineering*, 28(4):364–377, April 2002.

[GSX99]    Ravi K. Gedela, Sol M. Shatz, and Haiping Xu. Formal modeling of synchronization methods for concurrent objects in ada 95. In *Proceedings of the 1999 annual ACM SIGAda international conference on Ada*, pages 211–220. ACM Press, 1999.

[Hol00a]    G.J. Holzmann. Logic verification of ansi-c code with spin. pages 131–147. Springer Verlag / LNCS 1885, Sep. 2000.

[Hol00b]    G.J. Holzmann. Software verification at bell labs: one line of development. *Bell Labs Technical Journal*, 5(1):35–45, Jan-March 2000. Bell Labs 75th year anniversary issue.

[HPP03]    S. Haddad and J.F. Pradat-Peyre. New powerfull Petri nets reductions. Technical report, CEDRIC, CNAM, Paris, 2003.

[HS99]    G.J. Holzmann and Margaret H. Smith. Software model checking - extracting verification models from source code. pages 481–497, Kluwer Academic Publ., Oct. 1999. also in: Software Testing, Verification and Reliability, Vol. 11, No. 2, June 2001, pp. 65–79.

[HS00]    G.J. Holzmann and Margaret H. Smith. Automating software feature verification. *Bell Labs Technical Journal*, 5(2):72–87, April-June 2000. Issue on Software Complexity.

[ISO99]    ISO/IEC 15291. *Ada Semantic Interface Specification (ASIS)*, 1999.

[Jen91]    K. Jensen. Coloured Petri nets: A hight level language for system design and analysis. In Jensen and Rozenberg, editors, *High-level Petri Nets, Theory and Application*, pages 44–119. Springer-Verlag, 1991.

[KPP97]    C. Kaiser and J.F. Pradat-Peyre. Comparing the reliability provided by tasks or protected objects for implementing a resource allocation service: a case study. In *TriAda*, St Louis, Missouri, november 1997. ACM SIGAda.

[Mak02]    M. Makela. Maria user's guide. Technical report, Helsinki Univ. of Technology, Finland, 2002.

[MSS89]    T. Murata, B. Shenker, and S.M. Shatz. Detection of Ada static deadlocks using Petri nets invariants. *IEEE Transactions on Software Engineering*, Vol. 15(No. 3):314–326, March 1989.

[NM94]    M. Notomi and T. Murata. Hierarchical reachability graph of bounded Petri nets for concurrent-software analysis. *IEEE Transactions on Software Engineering*, Vol. 20(No. 5):325–336, May 1994.

[PPP00]    D. Poitrenaud and J.F. Pradat-Peyre. Pre and post-agglomerations for *LTL* model checking. In M. Nielsen and D Simpson, editors, *High-level Petri Nets, Theory and Application*, number 1825 in LNCS, pages 387–408. Springer-Verlag, 2000.

[TSM90]    S. Tu, S.M. Shatz, and T. Murata. Applying Petri nets reduction to support Ada-tasking deadlock detection. In *Proceedings of the 10th IEEE Int. Conf. on Distributed Computing Systems*, pages 96–102, Paris, France, June 1990.

# A    Packages of the Analysed Program

```
1   package Types is
2      N_Philosophers : constant := 5;
3      type Id is mod N_Philosophers;
4      type Status is (Eating, Thinking, Hungry);
5   end Types;
```

```
1   package Random is
2      function Value return Natural;
3   end Random;
```

```
1   with Types; use Types;
2   package Output is
3      procedure Log (I : Id; S : Status; T : Natural);
4   end Output;
```

```
1   with Ada.Text_IO, Ada.Integer_Text_IO; use Ada.Text_IO, Ada.Integer_Text_IO;
2   package body Output is
3      type Info is record
4         S : Status;
5         T : Natural;
6      end record;
7      Info_Table : array (Id) of Info := (others => (Hungry, 0));
8      protected Mutex is
9         entry P;
10        procedure V;
11     private
12        Free : Boolean := True;
13     end Mutex;
14     protected body Mutex is
15        entry P when Free is begin Free := False; end P;
16        procedure V is begin Free := True; end V;
17     end Mutex;
18     procedure Log (I : Id; S : Status; T : Natural) is
19     begin
20        Info_Table (I) := (S, T);
21        Mutex.P;
22        for P in Info_Table'Range loop
23           case Info_Table (P).S is
24              when Eating   => Put (" ___eating");
25              when Thinking => Put (" _thinking");
26              when Hungry   => Put (" ___hungry");
27           end case;
28           Put (Info_Table (P).T, Width => 4);
29        end loop;
30        New_Line;
31        Mutex.V;
32     end Log;
33  end Output;
```

```
1   with Types;               use Types;
2   package Chop is
3      type Ready_Table is array (Id) of Boolean;
4      protected Sticks is
5         entry Get_Left   (C : Id);
6         entry Get_Right  (C : Id);
7         procedure Free   (C : Id);
8      private
9         entry Get_Left_Q   (C : Id);
10        entry Get_Right_Q  (C : Id);
11        Ready    : Ready_Table := (others => True);
12        Updating : Boolean := False;
13     end Sticks;
14  end Chop;
```

```
1   package body Chop is
2      protected body Sticks is
3         entry Get_Left   (C : Id ) when not Updating is
4         begin
5            if Ready (C) then
6               Ready (C) := False ;
7               requeue Get_Right ;
8            else
9               requeue Get_Left_Q ;
10           end if ;
11        end Get_Left ;
12
13        entry Get_Left_Q (C : Id ) when Updating is
14        begin
15           Updating := EXP1 ;
16           requeue Get_Left ;
17        end Get_Left_Q ;
18
19        entry Get_Right   (C : Id ) when not Updating is
20        begin
21           if Ready (C + 1) then
22              Ready (C + 1) := False ;
23           else
24              requeue Get_Right_Q ;
25           end if ;
26        end Get_Right ;
27
28        entry Get_Right_Q   (C : Id ) when Updating is
29        begin
30           Updating := EXP2 ;
31           requeue Get_Right ;
32        end Get_Right_Q ;
33
34        procedure Free   (C : Id ) is
35        begin
36           Ready (C) := True ; Ready (C + 1) := True ;
37           Updating := EXP3 ;
38        end Free ;
39     end Sticks ;
40  end Chop ;
```

```
1   with Chop , Random , Output , Types ; use Types ;
2   procedure Diner is
3      Tick_Delay   : constant Duration := 1.0 ;
4      Max_Tick     : constant := 1 ;
5      Max_Times    : constant := 7 ;
6
7      task type Philosopher is
8         entry Init (N : Id );
9      end Philosopher ;
10
11     task body Philosopher is
12        Self    : Id ;
13        N_Tick  : Natural ;
14     begin
15        accept Init (N : Id ) do Self := N ; end Init ;
16
17        for Times in 1 .. Max_Times loop
18           Chop . Sticks . Get_Left ( Self );
19           N_Tick := Random . Value mod Max_Tick ;
20           Output . Log ( Self , Eating , Times );
21           delay Tick_Delay * N_Tick ;
22           Chop . Sticks . Free ( Self );
23           Output . Log ( Self , Thinking , Times );
24           delay (Max_Tick − N_Tick) * Tick_Delay ;
25           Output . Log ( Self , Hungry , Times );
26        end loop ;
```

```
27      end Philosopher;
28      Philosophers : array (Id) of Philosopher;
29 begin
30      for P in Philosophers'Range loop
31          Philosophers (P).Init (P);
32      end loop;
33 end Diner;
```

# B  Sliced Package Output

```
1 with Types_Simple; use Types_Simple;
2 package Output_Simple is
3      procedure Log;
4 end;
```

```
1 package body Output_Simple is
2      protected Mutex is
3          entry P; procedure V;
4      private
5          Free : Boolean := True;
6      end Mutex;
7      protected body Mutex is
8          entry P when Free is begin Free := False; end P;
9          procedure V is begin Free := True; end V;
10     end Mutex;
11     procedure Log is
12     begin
13         Mutex.P;
14         Mutex.V;
15     end Log;
16 end;
```

# C  A Part of a Deadlock Execution Trace

```
(...)                Updating := false        Chop.Philosophers(0).40,41
                                              Chop.Philosophers(0).42     Updating := true
Diner.Philosophers(0).25                      Chop.Philosophers(0).43
Chop.Philosophers(0).07      Get_Left         Diner.Philosophers(0).25
Chop.Philosophers(0).09,10,11                 Chop.Philosophers(3).34     Updating := false
Chop.Philosophers(0).23      Get_Right        Chop.Philosophers(3).35
Chop.Philosophers(0).25,26,30                 Chop.Philosophers(3).23     Get_Right
Diner.Philosophers(0).29                      Chop.Philosophers(3).25,28
Diner.Philosophers(1).25                      Chop.Philosophers(3).32     Get_Right_Q ---x
Diner.Philosophers(2).25                      Chop.Philosophers(2).09,10,11
Diner.Philosophers(3).25                      Chop.Philosophers(2).23     Get_Right
Diner.Philosophers(4).25                      Chop.Philosophers(2).25,28
Chop.Philosophers(4).07      Get_Left         Chop.Philosophers(2).32     Get_Right_Q ---x
Chop.Philosophers(4).09,10,11                 Chop.Philosophers(1).07     Get_Left
Chop.Philosophers(4).23      Get_Right        Chop.Philosophers(1).09,10,11
Chop.Philosophers(4).25,28                    Chop.Philosophers(1).23     Get_Right
Chop.Philosophers(4).32      Get_Right_Q ---x Chop.Philosophers(1).25,28
Chop.Philosophers(3).07      Get_Left         Chop.Philosophers(1).32     Get_Right_Q ---x
Chop.Philosophers(3).09,10,11                 Chop.Philosophers(0).07     Get_Left
Chop.Philosophers(3).23      Get_Right        Chop.Philosophers(0).09,10,11
Chop.Philosophers(3).25,28                    Chop.Philosophers(0).23     Get_Right
Chop.Philosophers(3).32      Get_Right_Q      Chop.Philosophers(0).25,28
Chop.Philosophers(0).38      Free             Chop.Philosophers(0).32     Get_Right_Q ---x
```

# A Graphical Environment for GLADE*

Ernestina Martel, Francisco Guerra, Javier Miranda, and Luis Hernández

IUMA, Institute for Applied Microelectronics
and Departamento de Ingeniería Telemática
University of Las Palmas de Gran Canaria
Canary Islands, Spain
{emartel,fguerra,jmiranda,lhdez}@iuma.ulpgc.es

**Abstract.** This paper presents a graphical environment for programming distributed applications by means of GLADE. This environment facilitates the configuration and execution of Ada distributed applications. The environment frees the user from learning the GLADE configuration language, and keeps the state and the configuration of the system in a database for reconfiguration purposes. In order to provide fault tolerance the database can be replicated; the failure is then automatically masked by the graphical environment.

**Keywords:** Distributed Configuration, Ada, GLADE, Group_IO

## 1 Introduction

The Distributed System Annex (DSA) of Ada95 [1] provides a model to program distributed applications. The implementation of such annex is found in ADEPT [2] and GLADE [3]. In this paper we focus on GLADE, the GNAT implementation of the Ada95 DSA, because ADEPT has not evolved with recent GNAT releases. GLADE's users must program the Ada units and then write a configuration file.

Configuration languages have been widely used for the construction of distributed applications in Conic [4], Durra [5], Regis [6], Olan [7] and Jaden [8]. Moreover, the design and development of Regis distributed programs are also provided by means of a visual programming environment named Software Architect's Assistant [9]. Furthermore, Piranha [10], a CORBA tool for high availability, includes some features to graphically manage the configuration and execution of CORBA applications.

One of our goals is to free the user from learning the GLADE configuration language by means of an user friendly graphical interface. In our approach, the user only needs to set the properties of the GLADE components, and program them according to the Ada 95 distributed model. Besides, some long-lived applications need to keep their configuration state to ease reconfiguration.

---

* This work has been partially funded by the Spanish Research Council (CICYT), contract number TIC2001–1586–C03–03

J.-P. Rosen and A. Strohmeier (Eds.): Ada-Europe 2003, LNCS 2655, pp. 182–195, 2003.
© Springer-Verlag Berlin Heidelberg 2003

In a previous paper [11], we presented a tailorable *Environment for Programming Distributed Applications* (EPDA) which provides an environment for configuration and execution of distributed applications. The environment is based on the *component-based model* which has been highly exploited in Durra, Darwin, Olan and LuaSpace [12]. According to this model, distributed applications are composed by components (with their properties) and the relationships between components. Components refer to software entities which must be located on nodes, whereas properties deal with component description. Relationships represent dependency between components. Therefore, distributed applications can be viewed as a hierarchy of single and composite components. The tailorable EPDA is customizable to different distributed platforms by means of the EPDA-Modeler tool. In this paper we present an instantiation of the tailorable EPDA for GLADE named GLADE EPDA, and therefore we focus on the configuration and execution of GLADE applications. In order to easily incorporate future features of GLADE to GLADE EPDA, GLADE EPDA does not modify GLADE sources.

This paper is structured as follows. In Sect. 2 some details of the GLADE EPDA are presented. Section 3 shows an example of both configuration and execution in the GLADE EPDA. Finally, we conclude and present the current work.

## 2   GLADE EPDA Introduction

GLADE EPDA follows the client/server model. On the *client* side (Fig. 1a), GLADE EPDA provides a GUI which enables the user to instantiate the allowed GLADE components, their properties and relationships. This GUI is structured in two sections: one for hardware components (*node section*) and another for software components (*application section*). Every component instantiation is sent to the GLADE EPDA server. The GUI has been developed by means of both GtkAda and Glade (the user interface builder for Gtk+); the communication between GLADE EPDA client-group of GLADE EPDA replicas and group of GLADE EPDA replicas-GLADE EPDA proxy is provided by the communication library Group_IO [13].

The GLADE EPDA *server* (see Fig. 1b) keeps the configuration of the system and the instantiated components in a database. Besides, the GLADE EPDA server processes the commands sent by the user through the GLADE EPDA GUI. This means that the GLADE EPDA server must issue requests to the physical nodes to know their current state and availability, or to execute/kill a component. Therefore each physical node must have a proxy (see Fig. 1c) which executes the requests issued by the group of GLADE EPDA replicas. In order to guarantee data availability in presence of node failures the GLADE EPDA server can be replicated [14]. In this case the proxy must also filter the duplicated requests issued by the replicas of the GLADE EPDA server. In order to provide a higher level of fault tolerance on the server side, the group of GLADE EPDA replicas is dynamic; that is, in case of replica failures, new replicas can be entered

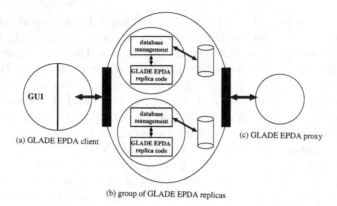

(a) GLADE EPDA client

(b) group of GLADE EPDA replicas

(c) GLADE EPDA proxy

**Fig. 1.** GLADE EPDA architecture

the group to keep the replication degree: the incoming replicas receive the state from the live replicas, and only then they start working as another member of the group [13].

In the following subsections, we centered on some details of the GLADE EPDA functionality: the configuration and execution of GLADE applications.

## 2.1   GLADE EPDA Configuration

The GLADE EPDA configuration has two main parts: the hardware and software configuration. The former allows us to define the nodes of the distributed system, whereas the latter focuses on the instantiation of the GLADE components (applications, partitions, channels and Ada units).

In the *hardware configuration*, GLADE EPDA distinguishes two node types: abstract and physical nodes. Ada partitions run on abstract nodes and abstract nodes must be assigned to physical nodes. This indirection level lets the user reconfigure Ada applications without changing the application configuration but only the association between abstract and physical nodes in the node configuration. GLADE EPDA prevents the user from assigning one abstract node to more than one physical node. The binding of application's abstract nodes can be modified at any time, but always before its execution. The tool avoids the execution of applications with abstract nodes which are not assigned to physical nodes.

In the *software configuration* GLADE EPDA allows us to instantiate the GLADE components: applications, partitions, channels and Ada units. Applications are formed by partitions and optionally channels: a partition is a collection of Ada units and the main distribution unit, whereas a channel describes the bidirectional communication between partitions. Two kinds of partitions are considered: active and passive partitions. The GLADE EPDA GUI represents the hierarchy of components by means of a graphical tree where applications are allowed in its first level, partitions and channels must be in the second level

**Table 1.** Properties of GLADE components in GLADE EPDA GUI

| PROPERTIES | GLADE COMPONENTS | | | |
|---|---|---|---|---|
| | Application | Partition | Channel | Ada Unit |
| Application name | X | | | |
| Partition name | | X | | |
| Channel name | | | X | |
| Ada unit name | | | | X |
| Application's main procedure | X | | | |
| Application's main procedure partition | X | | | |
| Partition's main procedure | | X | | |
| Starter | X | | | |
| Boot location | X | | | |
| Self location | | X | | |
| Passive | | X | | |
| Data location | | X | | |
| Task pool | X | | | |
| Version control | X | | | |
| Target host name | X | X | | |
| Target host function | X | X | | |
| Register filter | X | | | |
| Storage directory | X | X | | |
| Reconnection policy | | X | | |
| Termination policy | | X | | |
| Filters | X | X | X | |
| Command line | | X | | |
| First Partition | | | X | |
| Second Partition | | | X | |
| *Source Host* | X | X | | X |
| *Source Directory* | X | X | | X |

and Ada units in the third level. In this hierarchy, the GLADE components in a level constitutes the container of the components in the following immediate level. For each component instantiation, its properties and the relationship with other components must be introduced. The features of the GLADE configuration language (described in GLADE User's Guide [15]) are the properties of the GLADE components in our GUI (see Table 1). Some of the properties are associated exclusively to a specific GLADE component, whereas some properties are common to different components. When a property is common to different component instances in the same hierarchy, the property of the component in the lower level overrides the property of the component in the higher level. For instance, let us suppose the user defines the storage directory "temp" for a new GLADE application, which means that all the partitions will have the same value for the storage property. If the user redefines the storage property for one of the partitions of the application, every partition of the application will use "temp" as storage directory, except for the one whose property has been redefined.

Apart from the above mentioned properties, we add the source location property (showed in italics in Table 1) for applications, partitions and Ada units to specify the source host and directory where the source code is provided. Moreover, some of the properties are compulsory (e.g. the component names, the application target host, etc.), whereas others are optional. Furthermore, every property which involves a host must be fulfilled with an abstract node name (this abstract node is automatically added to the abstract node area in the node section of the GLADE EPDA GUI). The user does not have to fulfill all the configuration values because most of them are set to their default values. Besides, some properties are formed by a set of parameters which only are shown when the property is marked. For example, the source location property is formed by two parameters: the source host and the source directory. These parameters are shown when the user selects the source location property; otherwise they are not visible.

## 2.2   GLADE EPDA Execution

Once the configuration is completed, the hierarchy of components and their properties are kept in the GLADE EPDA server's database and the application is ready for execution or reconfiguration (any configuration change before the application execution). When the user executes an existing application by means of the menu bar option (in the GLADE EPDA GUI), the GLADE EPDA server carries out the following steps: (1) to generate the GLADE configuration file (.cfg) and install the source code; (2) to process the configuration file; (3) finally, to launch the application. GLADE EPDA also allows this execution step by step.

Every GLADE EPDA replica generates the *GLADE configuration file* for each application (from the configuration data in the GLADE EPDA replica's database) and installs the source code in its local disk. Some checks are made before the generation of the .cfg file: (1) all the abstract nodes must be assigned to one of the available physical nodes, that is, the physical nodes which have been added to the GLADE EPDA; (2) as the main procedure for a GLADE application must be specified along with a partition (the partition where it will be located), the GLADE EPDA checks whether this partition has been instantiated before the generation of the configuration file; (3) channels must be formed exactly by two partitions[1]. Moreover, for the generation of the configuration file, every abstract node which is used in the configuration of the application is replaced by the corresponding physical nodes. The generation of the configuration file only involves those attributes which have been considered in the GLADE release (the source location property is not considered in the configuration file but in the application installation).

The *application installation* implies installing the source code of all the components of the application in the target nodes. Every GLADE EPDA replica knows the allocation of this code and it is copied to its local disk. The source code, like the configuration file, is kept in the local disk of every replica in order to guarantee its availability in presence of node failures.

Both the configuration file and the source code are needed to generate the binaries by means of the GLADE *gnatdist* tool. These steps take place in the target host of the application (compulsory property for applications in the GLADE EPDA). In consequence, the GLADE EPDA replica must send both the configuration file and the Ada source code to the GLADE EPDA proxy on the target host.

---

[1] Although two partitions are defined during the instantiation of the channel, the user may have deleted one of them through the GUI utilities.

Finally, the application is launched. However, in GLADE there are different ways *to start an application* which are decided at configuration time: (1) if the user specifies the *Starter* property as *None*, the startup must be done manually; (2) if the *Starter* property is set to *Ada*, the startup is made from an Ada program; (3) if the *Starter* is set to *Shell*, the application is launched from a shell script. In the first case, at configuration time the user must indicate the startup order of the partitions (e.g. the partition server must be launched before the partition client) and the GLADE EPDA launches the application according to the order specified by the user. In the remaining cases, the GLADE EPDA proxy launches the application from the specified starter: an Ada binary or a shell script.

# 3   GLADE EPDA Usage Example

This section presents the configuration and execution of an Ada95 distributed application through the GLADE EPDA. It is structured as follows. Section 3.1 briefly summarizes the *Bank* example which is provided in the GLADE release. Section 3.2 introduces some specific features of the GLADE EPDA to the configuration of the *Bank* example. Section+3.3 describes the user actions to configure and execute the *Bank* example throughout the GLADE EPDA (client and server).

## 3.1   GLADE's Bank Example

The *Bank* example provided with the GLADE release is a client/server application. The *server* manages a simple database with the client accounts (the name, password and balance of the customers). The *clients* are provided with the following utilities: to present their balance, to withdraw or deposit money into their account and to transfer money from their account to the account of someone else. Several Ada units are developed for the *Bank* example: *Bank, Alarm,* etc. However, the main Ada units for this application are the following: the *Server* (the only RCI package), the *Manager* and the *Client*. The user must specify the partitions of the application at configuration time, and GLADE builds them from the provided Ada units.

The *Bank* application, named *Simcity* in the GLADE release, has the following properties: the starter method (*None*) indicates a manual startup of the application; there is no version control (*Version* is set to false); the boot server location is described by means of the protocol (*tcp*), host (*localhost*) and port (5557); the main procedure of the *Simcity* application (*Manager*) will be located into the *Bank_Server* partition. The *Simcity* application has two partitions: the *Bank_Server* partition, which contains the *Server* Ada unit, and the *Bank_Client* partition, with the *Client* main procedure, and a *Local* policy of termination.

### 3.2  Additional Properties for the Bank Example in the GLADE EPDA

Although the basic properties for the *Simcity* application have been presented, some additional properties must be provided to the GLADE EPDA. First, the location of the source code (source location) and the site where the components will be launched (target location) must be set. As we stated in Sect. 2.1, these locations are composed by an abstract node and a directory (in the form *abstract node:/directory*). Thus, the *Simcity*'s source location is set to *node1:/src*, whereas its target location is set to *node2:/bin*. Although the location nodes are specified in terms of abstract nodes, they must be bound to physical nodes before the execution of the *Simcity* application. Due to this fact, physical nodes of the distributed system must be setup (for example, *pcemartel*, *pcfguerra* and *pcjmiranda*), and then the user associates each abstract node to one of the available physical nodes. For example, we have bound *node1* to *pcemartel* and *node2* to *pcfguerra* for a distributed execution; for a centralized execution, all the abstract nodes can be bound to only one physical node, i.e. *localhost*.

### 3.3  User Actions in GLADE EPDA

The user actions are given in two stages: component configuration stage and application execution stage. Component *configuration* implies fulfilling its properties and its relationship with other components. This information is sent to the GLADE EPDA server to be stored in its database. The user actions to configure and execute applications are the following: (1) to create the application; (2) to create the partitions and channels which are contained into the application; (3) to create the partition's Ada unit; (4) to create the abstract nodes and the binding between the abstract nodes and the available physical nodes of the distributed system; (5) to execute the application. In the following, we describe the user actions for the *Bank* example throughout GLADE EPDA.

**Step 1: Application Creation.** For the *Bank* example, the user must create the *Simcity* application through the application section of GLADE EPDA GUI. At first, there is no component in the application section and the application creation is the only allowed option in the menu bar (see Fig. 2a). When the user selects this option, a property form is presented in the right area of the application section (see Fig. 2b). *Simcity* properties are fulfilled and sent to the GLADE EPDA server when the confirmation button of the form is clicked. After the application registration on the server side, *Simcity* is added to the application tree of the GLADE EPDA GUI (see Fig. 2c); otherwise, a warning message is presented to the user.

**Step 2: Partition and Channel Creation.** After the application creation, the user defines its partitions and channels. This implies registering its properties and its container application in the server's database. For the partitions two choices are presented: active and passive partitions.

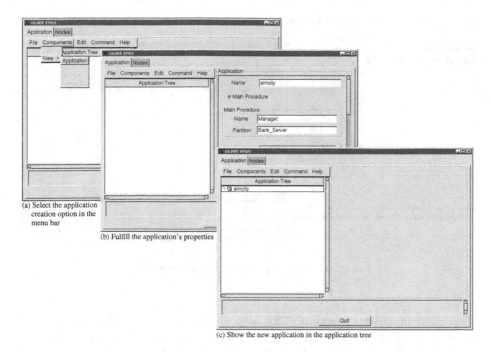

(a) Select the application creation option in the menu bar

(b) Fulfill the application's properties

(c) Show the new application in the application tree

**Fig. 2.** An example of application creation

For the *Simcity* application, the *Bank_Client* and *Bank_Server* partitions must be created (in this example, no channels are defined). For this, the user selects the container application (*Simcity*) in the application tree of the GLADE EPDA GUI, and accesses to the option in the menu bar which allows partition creation. Figure 3a shows the selection of the *Simcity* application in the left area. Because the *Simcity* application is selected, two choices are available in the menu bar: to add either a partition or a channel. If the operation to add a partition is selected, a property form is presented in the right area of the application section (see the right area of Fig. 3a). This form is used to introduce the properties of the *Bank_Client* partition. After a click in the confirmation button of the form, the properties and the relationship between the *Bank_Client* partition and its container (*Simcity*) are sent to the GLADE EPDA server. After the registration of these data in the server's database, the *Bank_Client* is added to the *Simcity*'s tree in the GLADE EPDA GUI (see Fig. 3b).

The creation of the *Bank_Server* partition is similar to the creation of the *Bank_Client* partition, except for its specific properties and the creation of its Ada unit (*Server*). Due to the fact the *Server* Ada unit is created in the following step, the user only has to introduce the name of the *Bank_Server* partition. Ada units are created in the following step because they are only allowed in the third level of the application hierarchy in the GLADE EPDA GUI.

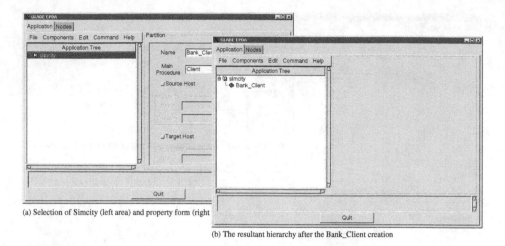

(a) Selection of Simcity (left area) and property form (right

(b) The resultant hierarchy after the Bank_Client creation

**Fig. 3.** An example of partition creation

**Step 3: Ada Unit Creation.** This step allows us to define the Ada units of partitions. For the *Bank* example, *Server* is the only RCI Ada unit contained in the *Bank_Server* partition. For the creation of this Ada unit, the user selects the container partition (*Bank_Server*) and accesses to the menu bar option to create an Ada unit. After this, a property form is presented in the right area of the application section. The only property for the *Server* Ada unit is its name, which is sent to the GLADE EPDA server along with the container name (*Bank_Server*). As always, the *Server* Ada unit is added to the application tree when the registration on the server side is completed. The resultant application hierarchy is showed in Fig. 4.

**Step 4: Physical and Abstract Node Creation and Binding.** Whenever a GLADE EPDA proxy is installed on a physical node of the distributed system, a message is sent to the group of GLADE EPDA replicas in order to register the physical node properties: name, IP address, CPU, operating system and the identity of its proxy. After the registration of the physical node, an icon is automatically added to the physical node section of the GUI. Therefore, the available physical nodes in GLADE EPDA are those which have been registered to the group of GLADE EPDA replicas. The system formed by these physical nodes is heterogeneous as each proxy will compile the partition for the target.

Abstract nodes are referenced during the creation of application components (for instance, *node1* and *node2* are used as properties of *Simcity*). GLADE EPDA adds automatically the abstract nodes which have been referenced in the component creation to the abstract node area in the GLADE EPDA GUI's node section (see *node1* and *node2* in Fig. 5). Any abstract node which is added to the GLADE EPDA GUI is sent to the GLADE EPDA server's database. GLADE

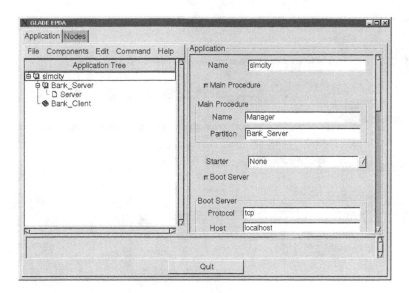

**Fig. 4.** Simcity hierarchy

EPDA also allows to create abstract nodes directly through the menu bar in the node section.

Every abstract node must be assigned to one of the available physical nodes in the node section of the GLADE EPDA GUI before the application execution. The abstract nodes which are referenced by the *Simcity* application (*node1, node2*) must be bound to the available physical nodes in the GLADE EPDA (*pcemartel, pcfguerra, pcjmiranda*): the user selects the set of abstract nodes to bind, the menu bar option for binding and the target physical node for the bind operation. In Fig. 5a the user has selected only one abstract node (*node1*) and also the menu bar option for binding. After this, the user is asked for the selection of the target physical node and *pcemartel* is selected (see Fig. 5b). Finally, *node1* is bound to *pcemartel* and added to the physical node tree in the node section. Figure 5c shows the binding between *node1* and *pcemartel*, and *node2* and *pcfguerra*.

**Step 5: Application Execution.** After the application configuration, the application is ready to be executed. The application execution depends on how the *Starter* property of the application has been configured. When this property is set to *None* the user will be asked for the execution order of the partitions.

For the *Bank* example, the user executes the *Simcity* application through the menu bar of GLADE EPDA GUI's application section. Because the *Starter* property is set to *None*, the user is asked for the execution order of the partitions: the server partition (*Bank_Server*) must be executed before the client partition (*Bank_Client*). This information along with the execution command is sent to the GLADE EPDA server (see Fig. 6-1). When the execution command is received

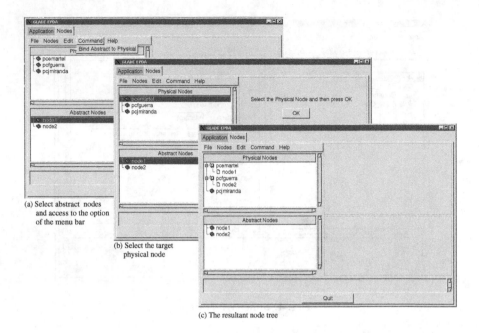

(a) Select abstract nodes
and access to the option
of the menu bar

(b) Select the target
physical node

(c) The resultant node tree

**Fig. 5.** An example of binding between the abstract and physical nodes

on the server side, the GLADE EPDA replicas do the following actions according to the configuration data in the GLADE EPDA replica's database:

- To generate the configuration file (*simcity.cfg*) and install the code from the source location to its local disk[2] (see Fig. 6-2). For the *Bank* example, this location is *node1:/src*, and because *node1* is bound to *pcemartel*, the GLADE EPDA replicas copy the source code from *pcemartel* to its local disk.
- To send the execution request to the proxy in the target location: *node2:/bin* (see Fig. 6-3). Because applications run on physical nodes, the GLADE EPDA replicas get the corresponding physical node for *node2* (*pcfguerra* because *node2* is bound to it). Therefore, the execution request is sent to the proxy in *pcfguerra* along with the configuration file, the source code and the execution order of the partitions.

Finally, the GLADE EPDA proxy processes the *simcity.cfg* by means of the *gnatdist simcity* (see Fig. 6-4) and launches the application according to the execution order: first, the *Bank_Server* partition, and then the *Bank_Client* partition. Both partitions are launched in the application's target host (*pcfguerra*) because no target host was specified for these partitions.

---

[2] The source code and the configuration file are stored in the local disk of every replica in order to guarantee their availability in presence of node failures.

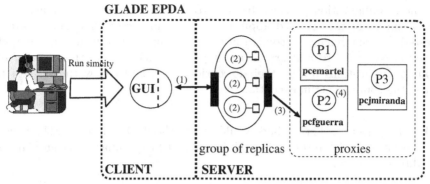

(1) GUI run command
(2) Genenerate the configuration file (simcity.cfg), install the source code from
    pcemartel (node1) to its local disk and get the proxy for pcfguerra (node2)
(3) Installs the source code and send the configuration file (simcity.cfg) to
    pcfguerra
(4) Execute 'gnatdist simcity' and start the application

**Fig. 6.** Execution of the *Simcity* application

## 4   Conclusions and Current Work

This paper has presented an environment to configure and execute GLADE
distributed applications. It is assumed that the user writes the code according to
the distributed model of Ada95 and knows the component properties of GLADE.
The GUI facilitates the GLADE configuration as the user does not have to get
used to the GLADE configuration language. The environment keeps the state and
data configuration to provide a basic level of static reconfiguration. Furthermore,
GLADE EPDA's functionality is provided without altering the GLADE release,
which facilitates the addition of future GLADE features to the current GLADE
EPDA. Therefore, each feature provided by GLADE is provided by GLADE
EPDA as long as the related configuration property is included in the GLADE
EPDA's GUI.

Nowadays, the main limitation of the GLADE EPDA is located on the client
side because software clients must be generated for different operating systems
(Windows, Linux, Solaris, etc.) and installed in every client node. In order to
gain flexibility, we are now moving to an environment where the configuration
and execution is made from a Web navigator. In this approach there is no need
to install specific software on client-side in order to provide configuration and
execution of distributed applications. We are working on that direction through
AWS (Ada Web Server) [16].

On the other hand, the interaction between the GLADE EPDA replica and
its database has been implemented by means of a binding to *postgresql*, without
support for *odbc*. We plan to use GNADE (GNAT Ada Database Development
Environment) [17] for this purpose.

The reconfiguration provided by the GLADE EPDA must take place before the execution of the involved application. In future research work, we are interested in some dynamic reconfiguration issues: migration, load balancing, etc. Moreover, we plan to add graphical support for monitoring the state of distributed applications during their execution, that is, to view the load of the nodes, the components in execution, etc.

**Ackowledgments.** The authors would like to thank David Shea of the Faculty of Translation and Interpretation (University of Las Palmas de Gran Canaria) for editorial assistance.

# References

1. Intermetrics. *Ada 95 Language Reference Manual.* International Standard ANSI/ISO/IEC-8652: 1995, January 1995.
2. Gargaro, Kermarrec, Nana, Pautet, Smith, Tardieu, Theriault, Volz, and Waldrop. *Adept (Ada 95 Distributed Execution and Partitioning Toolset).* Technical Report, April 1997.
3. L. Pautet, T. Quinot, and S. Tardieu. Building Modern Distributed Systems. In *2001 Ada-Europe International Conference on Reliable Software Technologies*, pages 122–135, Belgium, May 2001. Springer-Verlag.
4. G. Etzkorn. Change Programming in Distributed Systems. In *Proceedings of the International Workshop on Configurable Distributed Systems*, pages 140–151, United Kingdom, March 1992.
5. M. Barbacci, C. Weinstock, D. Doubleday, M. Gardner, and R. Lichota. Durra: a Structure Description Language for Developing Distributed Applications. *Software Engineering Journal*, 8(2):83–94, March 1993.
6. J. Magee, N. Dulay, S. Eisenbach, and J. Kramer. Specifying Distributed Software Architectures. In W. Schafer and P. Botella, editors, *Proc. 5th European Software Engineering Conf. (ESEC 95)*, volume 989, pages 137–153, Sitges, Spain, 1995. Springer-Verlag, Berlin.
7. R. Balter, F. Bellisard, Boyer, M. Riveill, and J.-Y. Vion-Dury. Architecturing ans Configuring Distributed Application with Olan. In *Proceedings IFIP International Conference on Distributed System Platforms and Open Distributed Processing (Middleware'98)*, pages 15–18, The Lake District, September 1998.
8. J. Bishop and L. Botha. Configuring Distributed Systems in a Java-based Environment. In *IEEE Proceedings of Software Engineering*, pages 65–74, http://citeseer.nj.nec.com/botha00configuring.html, April 2001. IEEE Computer Society.
9. K. Ng, J. Kramer, J. Magee, and N. Dulay. A Visual Approach to Distributed Programming. In *Tools and Environments for Parallel and Distributed Systems*, pages 7–31. Kluwer Academic Publishers, February 1996.
10. S. Maffeis. Piranha:A CORBA Tool for High Availability. *IEEE Computer*, pages 59–66, April 1997.
11. E. Martel, F. Guerra, and J. Miranda. A Tailorable Distributed Programming Environment. In *2002 Ada-Europe International Conference on Reliable Software Technologies*, pages 269–281, Vienna, June 2002. Springer-Verlag.

12. Thaís Batista and Noemi Rodriguez. Dynamic Reconfiguration of Component-based Applications. In IEEE, editor, *Proceedings of the International Symposium on Software Engineering for Parallel and Distributed Systems (PDSE 2000)*, volume 989, pages 32–39, Limmerick, Ireland, 2000.
13. F. Guerra, J. Miranda, J. Santos, E. Martel, L. Hernández, and E. Pulido. Programming Systems with Group_IO. In *Proceedings of the Tenth Euromicro Workshop on Parallel, Distributed and Network-based Processing*, pages 188–195, Canary Islands, Spain, January 2002. IEEE Computer Society Press.
14. F. Guerra, J. Miranda, J. Santos, and J. Calero. Building Robust Applications by Reusing Non-Robust Legacy Software. In *2001 Ada-Europe International Conference on Reliable Software Technologies*, pages 148–159, Belgium, May 2001. Springer-Verlag.
15. L. Pautet and S. Tardieu. *GLADE User's Guide. Version 3.14p*. GNAT, May 2001.
16. D. Anisimkov and P. Obry. *AWS User's Guide. Version 1.2*. ACT, Ada Core Technologies, October 2001.
17. M. Erdman. Gnat Ada Database Development Environment. In *2002 Ada-Europe International Conference on Reliable Software Technologies*, pages 334–343, Vienna, June 2002. Springer-Verlag.

# The Use of Ada, GNAT.Spitbol, and XML in the Sol-Eu-Net Project

Mário Amado Alves[1], Alípio Jorge[1], and Matthew Heaney[2]

[1] LIACC – Laboratório de Inteligência Artificial e Ciências de Computação (Universidade do Porto), Rua do Campo Alegre 823, 4150-180 PORTO, Portugal
{maa,amjorge}@liacc.up.pt, www.liacc.up.pt
[2] mheaney@on2.com, home.earthlink.net/~matthewjheaney

**Abstract.** We report the use of Ada in the European research project Sol-Eu-Net. Ada was used in a web mining subproject, mainly for data preparation, and also for web system development. Open source Ada resources e.g. GNAT.Spitbol were used. Some such resources were modified, some created anew. XML and SQL were also used in association with Ada.

## 1 Project Settings

> *Sol-Eu-Net is a European Union sponsored network of expert teams from academia and industry, offering tools and expertise designed to meet your Data Mining and Decision Support needs.*
>
> *(soleunet.ijs.si)*

In this article we describe the use of Ada in the Sol-Eu-Net project. The project effective period was January 2000 – March 2003. LIACC, the Portuguese partner, began using Ada on 1 June 2001, when the first author joined the project. This person became rapidly known in the project community as 'the Ada guy'. LIACC's tasks requiring programming were:
- data mining in the INE subproject;
- development of the SENIC web system.

We describe these tasks in detail in Sects. 2 and 3 respectively, and how Ada was used to support them, along with associated libraries, particularly GNAT.Spitbol, and other standard technologies namely SQL and XML. We reserve a special word on SQL for Sect. 4. In Sect. 5 we conclude, with a critical review of the reported approach, including an account of other programming languages used in the project.

The interested reader should accompany the commentaries with a view of the source code. The used Ada packages come from a variety of sources. Indexed_IO, for example, is on *AdaPower dot com*. The packages by Mário A. Alves are available on *www.liacc.up.pt/~maa*, or upon request to the author. In general, the packages are gratis for non-commercial use, and licensable at low cost for commercial projects. All packages are open source, or 'quasi' (see *groups.yahoo.com/group/softdevelcoop*).

The second author of this article features mainly as the coordinator of LIACC's participation in the project. The third author features as the author of Indexed_IO

J.-P. Rosen and A. Strohmeier (Eds.): Ada-Europe 2003, LNCS 2655, pp. 196–207, 2003.
© Springer-Verlag Berlin Heidelberg 2003

(Sect. 2.4) and other used or reviewed Ada packages (Sect. 5). The first author alone is accountable for any opinions expressed, even where the plural form is used.

# 2 The Use of Ada and GNAT.Spitbol in the INE Project

INE_Infoline, or simply INE, was a web mining project. *INE* stands for *Instituto Nacional de Estatística = National Institute for Statistics*, which is the Portuguese government office in charge of producing and making public statistical data about the country. Publication is done in part through INE's website *Infoline* (*www.ine.pt*). INE wanted to mine the record of accesses to *Infoline* in order to discover usage patterns and ultimately improve site usability.

## 2.1 Data Understanding and Preparation

The main source data are the web access logs produced—independently—by INE's two HTTP servers (located in Lisbon and Oporto). Complementary source data are a database of registered users and a mirror of the web site on disk. The whole data in compressed form occupy 5 CD-ROMs (c. 3 Gigabytes).

The logs were packaged periodically by INE, with varying periodicity, and also varying in physical as well as conceptual structure, both along time and between servers.

Figures 1 and 2 show example log files. These are text files. The first three lines of each file are shown. Characters carrying sensitive information (IP and User_Id) are hidden behind '☐' here. The space character is represented by the middle dot '•'. The end of line is represented by '↵'. White space means (literally) nothing.

```
xxx.xx.xxx.xxx,•-
,•1/1/00,•0:06:05,•W3SVC1,•INFOLINE,•194.65.84.194,•0,•285,•141,•304,•0,•GET,•/verdem
o/subsistema.htm,•-,•↵
xxx.xx.xxx.xx,•-
,•1/1/00,•0:13:06,•W3SVC1,•INFOLINE,•194.65.84.194,•0,•194,•1348,•200,•0,•GET,•/erros/
erro.asp,•404;http://infoline.ine.pt/si/apresent/senanual/cap3a.html,•↵
```

**Fig. 1.** Example log file

```
"xx.xx.xxx.xxx";"xxxxxxx";2001-01-08•0:00:00;1899-12-
30•17:11:47;5860;595;23721;200.00;"GET";"/inf/prodserv/quadros/tema16/sb1601/htm/0010
0099.htm"↵
"xx.xx.xxx.xxx";"xxxxxxx";2001-01-08•0:00:00;1899-12-
30•17:13:18;15266;595;31090;200.00;"GET";"/inf/prodserv/quadros/tema16/sb1601/htm/002
00099.htm"↵
```

**Fig. 2.** Another example log file

The whole time period under analysis was 1 January 2000–31 December 2001, so the whole set of logs was multitudinous and variegated. These conditions imposed a substantive effort in understanding, collecting and pre-processing the data. Namely, all the different file formats and fields had to be reverse-engineered. Mainly two software items were used to effect this task: a standard set of field definitions (Table 1) and a standard database system (SQL). The varying source fields were mapped onto the standard set, and corresponding transformation procedures were developed and effected to populate the database.

**Table 1.** Standard web access fields

| Name | Type (SQL) |
|------|-----------|
| Id | int(11) |
| Server_Id | char(1) |
| Date | varchar(19) |
| IP | text |
| User_Id | text |
| Method | text |
| URI | text |
| Status | decimal(3,0) |
| Request_Volume | decimal(10,0) |
| Response_Volume | decimal(10,0) |
| Processing_Time | decimal(10,0) |
| Referer_URI | text |

These procedures were directly done with Ada, using the GNAT.Spitbol library for character string pattern matching. The programs read the log files and generate the appropriate SQL statements to populate the database. Figure 4 on next page lists one such program designed to operate on input formatted as in Fig. 2. Figure 3 below shows the corresponding output.

Note the date in ISO format (5th field of the values clauses on Fig. 3). The usual tools (Microsoft Access, Microsoft Excel, MySQL, etc. – add © and ® marks at will) always have problems with date representation. Just take a look at how the date and time was represented in Figs. 1 and 2. One lesson we learned and now share: use only ISO format and Ada.Calendar and 'friends' for date and time – and never any 'proprietary' format. To this effect package Datetime has proved of great help, as will be further demonstrated in Sect. 2.2.

```
use•ine_2b↵
in-
sert•into•accesses•(id,•server_id,•ip,•user_id,•date,•response_volume,•request_volume,•status,
•method,•uri)•values•(•1,•'o',•'xx.xx.xxx.xxx',•'xxxxxxxx',•'2001-01-
08•17:11:47',•595,•23721,•200,•'GET',•'/inf/prodserv/quadros/tema16/sb1601/htm/00100099.
htm');↵
in-
sert•into•accesses•(id,•server_id,•ip,•user_id,•date,•response_volume,•request_volume,•status,
•method,•uri)•values•(•2,•'o',•'xx.xx.xxx.xxx',•'xxxxxxxx',•'2001-01-
08•17:13:18',•595,•31090,•200,•'GET',•'/inf/prodserv/quadros/tema16/sb1601/htm/00200099.
htm');↵
```

**Fig. 3.** Example generated SQL statements

```
with Ada.Text_Io; use Ada.Text_Io;                      procedure Tell_Line is
with Ada.Exceptions; use Ada.Exceptions;                begin
with Gnat.Spitbol; use Gnat.Spitbol;                       if Line_Count mod Chunk = 0 then
with Gnat.Spitbol.Patterns; use Gnat.Spitbol.Patterns;        Put(Current_Error, ".");
with Mysql;                                                end if;
with Arg;                                                  if Line_Count mod (10 * Chunk) = 0 then
                                                              Put_Line(Current_Error, Natural'Image(Line_Count));
procedure Oporto2sql is                                    end if;
                                                        end Tell_Line;
   IP_VS              : VString;
   User_Id_VS         : VString;
   Date_1_VS          : VString;                        begin
   Date_2_VS          : VString;
   Response_Volume_VS : VString;                           Put_Line ("use " & Arg (1, "ine_2b"));
   Request_Volume_VS  : VString;                           Id := Integer'Value (Arg (2, "1"));
   Status_VS          : VString;                           Chunk := Natural'Value (Arg (3, "1000"));
   Method_VS          : VString;                           loop
   URI_VS             : VString;                              begin
                                                                Get_Line(Line_Buffer, Line_Length);
   Line_Pattern : Pattern :=                                    Line_Count := Line_Count + 1;
      """" & Break("""") * IP_VS           & """" & ";" &       Tell_Line;
      """" & Break("""") * User_Id_VS      & """" & ";" &    exception
             Break(";" ) * Date_1_VS       & ";" &             when End_Error => exit;
             Break(";" ) * Date_2_VS       & ";" &          end;
             Break(";" )                   & ";" &          if Line_Length > 0 then
             Break(";" ) * Response_Volume_VS & ";" &          begin
             Break(";" ) * Request_Volume_VS  & ";" &            if Match
             Break(";" ) * Status_VS       & ";" &                  (Line_Buffer (1 .. Line_Length),
      """" & Break("""") * Method_VS       & """" & ";" &          Line_Pattern)
      """" & Break("""") * URI_VS          & """"    ;         then
                                                                   Put_Line (SQL_Statement);
   Line_Buffer : String(1 .. 10_000);                           Id := Id + 1;
   Line_Length : Natural range 0 .. Line_Buffer'Last;           Stmt_Count := Stmt_Count + 1;
   Line_Count  : Natural := 0;                               else
   Chunk       : Natural;                                       raise Error;
   Stmt_Count  : Natural := 0;                               end if;
   Error_Count : Natural := 0;                            exception
   Id          : Integer;                                    when E : others =>
   Error       : exception;                                     Error_Count := Error_Count + 1;
                                                                Put_Line
function Good_Date return String is                                ("-- ERROR!");
begin                                                            Put_Line
   return S(Date_1_VS)(1..10)&" "&S(Date_2_VS)(12..19);           ("-- Exception: " & Exception_Name (E));
exception                                                        Put_Line
   when Constraint_Error =>                                          ("-- Input line: [" &
      return S(Date_1_VS)(1..10)&" 0"&S(Date_2_VS)(12..18);          Line_Buffer (1 .. Line_Length) & "]");
end Good_Date;                                              end;
                                                        end if;
function SQL_Statement return String is begin return    end loop;
   "insert into accesses " &                            New_Line
   "(id, server_id, ip, user_id, "&                       (Current_Error);
   "date, response_volume, request_volume, "&           Put_Line
   "status, method, uri) values (" &                      (Current_Error, Natural'Image(Line_Count) &
      Integer'Image(Id)        & ", "&--Id                 " input lines processed");
   "'o'"                        &"'", "&--Server_Id      Put_Line
   '''&S(IP_VS)                 &"'", "&--IP               (Current_Error, Natural'Image(Stmt_Count) &
   '''&Mysql.Escape(S(User_Id_VS)) &"'", "&--User_Id      " statements generated");
   '''&Good_Date               &"'", "&--Date           Put_Line
      S(Response_Volume_VS)    & ", "&--Response_Volume    (Current_Error, Natural'Image(Error_Count) &
      S(Request_Volume_VS)     & ", "&--Request_Volume     " errors");
      S(Status_VS)(1 .. 3)     & ", "&--Status
   '''&S(Method_VS)            &"'", "&--Method       end Oporto2sql;
   '''&Mysql.Escape(S(URI_VS)) &"');";--URI
end SQL_Statement;
```

**Fig. 4.** Example Ada program to transform a log file into SQL statements

## 2.2  Data Enrichment: Sessions

The total number of accesses for the period 2000-2001 was circa five million. This does *not* include accesses to GIF images and similar stuff, which were deemed irrelevant for the data mining objectives.

Now, certain data mining models required *sessions*, not simply accesses. A session is a sequence (in time) of accesses with the same Session_Owner, which is either the IP or the User_Id. Not all accesses have User_Id.

To identify and represent sessions, five additional fields were used (Table 2). Unix_Time is the time in seconds elapsed since 1970-01-01 00:00:00.

**Table 2.** Additional access fields, for sessions

| Name | Type |
|------|------|
| Unix_Time | decimal(10,0) |
| IP_Session_Id | int(11) |
| Order_Number_In_IP_Session | int(11) |
| User_Session_Id | int(11) |
| Order_Number_In_User_Session | int(11) |

Sessions were identified by traversing all accesses ordered by the composite key (Session_Owner, Unix_Time): each new Session_Owner or a Unix_Time more than 30 minutes greater than the previous starts a new session.

This computation was also specified in Ada, in order to overcome two problems:
- the poor and/or incorrect treatment of dates and time provided by the usual tools, as noted above;
- the sheer kind of computation necessary, not expressible in SQL.

Namely, Ada programs *correctly* computed the values of Unix_Time, and identified the sessions based on the aforementioned criterion. For time computations, package Datetime was used, which converts between Ada.Calendar.Time and dates in ISO format. The conversion to Time is robust in the sense that it accepts any 'stretch' of an ISO date/time string between year only and up to the micro-second (the missing parts default to month 1, day 1, micro-second 0).

## 2.3  Data Enrichment: Categories

> *SNOBOL is both easier to use and more powerful than regular expressions. NO backslashes to count.*
>
> (*sourceforge.net/projects/snopy/*)

Each accessed page in the *Infoline* web site, either static or dynamic, has associated categories e.g. Tema (theme), Subtema (sub-theme), Tópico (topic), which are organizing concepts of the site—and which were the attributes for most data mining modelling done in the project. A particularly important category is Visualization, which is a flag marking a page as a *visualization = conversion = target*, as opposed to being merely a *click-through = active = navigational* page.

In order to enrich the data with these categories, a function of Knowledge About Accesses, or KAA, must be defined, that maps each access entry onto the corresponding categories. Such a function is mainly based on the character string pattern of the accessed URI.

A document prepared by INE provided most of this knowledge, in semi-formal style. Other parts of the KAA were obtained in meetings and from data exploration by the data mining team.

The KAA was formalised as an Ada package and applied to the database. Again, the superior string pattern matching capabilities of GNAT.Spitbol were explored.

The current state of the KAA package covers (gives some category to) one third of the visited pages. The unmatched URIs were assumed to be irrelevant. A table Pages in the database held the results of the KAA. Sessions are also enriched, with information derived from accesses and the KAA (Table 3).

**Table 3.** Enriched sessions

| Name | Type |
|------|------|
| Owner | enum('ip','user') |
| Id | int(11) |
| Date | char(19) |
| Number_Of_Accesses | int(11) |
| Duration | bigint(13) |
| Volume | int(11) |
| Number_Of_Visualizations | int(11) |

## 2.4  A Clustering Algorithm and Its Implementation

To profile users – a stated objective of the INE project – a clustering model is appropriate. Early experiments utilized the commercial tool Clementine, for exploration. For deployment, we planned to integrate more tightly. To this effect, an Ada package was developed that implements the simple but effective clustering algorithm in Fig. 5.

| 1. Start with one cluster per item. |
|---|
| 2. While not satisfied, unite the two most similar clusters. |

**Fig. 5.** A clustering algorithm

The implementation difficulty resides in finding the two most similar clusters. A similarity matrix, holding all distances between each pair, has space complexity $O(n^2)$. When $n$ is high—as is the case in the INE project—, the computation becomes impracticable.

A trick is to work with a *sparse* matrix, by eliminating, for example, all pairs of distance greater then a given value. Also, as the number of clusters decreases by one on each iteration, the matrix gets smaller and smaller.

This approach was taken in Ada package Clustering. Clustering.Disk_Based uses package Indexed_IO to implement the required data structures, mostly ordered sets.

Package Indexed_IO is free software authored by Matthew Heaney. The version in use by Clustering.Disk_Based was slightly modified by Mário A. Alves.

This clustering algorithm has been proposed as a test case for the emerging Ada Standard Containers Library [1], because of the conciseness of its mathematical formulation, in contrast with the complexity of data structures required for its implementation. For example, Clustering.Disk_Based.Find_Clusters works with the six data structures described in Fig. 6.

---

Items are identified with a derived type from Positive, for compatibility with file positions.

In what follows *direct file* and *indexed file* denote objects of type File_Type of instances of Ada.Direct_IO and Indexed_IO respectively.

1. The set of items is passed as a direct file open for reading. The item position is its Id.

The clustering results are held in a complex of 3 files:

2. The set of clusters is held in an indexed file of triples (Cluster_Id, Cluster_Centroid, Cluster_Size), with key Cluster_Id.

3. The (many-to-one) mapping of items to clusters is held in a direct file of Ids. The position is the item Id. The contained element is the cluster Id.

4. The (one-to-many) mapping of clusters to items is held in an indexed file of pairs (Cluster_Id, Item_Id) with key (Cluster_Id, Item_Id). To get all items of a cluster X the trick of starting a visit with (X, Id_Type'First) is used.

The similarity matrix is held in a complex of 2 files, with an order relation defined between pairs of clusters:

5. The set of admissible pairs of clusters is held in an indexed file of scheme (Distance (A, B), (A, B)), where A, B are cluster ids. An admissible pair (A, B) is one with Distance (A, B) <= Maximum_Distance. This file provides access to the pair with the smallest distance (greatest similarity), using the trick of starting the visit with (Distance_Type'First, (1, 1)).

6. The (one-to-many) mapping of clusters to pairs is held in an indexed file of scheme ((A, B), (Distance (A, B), (A, B))). A variation of the usual trick gets all pairs of a cluster i.e. all pairs of which the cluster is an element.

---

**Fig. 6.** The data structures of Clustering.Disk_Based.Find_Clusters

The distance between two clusters is the Euclidean distance between their centroids. Other distance methods can be substituted easily.

Package Clustering.Disk_Based has been tested for correctness and time performance. On a test set of 100 random points in a planar real space of 1.0 * 1.0, with Maximum_Distance = 0.5 and Minimum_Number_Of_Clusters = 5, processing time is a couple of seconds on the 'vanilla' PC.

The package can be improved in many ways. Unfortunately the Sol-Eu-Net project has come to its end before the package was utilized on real data. So further package development is on wait. Ideas: a method to determine Maximum_Distance, namely as a fraction (e.g. 1/2) of the greatest distance; develop Clustering.RAM_Based with the same 'interface'; methods to support non-real attributes (Boolean, discrete, categories).

## 3   The Use of Ada and XML in SENIC

SENIC (Sol-Eu-Net Information Collector) is a web system designed to support the Sol-Eu-Net project task of collecting information about the tools and case studies of the project. Figures 7–9 are screenshots of interesting moments in a SENIC session. SENIC 3 implements the Unified Schema for the description of problem solving projects involving both Data Mining and Decision Support sub-parts [2].

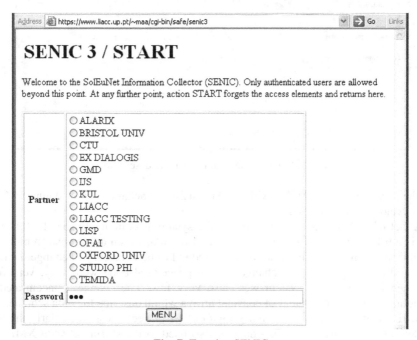

**Fig. 7.** Entering SENIC

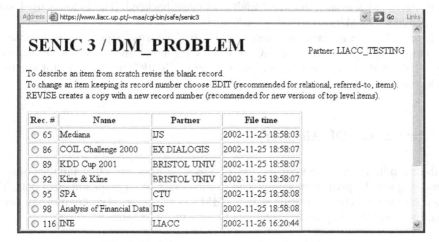

**Fig. 8.** Listing data mining problem descriptions

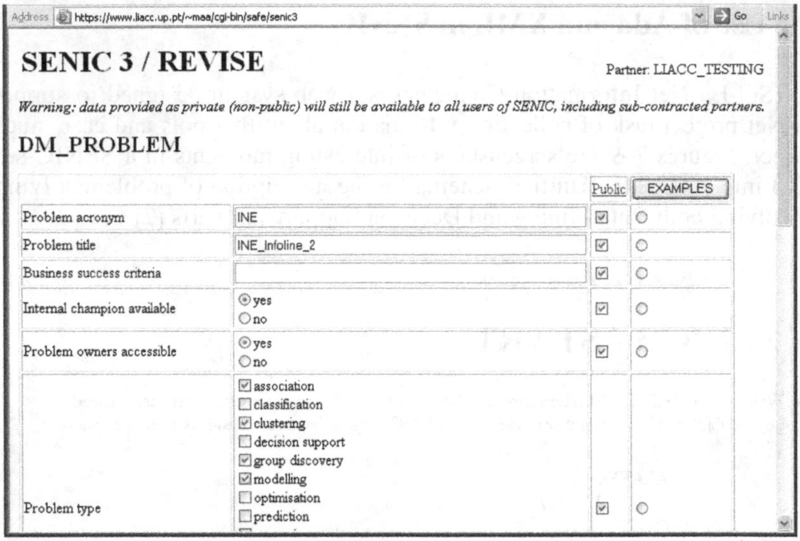

**Fig. 9.** Filing a data mining problem description

The system was built with Ada, using packages like Generic_CGI, XML_Automaton, and Pgsql.

A fact of notice is that the structure of the system is completely specified in Ada. This is a tribute mostly to the rich type system of Ada, which allows one to build the model directly in the source code. It has happened to be necessary to change the conceptual structure greatly. The changes to the procedural code were very, very few. The changes in the type system were also very few. The changes concentrated in enumeration and in the content of constant arrays which describe the data scheme.

XML [4] was used to communicate with other project sites, particularly the Sol-Eu-Net website (*soleunet.ijs.si*). Ada was also used to transform between XML and SENIC structures. Package XML_Automaton was used to transform from XML.

Postgres [5] was used for the SENIC database backend. The binding used was package Pgsql by Mário A. Alves – see Sect. 4.

Lately SENIC was reused to create DATIC [6]. The reuse experience was successful enough to suggest reusing SENIC (or DATIC) again to implement the *knowledge-map* of the Sol-Eu-Net project, and even creating GENIC, the Generic Information Collector.

# 4   The Use of SQL All Over

Both data mining and web systems – as well as a multitude of other families of systems – require the permanent storage and manipulation of large quantities of data. To attain efficiency here, integration with a DBMS is the most taken engineering decision. Existing DBMSs provide essentially two means of integration:

- a native programming language, tightly bound with the database module, in which the whole system is supposed to be written; however this language is often of not enough expressive power;
- an API, usually written in C, to the SQL processing module (APIs to the internal modules are seldom provided).

When writing the system in another language (e.g. Ada) ultimately it is necessary to link with the API, adapting the two languages (e.g. Ada and C). A standard means of adaptation is embedded SQL. This in fact introduces a new (small) language, which is neither Ada nor C, and requires pre-processing of the source code. Another means is to use the API directly. A lot of proposals – bindings – exist for this. The most followed one is perhaps ODBC (Open Database Connectivity, actually SQL/CLI [7]). Also of note is ADBC, associated with GNADE [8].

However in our projects we have been using our own bindings. In part because they started developing a long time ago when hardly anything else was available, in part because they simply seem better fit for the job. Packages Mysql and Pgsql bind to the Mysql and Postgres APIs respectively [5, 9]. They are much simpler to understand and operate than ODBC and GNADE, and the trade-off in functionality is not much – if any at all. These packages allow the programmer to communicate with the database server in SQL and provide ADTs to give access to the data resulting from such queries.

Mysql and Pgsql differ only very slightly. It has happened to be necessary to re-target an application (INE) from Pgsql to Mysql: the required code changes were easy and few.

## 5  Ada and the Others: Lessons Learnt—Empower the Future

Ada was used in the Sol-Eu-Net project for data preparation and web system development. The results were satisfactory. This may be unsurprising, since Ada is a general purpose language. However, perhaps the last myth about Ada is that it does not excel in programming in the small. May the current report help dissipate that myth. What deserves emphasizing is the decision to use Ada instead of the multitude of technologies some organizations still use to accomplish the kind of tasks reported here.

The first author of this article calls the principle behind this decision the *Reduced Technology Mix* principle: essentially, choose fewer – but better – components.

It was nice to find, from recent threads on CLA [10], that Ada-based organizations also do all their 'scripting' in Ada.

With the work reported in this article we have also strengthen our experience that the rich type system of Ada lets the problem express itself as source code. With Ada, the program truly is the documentation. Especially if the design patterns utilized are recognized.

With respect to comparison with other languages, the first author is convinced the right way to go is experimental software engineering. However there is a data famine in this area, so we can only offer a qualitative review.

Other programming languages used in the Sol-Eu-Net project were C++, Java, Prolog, Sumatra, and R.

**C++** was used for text mining and produced nice graphical visualizations of clusters [11]. The C++ team reduced pointer bugs by using SmartPointers. We consider this a lesson learnt: C++, only with SmartPointers.

**Java** was also used for graphical visualization. Some compatibility problems were felt here. It seems no computer has exactly the same Java machine installed. This is curious enough, because perhaps the main purported advantage of Java is precisely its interoperability.

Java is also the implementation language of Caren, an association rule learner used in the INE subproject. Caren works with text input and output. Probably because of this, no compatibility problems were felt here.

Java is also the implementation language of Zeno (*zeno8.gmd.de/zeno*), the official Internet groupware of the Sol-Eu-Net project. No implementation problems were felt here.

**Prolog** was used, by the second author, to format the data for Caren, invoke it, and create reports of these operations, along with their results. It took some, but not much, debugging to get the system work correctly.

**Sumatra** is a specific domain language for data transformation. It is the language of SumatraTT, a visual data transformation tool. SumatraTT was the recommended data pre-processing tool in the Sol-Eu-Net project, as a way to achieve uniformity in data transformation specifications. As such, it was used in one or two occasions to an advantage, but naturally most times teams preferred to use their 'own' languages and tools. Also, the poor English of the manual did not help its adoption.

**R** is being taught and used at LIACC as a data analysis tool. A nice feature of R is its graphical output capabilities.

In summary, what the other languages seem to offer advantageously over Ada is not any 'core' language features but 'external' traits like nice graphical output and Internet gadgetry. However this issue should not be overlooked.

In the opinion of the first two authors of the current article, Ada and Prolog are unsurpassed in readability. In the opinion of the first and third author, Ada is unsurpassed in many other factors as well. We will not preach to the choir here: the expressed opinions are extensively backed up – search the Internet. Of all Ada intrisic properties, let us just single one out for the ongoing discussion: *rigour*.

However we think the 'external' capabilities associated with a language are equally important.

Increasingly a need for a standard data structures library for Ada has been recognized. Currently several libraries seem to be competing for a *de facto* status, notably Charles (by the third author), the Booch Components, Mats Weber's library, PragmARC, etc. (see AdaPower; see also *groups.yahoo.com/group/asclwg*).

In the areas of graphical output, graphical user interface, and networking, and as to whether there should be a standard specification (an annex) for each of these areas as well, opinions in the community diverge. We think that results like the currently reported motivate standardization.

All this relates to the issue of programming *language* vs. programming *system*. Programming systems, like Smalltalk (*minnow.cc.gatech.edu/squeak*), are appealing. As is a language with a (*de facto* or otherwise) standard library, complete with respect to the necessities of the current computing age (e.g. Python, PERL). The advantage is obvious: power and uniformity. We advocate pushing Ada in this direction. We dream of an Ada 2007 with well designed annexes and package specifications for data structures, graphical user interface, and perhaps networking. The first author would also very much like to see *persistence* added to this lot. Ada 2007: rigour *and* power?

**Acknowledgements.** Sol-Eu-Net was EU project IST-1999-11495. The first and second authors wish to thank all EU and project people. The first author wishes to thank the Foundation for Science and Technology of the Portuguese government, for doctoral grant SFRH/BD/11109/2002 and for support to attend Ada-Europe 2003. And we all wish to thank the Ada-Europe 2003 reviewers for their insightful comments, which were invaluable in shaping this article to its final state, and Programme co-chair Jean-Pierre Rosen for that extra help.

# References

1.  ASCL. Workshop: Standard Container Library for Ada. In: Ada-Europe 2002. *www.auto.tuwien.ac.at/AE2002*. See also *groups.yahoo.com/group/asclwg*.
2.  Web Support to E-Collaboration: from Knowledge Management to Organizational Learning / Alípio Jorge; Damjan Bojadziev; Dunja Mladenic; Olga Stepankova; Jiri Palous; Mário Amado Alves; Johann Petrak; Peter Flach. –12 p. – Chapter 19 of [3]
3.  Data Mining and Decision Support: Integration and Collaboration / Dunja Mladenic; Marko Grobelnik; Steve Moyle; Nada Lavrac (Eds.) – Foreword by Maarten van Someren. – IV Parts, 22 Chapters, c. 300 p. – Kluwer Academic Publishers, to appear in 2003.
4.  Extensible Markup Language (XML). *www.w3.org/XML*
5.  PostgreSQL. *www.postgresql.org*
6.  The DATIC Initiative. *www.liacc.up.pt/datic*
7.  ISO/IEC 9075-3:1995 Information technology: Database languages: SQL: Part 3: Call-Level Interface (SQL/CLI)
8.  GNAT Ada Database Development Environment / Michael Erdmann. – p. 334 ff. – In: Reliable Software Technologies: Ada-Europe 2002: 7th Ada-Europe International Conference on Reliable Software Technologies, Vienna, Austria, June 17-21, 2002: Proceedings / J. Blieberger; A. Strohmeier (Eds.) – Springer-Verlag. – (LNCS 2361)
9.  MySQL. *www.mysql.com*
10. Comp.lang.ada news group. *groups.google.com/groups?group=comp.lang.ada*
11. Web Site Access Analysis for a National Statistical Agency / Alípio Jorge; Mário A. Alves; Marko Grobelnik; Dunja Mladenic; Johann Petrak. – 11 p. – Chapter 12 of [3]

# Transactions and Groups as Generic Building Blocks for Software Fault Tolerance*

Marta Patiño-Martínez[1], Ricardo Jiménez-Peris[1], and Alexander Romanovsky[2]

[1] Technical University of Madrid (UPM), Spain
{mpatino,rjimenez}@fi.upm.es
[2] School of Computing Science, University of Newcastle upon Tyne, UK
Alexander.Romanovsky@newcastle.ac.uk

**Abstract.** During the last decades several mechanisms for tolerating errors caused by software (design) faults have been put forward. Unfortunately only few experimental programming languages have incorporated them, so these schemes are not available in programming languages and systems that are used in developing modern applications. This is why programmers must either implement these mechanisms themselves or follow very complicated guidelines. It is not the case for software mechanisms developed for tolerating hardware faults (site crashes). Many programming languages and development systems provide mechanisms to cope with site failures. For instance, transactions are defined as one of the basic services in CORBA, Enterprise JavaBeans and Jini. In this paper we demonstrate how to implement mechanisms to tolerate software faults on the top of the mechanisms proposed for tolerating hardware errors.

## 1 Introduction

Many schemes and techniques have been developed for dealing with both hardware and software faults. With the increasing use of distributed systems, many of the software techniques for tolerating (site) hardware faults like transactions, group communication, and exception handling have been incorporated into a number of software development kits. In contrast, the use of software fault tolerance in order to detect errors caused by design faults is not particularly widespread. There are two important factors that make software fault tolerance more and more important in the development of new applications. On one hand the complexity of software increases more and more. The development of new distributed applications, such as web services, have an inherent complexity that centralized applications do not exhibit. On the other hand, these new applications are not developed from scratch. Many software components are (freely) available ready to be used in order to build new applications. Due to this fact, the cost of

---

* This research has been partially funded by the Spanish National Research Council, CICYT, under grant TIC2001-1586-C03-02, and by European IST DSoS project (IST-1999-11585).

J.-P. Rosen and A. Strohmeier (Eds.): Ada-Europe 2003, LNCS 2655, pp. 208–219, 2003.

software fault tolerance, which is generally based on software diversity, is not prohibitive any more. However, software fault tolerance techniques are not easy to implement nor available in current software development kits. In this paper we explore similarities in some of the software and hardware fault tolerance techniques. The contribution of the paper is that it proposes ways of implementing software fault tolerance on top of hardware fault tolerance techniques. The applicability of these ideas is demonstrated with a distributed programming language, *Transactional Drago*, supporting transactions and group communication. Although we use a particular programming language, our analysis shows that minimal modifications will be required to allow the same approaches to be used in building software fault tolerance on top of current middleware systems. For example, support for transactions is available in CORBA, Jini, and Enterprise Javabeans. Moreover, a group service has been recently included in Fault Tolerant CORBA specification [OMG00]. Most of the behaviour we assume *Transactional Drago* provides for groups is part of this specification or will soon be included. We assume that other middleware technology will follow a similar approach to that of the Fault Tolerant CORBA.

## 2   Model and Definitions

The system consists of a set of sites (nodes) $S = \{S_1, S_2, ..., S_N\}$ provided with stable memory that communicate by exchanging messages through reliable channels. We assume a partially synchronous system where nodes fail by crashing (no Byzantine failures). Failed sites may recover with its stable memory intact. Sites are provided with a group communication system (which provides reliable multicast and group membership services) supporting strong virtual synchrony [FvR95]. Group membership services provide the notion of view (current connected and active processes). Changes in the composition of a view are delivered to the application. We assume a primary component membership [CKV01]. Strong virtual synchrony ensures that messages are delivered in the same view they were sent and that two processes transiting to a new view have delivered the same set of messages in the previous view. Regarding ordering guarantees of multicast messages, we will use total order multicast, which ensures that messages are delivered in the same order at all group members.

A transaction is a sequence of operations that are executed atomically, that is, they are all executed (the transaction commits) or the result is as if none of them had been executed (it aborts). Two operations on the same data item conflict if they belong to different transactions and at least one of them modifies the data item. Concurrent transactions with conflicting operations must be serialized [BHG87]. That is, the final effect of the concurrent execution of conflicting transactions is equivalent to some sequential execution. Transactions can be nested (*subtransactions*). Transactions that are not nested are called *top-level transactions*. If a transaction aborts, all its subtransactions and their descendants will also abort. However, a subtransaction abortion does not compromise the result of its parent transaction (the enclosing one).

## 3  Transactional Drago

*Transactional Drago* [PJA98] is an extension to Ada [Ada95] and *Drago* [MAAG96] for programming distributed transactional applications. *Transactional Drago* implements *group transactions* [PJA02], a transaction model that supports multithreaded transactions.

Programmers can start transactions using the begin-end transaction statement or *transactional block*. Transactional blocks are similar to block statements in Ada. They have a declarative section, a body and can have exception handlers. The statements inside a transactional block are executed within a transaction. A transactional block can be nested leading to a nested transaction structure. Any unhandled exception in a transaction causes transaction abort. The exception is signalled in the transaction enclosing scope.

All data used in a transaction are subject to concurrency control (locks) and are recoverable. If a transaction aborts, data will be restored to the value they had before executing that transaction. Concurrency control is implicitly handled by the run-time system [PJKA02]. Programs access transactional data just as regular non-transactional data.

*Transactional Drago* inherits the process and distribution model of *Drago* [MAAG96]. Processes are the unit of distribution and belong to groups. Processes belonging to the same group share a common interface, which is a description of remotely callable services. A request to a group is multicast in total order to all the group members. There are two kinds of groups, *replicated* and *cooperative* ones, according to the state and behavior of its members. *Replicated groups* implement the active replication model. All the group members are identical deterministic replicas. They have the same state and code, and should run on failure-independent sites. Since all group members receive the same requests in the same order, they will produce the same answers. The results of the replicas are filtered at the client side, so that the results of a single replica are returned to the client (transparent replication). A replicated group can act as a client of another group. The underlying communication system, *Group_IO* [GMAA97], filters the replicated requests so that a single message is issued. Members of a *cooperative group* do not need to have either the same state or the same code. They are intended to divide data among its members and/or to express parallelism taking advantage of multiprocessing or distribution capabilities. Invocations to or from a cooperative group are independent so they are not filtered by the communication system. The client receives a reply per group member.

*Transactional Drago* provides support for exception handling in both kinds of groups [PJA01]. The interface of a group can include the exceptions the group can signal. In replicated groups, all members are supposed to signal the same exception, if it is not the case, members that disagree are considered faulty. Members of a cooperative group can finish a request signalling different exceptions. In such a situation, by default, the predefined exception *several_exceptions* is signalled to the client. The group programmer can overwrite this behaviour defining an exception resolution function for each group service. Whenever there is no primary component (majority) in a group, the predefined exception *group_error* is signalled to the client. Failed group members will join the group their group when they are available again.

# 4  Recovery Blocks

## 4.1  The Basic Mechanism

The recovery block [Ran75] scheme is based on software redundancy and rollback to recovery points. A number of variants (*alternates*) are independently produced from the same specification by different designers (maybe by using different algorithms). Besides, an acceptance test is to be implemented to check the correctness of the alternate results. A recovery point is established when the program enters the recovery block. The primary alternate is executed and the acceptance test checks its correctness. If it fails, the program is restored to the recovery point, and the next alternate is tried. This sequence continues until either the acceptance test is ensured or all alternates have been tried and failed. In the first case, the recovery point is discarded, and the block is exited. In the second one, the recovery point is restored (to keep the program in a predefined known state), and a failure exception is signalled. Recovery blocks can be nested, in which case the results of the nested recovery block must be flushed away if the alternate containing it has not passed its acceptance test. Note that recovery blocks and transactions belong to backward error recovery schemes, and that there are certain similarities in their behaviour when the acceptance test fails and a transaction aborts because the system is rolled back to a previous state.

## 4.2  Extensions of the Mechanism for Exception Handling

The original idea of the recovery blocks does not include a full-fledged exception handling mechanism: the use of exception handling here is restricted to only signalling the failure exception when a recovery block fails to produce the results, and to handling any exception by interrupting the current alternate, restoring the previous state and trying another alternate. However, exception handling is widely used and most modern programming languages have features for declaring exceptions in different scopes and associating handlers with them. In general, exceptions of each scope can be divided into *external* and *internal* [Rom00a], depending on whether they are propagated out of the scope where they are signalled or not. Exceptions handled in the scope where they are *raised* (Ada terminology) are classified as internal exceptions. Exceptions that are propagated outside the scope in which they are raised are called external exceptions. The recovery block scheme can benefit from introducing a general exception handling mechanism. First of all, this will allow any alternate to have a number of internal exceptions declared by the alternate developer (with the corresponding handlers), so that when such an exception is signalled the corresponding handler tries to handle it locally. If this succeeds, the acceptance test is checked and the execution of the recovery block proceeds. If an external exception is signalled outside the alternate, the system state is restored by the recovery block support and another alternate is tried. This approach seems to be the most practical one and in the following subsections we discuss solutions based on it, although it may raise some doubts because all changes in the system state are always aborted irrespective of the external exception signalled and the exception itself is ignored. One of the advanced solutions here is to allow recovery blocks to have several outcomes: one normal, several exceptional and a failure. The exceptional outcomes will correspond to the external exceptions and each of those outcomes will have a special acceptance test associated with it. If the acceptance test is passed after the external exception has been signalled, the alternate succeeds and the exception is signalled. If the acceptance test is not passed, the state is restored and the next alternate is tried.

### 4.3   Implementation of Recovery Blocks with Transactions

We will first show how to implement a recovery block in which external exceptions cause state restoration (Fig. 1a). In order to simplify the code, each alternate is encapsulated in a procedure, `alternateX`, which includes the handlers for its internal exceptions. The recovery block tries sequentially its constituent alternates. This behaviour can be modelled by using a loop that tries one alternate at a time until either an alternate succeeds or all the alternates have been executed. The execution of an alternate must be such that if the acceptance test fails, the state is restored. This behaviour is automatically provided by enclosing both the alternate and the acceptance test in a transaction. The loop contains the transaction that encapsulates the execution of the current alternate. If the acceptance test fails, an exception is signalled to cause transaction abort and therefore, rollback of the state. The transaction must be enclosed within a block statement to capture any exception that is propagated out of an alternate or is signalled in the acceptance test (aborting the transaction and causing state restoration). The exception handler (associated to the block statement) will capture the exception (`when others`) and force the execution of the next alternate, if any. If there are no more alternates the recovery block ends by signalling a `failure` exception.

   If an alternate contains a recovery block, the alternate must contain the previous code, resulting in a nested recovery block (transaction) structure.

   The previous code must be modified to consider external exceptions as a valid result of an alternate (Fig. 1b). If an alternate signals some external exception, and the associated acceptance test is passed, the alternate succeeds and that exception is propagated out of the recovery block. To achieve this behaviour, those exceptions must be handled inside the transaction, otherwise the transaction will abort and the state rolled back. Therefore, we need an additional exception context (block statement) within the transaction to capture those exceptions. Otherwise, an (unhandled) external exception will abort the transaction and therefore, the alternate. The block has a handler for each external exception, which contains the associated acceptance test. If the acceptance test succeeds, the exception is saved to be signalled again after the transaction has committed (the alternate has succeeded). If the acceptance test fails, the `failure` exception is raised, the transaction aborts and the next alternate is executed.

```
ok:= false;
alternate:= 1;
while alternate <= N and not ok loop
 begin -- capture unhandled
 -- exceptions in the transaction
  begin transaction
   case alternate is
    when 1 => alternate1(params);
    when 2 => alternate2(params);
     ...
   end case;
   ok:= acceptance_test(params);
   if not ok then
   -- the acceptance test has failed,
   -- abort the transaction
    raise failure;
   end if;
  end transaction;
  exception -- current alternate has
  -- failed, try the next one
   when others =>
    if alternate < N then
     alternate:= alternate +1;
    else
     raise failure;

    end if;
 end;
end loop;
```

```
raise_exception:= false;
ok:= false;
alternate:= 1;
while alternate <= N and not ok loop
 begin-- capture unhandled
 -- exceptions in the transaction
  begin transaction
   begin
    case alternate is
     when 1 => alternate1(params);
     when 2 => alternate2(params);
      ...
    end case;
    ok:= acceptance_test(params);
    if not ok then
     raise failure;
    end if;
    exception -- execute an
    -- acceptance test for
    -- external exceptions
     when excep: external1 =>
      ok:= acceptance_test1(params);
      if not ok then
       -- abort transaction
       raise failure;
      else
       raise_exception:= true;
       Save_Ocurrence(external, excep);
      end if;
     when excep: external2 =>
      ok:= acceptance_test2(params);
      if not ok then
       raise failure;
      else
       raise_exception:= true;
       Save_Ocurrence(external, excep);
      end if;
     end;
   end transaction;
   exception -- try the next alternate
    when others =>
     if alternate < N then
      alternate:= alternate +1;
     else
      raise failure;
     end if;
   end;
 end loop;
 if raise_exception then
     Reraise_Exception(external);
 end if;
```

(a)                                              (b)

**Fig. 1.** Recovery block. External exceptions (a) cause state restoration (b) are a valid result

## 4.4 Distribution and Replication

Recovery blocks were originally proposed as a general means for tolerating design faults. The author was deliberately not specific about how to apply the scheme in the context of distributed systems. In such a context any application willing to tolerate design faults is usually interested in tolerating hardware faults. Paper [LABK90] proposes several ways in which hardware and software fault-tolerance mechanisms can be used in combination to tolerate effectively faults of any nature. In particular, in order to tolerate hardware faults during the execution of a recovery block the paper proposes to replicate a recovery block and to concurrently execute replicas on different sites: when a site fails the rest of sites can continue execution and provide a correct result in spite of software design faults.

To apply these ideas in the context of distributed systems we have to distribute the server component implemented with software diversity into several sites. The simplest way of employing the recovery block technique for implementing a server component diversely is by designing each of its services as a recovery block. This completely hides diversity from the clients, makes it easier to implement recovery points, which have to only deal with the state of the server and facilitates the run time support. We call this approach service diversity as opposed to another approach which is called server diversity in which the entire server is implemented diversely.

In order to tolerate site failures two replication techniques can be used: active replication and the primary-backup scheme. Active replication avoids the delays in the execution of a recovery block when a site crashes (although the recovery blocks by their very nature can suffer from delays when errors caused by design faults are detected and tolerated) because in this case all (replicas) servers execute the same service concurrently. Group communication can be used to ensure that all replicas receive the same requests in the same order (total order multicast). If there are no hardware faults then, due to replica determinism and total order multicast, all replicas will produce the same results. Either all fail to ensure the acceptance test or ensure it. If due to hardware faults some of the replicas do not complete their execution of the alternate, the remaining replicas proceed with checking the acceptance test and either report the result or rollback and try the next alternate.

In contrast to this approach in the primary-backup scheme only one replica (the *primary*) is running; it periodically multicasts a checkpoint to the rest of replicas (the *backup* replicas). When the primary crashes, one of the backups takes over and continues the execution of the same alternate from the last checkpoint. Therefore, if the primary crashes, some part of the work could be lost and system execution can experience some delays.

Both replication techniques have to deal with the situation when a running replica is allowed to invoke other servers. If such invocations are allowed, in the case of active replication, they should be filtered so that only one invocation is performed. Special care must be taken in the primary-backup approach to inform backups about the success or failure of such invocations.

# 5    N-Version Programming

## 5.1    The Basic Mechanism

N-version programming (NVP) [Avi85] is another technique that uses redundancy to mask software faults. In this approach, N versions of a program (a module) are developed independently by different programmers, to be run concurrently. Their results are compared by an adjudicator. The simplest adjudicator uses the majority voting: the results produced by the majority of versions are assumed to be correct, the rest of the versions are assumed to be faulty. This technique requires a special support which controls the execution of versions and the adjudicator and passes the information among them. In particular, it synchronizes execution of N versions to obtain their results and to pass them to the adjudicator.

In order to tolerate site crashes, each version of an N-version program can be executed at a different site: if any site crashes the remaining sites can continue their execution and provide the required service. For the majority voting this means that this scheme is capable of tolerating K design faults or site crashes where $K < N/2$.

A group of processes can be used to implement N-version programming in distributed systems. To do this the processes should provide the same set of services which have diverse implementations. Such a group is built out of N versions as a diversely-designed server, so that each group member executes a corresponding version. It is possible to designate one member (for instance, the one with the smallest identifier) to acting as a group coordinator to be in charge of voting and returning the results to the client. Another approach is to return all version results directly to the client and to perform voting at the client side. Although the former solution hides the voting function from the client, the code becomes complex because it needs to take into account coordinator crashes. In the following section we will demonstrate how the latter approach can be implemented.

Recovering faulty versions is known to be a very difficult problem because of diversity of the internal version states. Several techniques have been developed; for example, the community error recovery [TA87] mainly relies on the assumption that all versions have a common set of internal data, whereas approach in [Rom00b] is based on developing an abstract version state and mapping functions from this state to a concrete state of each version and back. The same techniques can be used for recovering versions after site crashes.

## 5.2    Implementation of N-Version Programming Using Cooperative Groups

Cooperative groups in *Transactional Drago* provide all the functionality needed to implement N-version programming. Group specification (e.g. given in Fig. 2a) describes the services the group provides. Each group member implements that specification (Fig. 2b). To apply N-version programming the number of group members should be equal to the number of versions, so that each member implements one version. Clients invoke services of such diversely-designed server using multicast and the group members execute the request in parallel. When a member finishes, it returns its results to the client. The

```
group specification N_Version is              for group N_Version;
-- group services                             agent Version1 is
 entry Service1(parameters1);                 -- group member data declaration
 entry Service2(parameters2);                 begin
end N_Version;                                  select
                                                  accept Service1 (parameters1) do
                                                  -- version1 for service Service1
                                                  end Service1;
                                                  or
                                                  accept Service2 (parameters1) do
-- client code                                    -- version1 for service Service2
Service1 (parameters1, results);                  end Service2;
Voting(results, resultsService1);               end select;
                                              end Version1;
```

(a)                                            (b)

**Fig. 2.** N-version programming. (a) group specification and invocation (b) a group member

client is blocked until all the available members have replied, after that a voting function is called. For example, in Fig. 2a the Voting procedure is in charge of returning the majority result and the results parameter is an array with the individual results of all N versions.

### 5.3   N-Version Programming and Exception Handling

Allowing servers to signal external exceptions to the clients is an important feature that supports recursive system structuring and promotes disciplined approaches to tolerating faults of many types. Such exceptions are used in all situations when the required results cannot be delivered (for example: illegal input parameters, environmental faults, delivering partial results, server failure). Paper [Rom00a] proposes a scheme that allows diversely-designed servers to have external exceptions in their interfaces. Clearly all of the versions have to have the same set of external exceptions because such exceptions have to be treated as an immanent part of server specification. The scheme requires an adjudicator of a special kind to allow interface exception signalling when a majority of versions have signalled the same exception.

This feature can be introduced into the distributed setting when N versions of a server are structured as a group to tolerate hardware faults. The only extension which is needed is an ability to pass an identifier of an exception that a group member has signalled to the group coordinator (it can be either one of the group members or the client itself) and an extended majority voting that adjudicates not only normal results but exception identifiers as well.

### 5.4   Implementation of N-Version Programming with Exceptions

External exceptions can be declared in the group specification. Those exceptions are propagated to the client when they are not handled in the group during the execution of a service. Cooperative group members might have different code and therefore, each

```
group specification N_Version is
-- exceptions signalled by the group
exception1, exception2 : exception;
-- group services
entry Service1(parameters1);
entry Service2(parameters2);
private
 use  Ada.Exceptions;
 -- predefined type Exception_ID_List_Type is
 --      array (Positive <>) of Exception_ID;
 function MajorityResolution
  (exceptionList: Exception_ID_List_Type): Exception_ID is
 begin
  --select the majority exception
 end  Service1Resolution;

 for Service1'resolution use MajorityResolution;
 for Service2'resolution use MajorityResolution;
end N_Version;
```

**Fig. 3.** Exception resolution

member can signal a different exception for the same service. In this case, by default, the predefined exception several_exceptions is propagated to the client. This behaviour can be overwritten by defining an exception resolution function in the group specification for each group service (Fig. 3). This function is invoked by the run-time when two or more group members finish the execution of that service signalling different exceptions. This function can be programmed so that the majority exception is signalled.

# 6   Discussion and Conclusions

Our approach demonstrates how the mainstream languages (such as Ada) can be used for providing software fault tolerance. We have to rely on a number of constraints for the application programmers to follow, otherwise it is impossible to guarantee the properties of such techniques [RS95]. These restrictions mainly concern the absence of information flows between the versions and the rest of the system, non-interference of the tasks in which such techniques are used, etc.

The solutions presented could be implemented using any toolkit that supports group communication and transactions. As an example, CORBA supports transactions and the FT CORBA specification [OMG00] supports groups of objects including an active replication style. CORBA active replication and our replicated groups share the same properties: replication transparency, determinism and total order. In the FT CORBA *active_with_voting* replication style both requests addressed to and replies from a replicated group are voted and majority results are delivered. This style can be used for implementing N-version programming.

In general, programmers do not get much support when they are faced with the task of applying N-version programming or recovery blocks. By the contrast, software techniques for tolerating site failures (transactions, group communication, and replication) have become very popular and are part of most middleware platforms used for developing distributed applications. In this paper we have shown how recovery blocks and N-version programming can be programmed using these two building blocks. Besides, the proposed techniques allow site failures and design faults to be treated uniformly.

# References

[Ada95]    *Ada 95 Reference Manual, ISO/8652-1995*. Intermetrics, 1995.

[Avi85]    A. Avizienis. The n-version approach to fault-tolerant software. *IEEE Transactions on Software Engineering*, 11(12):1491–1501, 1985.

[BHG87]    P. A. Bernstein, V. Hadzilacos, and N. Goodman. *Concurrency Control and Recovery in Database Systems*. Addison Wesley, Reading, MA, 1987.

[CKV01]    G. V. Chockler, I. Keidar, and R. Vitenberg. Group Communication Specifications: A Comprehensive Study. *ACM Computer Surveys*, 33(4):427–469, December 2001.

[FvR95]    R. Friedman and R. van Renesse. Strong and Weak Virtual Synchrony in Horus. Technical Report TR95-1537, CS Dep., Cornell Univ., 1995.

[GMAA97]  F. Guerra, J. Miranda, A. Alvarez, and S. Ar valo. An Ada Library to Program Fault-Tolerant Distributed Applications. In K. Hardy and J. Briggs, editors, *Proc. of Int. Conf. on Reliable Software Technologies*, volume LNCS 1251, pages 230–243, London, United Kingdom, June 1997. Springer.

[LABK90]   J. Laprie, J. Arlat, C. B ounes, and K. Kanoun. Definition and Analysis of Hardware- and Software-Fault-Tolerant Architectures. *IEEE Computer*, 23(7):39–51, 1990.

[MAAG96]  J. Miranda, A. Alvarez, S. Ar valo, and F. Guerra. Drago: An Ada Extension to Program Fault-tolerant Distributed Applications. In A. Strohmeier, editor, *Proc. of Int. Conf. on Reliable Software Technologies*, volume LNCS 1088, pages 235–246, Montreaux, Switzerland, June 1996. Springer.

[OMG00]    OMG. *Fault Tolerant CORBA*. Object Management Group, 2000.

[PJA98]    M. Pati o-Mart nez, R. Jim nez-Peris, and S. Ar valo. Integrating Groups and Transactions: A Fault-Tolerant Extension of Ada. In L. Asplund, editor, *Proc. of Int. Conf. on Reliable Software Technologies*, volume LNCS 1411, pages 78–89, Uppsala, Sweden, June 1998. Springer.

[PJA01]    M. Pati o-Mart nez, R. Jim nez-Peris, and S. Ar valo. Exception Handling in Transactional Object Groups. In *Advances in Exception Handling, LNCS-2022*, pages 165–180. Springer, 2001.

[PJA02]    M. Pati o-Mart nez, R. Jim nez-Peris, and S. Ar valo. Group Transactions: An Integrated Approach to Transactions and Group Communication. In *Concurrency in Dependable Computing*, pages 253–272. Kluwer, 2002.

[PJKA02]   M. Pati o-Mart nez, R. Jim nez-Peris, J. Kienzle, and S. Ar valo. Concurrency Control in Transactional Drago. In *Proc. of Int. Conf. on Reliable Software Technologies, LNCS-2361* , pages 309–320, Vienna, Austria, June 2002. Springer.

[Ran75]    B. Randell. System Structure for Software Fault Tolerance. *IEEE Transactions on Software Engineering*, 1(2):220–232, 1975.

[Rom00a]   A. Romanovsky. An Exception Handling Framework for N-Version programming in Object-Oriented Systems. In *Proc. of the 3rd IEEE Int. Symp. on Object-oriented Real-time Distributed Computing (ISORC'2000)*, pages 226–233, Newport Beach, USA, March 2000. IEEE Computer Society Press.

[Rom00b]    A. Romanovsky. Faulty Version Recovery in Object-Oriented Programming. *IEE Proceedings –Software* , 147(3):81–90, 2000.

[RS95]    A. Romanovsky and L. Strigini. Backward error recovery via conversations in Ada. *Software Engineering Journal*, pages 219–232, November 1995.

[TA87]    K.S. Tso and A. Avizienis. Community error recovery in N-version software: a design study with experimentation. In *Proc. of 17th IEEE FTCS*, pages 127–133, Pittsburgh, PA, 1987.

# Getting System Metrics Using POSIX Tracing Services*

Agustín Espinosa Minguet[1], Vicente Lorente Garcés[2], Ana García Fornes[1], and
Alfons Crespo i Lorente[2]

[1] Departamento de Sistemas Informáticos y Computación
Universidad Politécnica de Valencia, Spain
{aespinos,agarcia}@dsic.upv.es
http://www.dsic.upv.es
[2] Departamento de Informática de Sistemas y Computadoras
Universidad Politécnica de Valencia, Spain
{vlorente,alfons}@disca.upv.es
http://www.disca.upv.es

**Abstract.** One of the possible applications of the POSIX tracing services is to
obtain system metrics from trace streams. A trace stream stores a sequence of
events generated by the system during the application execution. By interpreting
this sequence of events, interesting system metrics can be obtained. Unfortunately,
the interpretation of these sequences of events may be very difficult for a programmer who does not know the system implementation in detail. In order to solve this
problem, we present an interface which is implemented on top of the POSIX tracing services. This interface allows the programmer to obtain predefined system
metrics and user-defined metrics from trace streams without having to know the
system implementation.

## 1 Introduction

The POSIX tracing services [1] specify a set of interfaces to allow for portable access
to underlying trace management services by application programs. Programmers may
use these services to get a sequence of trace events generated by the system during the
execution of their application. These trace events are kept in a POSIX trace stream. The
contents of a trace stream can be analyzed while the tracing activity takes place or it
can be analyzed later, once the tracing activity has been completed. A trace event is
generated when some action takes place in the system and this trace event may be stored
in one or several trace streams. Each trace event contains data which is relative to the
action that has generated it. The POSIX tracing services require that event type, time
stamp, process identifier and thread identifier be associated to each trace event. Using
these data, we can get time related system metrics such as the execution time of a given
system call, the duration of an interrupt handler, etc.

Unfortunately, the interpretation of the events which are stored in trace streams may
be difficult for programmers who do not know the system implementation in detail.

* This work has been funded by the *Ministerio de Ciencia y Tecnologia* of the Spanish Government under grants DPI2002-04434-C04-02 and TIC2002-04123-C03-03 and by the Generalitat
Valenciana under grant CTIDIB/2002/61

J.-P. Rosen and A. Strohmeier (Eds.): Ada-Europe 2003, LNCS 2655, pp. 220–231, 2003.
© Springer-Verlag Berlin Heidelberg 2003

Events stored in a trace stream represent system actions such as context switches, hardware interrupts, state changes, etc. In order to extract metrics from these events it is necessary to know how the execution of the system generates these events. Usually only the programmer who has implemented the system knows this information. In order to solve this problem, we present an interface which is implemented on top of the POSIX tracing services. This interface allows the programmer to obtain predefined system metrics and user-defined metrics from trace streams without having to know the system implementation.

This metrics interface has been implemented in the real-time MaRTE OS operating system [7]. MaRTE OS is a real-time kernel for embedded applications that follows the Minimal Real-Time POSIX.13 subset [2]. Most of its code is written in Ada with some C code and assembler parts. It allows for the software development of Ada and C embedded applications. We implemented a subset of the POSIX tracing services in MaRTE OS and an Ada binding to these services in a previous work [5]. The metrics interface has been built on top of this Ada binding. The MaRTE OS operating system and the GNAT run-time system have been modified so that these systems generate events that are appropriate for a set of useful metrics.

The metrics interface is discussed in the remainder of the paper. Section 2 describes the Metrics package, which implements the metrics interface. Section 3 shows how this interface has been implemented. Section 4 presents how a set of system metrics which is suitable for real-time applications may be obtained using this interface.

## 2   The Metrics Package Specification

A metric is a temporal property of the system, just like the duration of a context switch or the execution time of a task. The Metrics package offers an interface that allows the programmer to obtain metrics from the events stored in a POSIX trace stream.

```
with Ada.Real_Time;
with Ada.Task_Identification;
with Posix.Trace;

package Metrics is

    type Metric                      is private;
    type Metric_Result (Size : Natural) is private;
    type Metric_Result_Element       is private;

    function System_Metric (Metric_Name   : String)
          return Metric;

    function New_User_Metric (Metric_Name    : String;
                              Initial_Event  : POSIX.Trace.Event_Identifier;
                              Final_Event    : POSIX.Trace.Event_Identifier;
                              Include_System : Boolean := True)
          return Metric;

    function Events_Required_By (M    : Metric;
                                 Trid : POSIX.Trace.Trace_Identifier)
          return POSIX.Trace.Event_Set;

    procedure Get_Metric_Result (From   : in Metric;
                                 Trid   : in POSIX.Trace.Trace_Identifier;
                                 Result : in out Metric_Result);
```

```
function Metric_Result_Element_Count (From : Metric_Result)
         return Natural;

function Get_Metric_Result_Element (From : Metric_Result;
                              I    : Natural)
         return Metric_Result_Element;

function Begin_Time    (From : Metric_Result_Element)
         return Ada.Real_Time.Time;

function End_Time      (From : Metric_Result_Element)
         return Ada.Real_Time.Time;

function Length        (From : Metric_Result_Element)
         return Ada.Real_Time.Time_Span;

function Task_Id       (From : Metric_Result_Element)
         return Ada.Task_Identification.Task_Id;

function Intermediate_Events (From : Metric_Result_Element)
         return Natural;

end Metrics;
```

In this interface, metrics are identified by names and these metrics are used through objects of the type Metric. There are two types of metrics, system metrics and user metrics.

System metrics are implementation-defined and usually correspond to temporal properties of the run-time or operating system. These metrics are used by calling the function System_Metric.

```
M1 : Metric;
M2 : Metric;

M1 := System_Metric ("Delay_Until_Wakeup");
M2 := System_Metric ("Schedule_And_Dispath");
```

Programmers can define their own user metrics, which allow them to measure the execution time between two points of their programs, with these points being defined by two trace events. These metrics may or may not include the system execution time and are created using the function New_User_Metric.

```
M3 : Metric;
Initial_Event : POSIX.Trace.Event_Identifier;
Final_Event   : POSIX.Trace.Event_Identifier;

Initial_Event := POSIX.Trace.Open ("Initial_Event");
Final_Event   := POSIX.Trace.Open ("Final_Event");

M3 := New_Metric (
      "My_Metric", Initial_Event, Final_Event, Include_System => False);
```

Each metric is obtained from a set of events that should be in the trace stream that is being analyzed. It is important that these events be stored in the trace stream when these events are generated. When programmers prepare the tracing of their applications, they can use the function Events_Required_By to obtain the set of events on which a metric depends, indicating to the trace system that the events of this set should be stored in the trace stream. The following code fragment illustrates how an application can be coded to create a trace stream so that programmers can obtain the metrics that they require.

```
Attr : POSIX.Trace.Attributes;
Trid : POSIX.Trace.Trace_Identifier;
File : POSIX.IO.File_Descriptor;

Interesting_Events : POSIX.Trace.Event_Set;
All_Events         : POSIX.Trace.Event_Set;

POSIX.Trace.Initialize (Attr);
File := POSIX.IO.Open("Logfile",POSIX.IO.Read_Write);
Trid := POSIX.Trace.Create_With_Log
            (POSIX.Process_Identification.Null_Process_ID,
             Attr,
             File);

POSIX.Trace.Add_Event_Set (Interesting_Events, Events_Required_By (M1, Trid));
POSIX.Trace.Add_Event_Set (Interesting_Events, Events_Required_By (M2, Trid));
POSIX.Trace.Add_Event_Set (Interesting_Events, Events_Required_By (M3, Trid));

POSIX.Trace.Add_All_Events (All_Events);
POSIX.Trace.Set_Filter (Trid, All_Events);
POSIX.Trace.Subtract_From_Filter (Trid, Interesting_Events);

POSIX.Trace.Start (Trid);

Start_Aplication;

POSIX.Trace.Shutdown (Trid);
```

In the previous code fragment, a trace stream is created by calling POSIX.Trace.Create_With_Log. Next, the trace stream is configured so that it filters all the events that are generated, except the events on which metrics M1, M2 and M3 depend. Once the trace stream is configured, this is activated by calling POSIX.Trace.Start and the application begins. From this instant, the events which are not filtered will be stored in the trace stream. Once the application ends, it stops the trace stream by calling POSIX.Trace.Shutdown. After this action, the trace stream is stored in the file logfile. This trace stream can be analyzed later to obtain the results of the metrics M1, M2 and M3.

A metric result is an object of the type Metric_Result. Each one of these objects is a list which is formed by all the measurements (objects of the type Time_Metric_Result_Element) which have been detected in the trace stream that is being analyzed. An individual measurement registers the following five values:

| Value | Description |
| --- | --- |
| Task_Id | Task identifier |
| Length | Measured time |
| Begin_Time | Instant in which the beginning of the measure is detected |
| End_Time | Instant in which the finalization of the measure is detected |
| Intermediate_Events | Number of trace events detected while making the measure |

Task_Id and Length are the main values obtained from a measurement. Begin_Time and End_Time allow the user to detect when the measurement has begun in the trace stream. For example, it can be useful to analyze a measurement with a strange Length value. Intermediate_Events allows the programmer to estimate the overhead caused by the tracing system. Sometimes different measurements from the same metric have a different number of trace events and so it is important to adjust the Length value by using the Intermediate_Events value. Of course, to make this adjustment it is necessary to known how much time is consumed in storing a trace event in a trace stream.

A metric result is obtained by calling the procedure Get_Metric_Result. When being called, this procedure scans the trace stream indicated in the parameter Trid searching for measurements that correspond to the metric indicated in the parameter From. These measurements are stored in the parameter Result. The number of measurements found can be consulted by calling function Metric_Result_Element_Count. The function Get_Metric_Result_Element is used to obtain one of these measurements from a result. The values of each measurement are consulted using the functions Begin_Time, End_Time, Length, Task_Id and Intermediate_Events. The following code fragment illustrates how programmers obtain the results of a metric from a previously created trace stream. Figure 2 shows the output of this example.

```
File := POSIX.IO.Open("Logfile",POSIX.IO.Read_Only);
Trid := POSIX.Trace.Open (File);

Get_Metric_Result (M1, Trid, Result);

N := Metric_Result_Element_Count (Result);

for I in 1 .. N loop
   Measure := Get_Metric_Result_Element (Result, I);
   L := Length (Measure);
   T := Task_Id (Measure);
   ....
   Show_Result_Element;
end loop;
```

| Task | Length | Begin_Time | End_Time | Int_Events |
|---|---|---|---|---|
| 8 | 44 | 14393 | 14445 | 5 |
| 9 | 51 | 24392 | 24495 | 5 |
| 10 | 47 | 24409 | 24521 | 5 |
| 6 | 45 | 24424 | 24545 | 5 |
| 7 | 43 | 24438 | 24569 | 5 |
| 11 | 44 | 45393 | 45445 | 5 |
| 6 | 50 | 49391 | 49464 | 5 |
| 7 | 45 | 49409 | 49489 | 5 |
| 12 | 51 | 54393 | 54482 | 5 |
| 13 | 46 | 54410 | 54507 | 5 |
| .. | .. | ..... | ..... | . |

**Fig. 1.** Metric result (time in $\mu s$)

## 3  The Metrics Package Implementation

The results of a metric are obtained by scanning sequences of events stored in a pre-recorded trace stream that correspond to the metric the programmer wants to obtain. Each metric provided by the Metrics package is analyzed by an automaton, or more precisely, by a group of equivalent automata, each of which is associated to a different task identifier. As these automata are internal to the implementation of the Metrics package users of this package do not use them directly.

We decided to use automata because they adapt very well to the nature of the information which is a sequence of trace events. Another advantage of using automata is they facilitate the implementation of new metrics, which is specially important if we want to port the implementation of the Metrics package to a system other than MaRTE OS.

The alphabet used by these automata is formed by tuples of three elements (Event_Type, Task_Id, Timestamp). These tuples are obtained from the events which are stored in a pre-recorded trace stream. The elements Event_Type and Task_Id are used to determine when a transition is applied, while the element Timestamp is used to obtain the measurement.

A transition between two states is applied in accordance with both the input tuple and the transition label according to the following rules:

| Label | The transition is applied if... |
|---|---|
| Event_Type, i | the input event is equal to Event_Type and the input task is the task associated to the automaton. |
| Event_Type, not i | the input event is equal to Event_Type and the input task is not the task associated to the automaton. |
| Event_Type | the input event is equal to Event_Type. |
| $\lambda$ | the label is $\lambda$. |
| * | the input tuple does not match any of the previous rules. |

The states of an automaton can be one of the following types: *begin, end, cancel, in, out* and *cond*. The calculation of a measurement takes place according to the following rules:

1. When entering in a *begin* state, a new measurement begins.
2. When entering in a *cancel* state, the current measurement is discarded.
3. When entering in an *end* state, the current measurement is completed and the values associated to this measurement are determined as follows:

| Value | Calculation |
|---|---|
| Task_Id | Identifier of the task associated to the automaton. |
| Begin_Time | Timestamp of the input tuple that caused the application of rule 1. |
| End_Time | Timestamp of the input tuple that caused the application of rule 3. |

Length                          The sum of the duration of the segments detected by the
                                automaton. A segment is formed by a sequence of states
                                which are exclusively of type *in*, or by a sequence of
                                states that begins with states of the type *cond* and which
                                ends with states of type *in*. The duration of a segment is
                                calculated as the difference between the Timestamp values
                                of the input tuples that have determined the initial and final
                                states of the segment.
Intermediate_Events             The sum of the sizes of the segments which are detected
                                by the automaton. The size of a segment is the number of
                                states that compose the segment minus one.

In other words, a measurement begins when the automaton reaches a state of type
*begin*. From this point the automaton looks for segments that belong to the metric that
the programmer wants to obtain. These segments are detected in accordance with the
type of the automaton states *in*, *out* and *cond*. *In* and *out* states are states that are inside
or outside of a segment, respectively. *Cond* states are states whose membership to a
segment is conditioned to later states of the automaton. If an *in* state is reached after a
sequence of *cond* states , the sequence of *cond* states belongs to the segment. If an *out*
state is reached, the sequence of *cond* states is discarded as belonging to a segment. In
this way, *cond* states allow the automaton to detect segments whose membership to the
measurement is conditioned by events that have not yet been obtained from the trace
stream. A measurement ends when an *end* or *cancel* state is reached. If an *end* state
is reached, the length of the measurement is the sum of the duration of the segments
detected. Conversely, if a *cancel* state is reached, the current measurement is discarded.
*Cancel* states are used when the automaton detects that the current situation does not
really correspond to the metric which is being analyzed.

An example of these automata is shown in Fig. 2. This is the automaton which analyzes user-defined metrics in which the system execution time is included. These metrics
measure the execution time performed by a task between two points of its execution.
These points are marked by two trace events which are user-defined. It is important take
into account that the hardware may interrupt the task execution, and so the execution time
consumed by the interrupt handler must not be part of the measurement. If the running

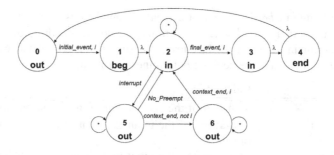

**Fig. 2.** User metric automaton

task is preempted by a higher priority task as a consequence of an interrupt, the execution time performed by this higher priority task must not be part of the measurement. In this way, a measurement of a user defined metric is formed by a sequence of segments in which the task is using the processor. The first segment is detected by the automaton (in the state 0) when the initial event defined by the user is detected. The automaton then reaches state 1 and the measurement begins. By the $\lambda$ transition, the automaton reaches state 2. With the automaton in state 2, if the task is interrupted, the current segment is completed and the automaton reaches state 5. If the running task is preempted as a consequence of the interrupt the automaton reaches state 6. States 5 and 6 are of type *out*, and so they are no part of the measurement. When the task again becomes the running task state 2 is reached and a new segment begins. When the final event is detected, the current segment is completed. The automaton reaches state 4 and the measurement ends. Finally, the automaton returns to state 0 and waits for a new measurement.

## 4   Usage Example

In this section, we describe how the Metrics package can be used to know the system overload that takes place in a real-time application formed by a set of periodic tasks. To carry out this study, we began with the following equation [3] which calculates the worst-case response time for a task $\pi_i$.

$$R_i = C_i + B_i + \sum_{\forall j \in hp(i)} \left\lceil \frac{R_i}{T_j} \right\rceil C_j \tag{1}$$

$R_i$   worst-case response time of $\pi_i$.
$C_i$   worst-case execution time of $\pi_i$.
$B_i$   blocking of $\pi_i$.
$T_j$   period of $\pi_i$.
$hp(i)$ set of all tasks of higher priority than $\pi_i$.

In order for equation 1 to be correct, it is necessary to incorporate the system overload to the terms of the equation. All the activities of the system caused by the task, such as its activation or system calls should be included in $C_i$. $B_i$ should be calculated keeping in mind the critical sections implemented by the program, as well as the system sections whose activity is in charge of lower priority tasks. These sections can delay the execution of $\pi_i$.

Our system is formed by the real-time operating system MaRTE OS and the run-time system of the GNAT compiler. We want to analyze the overload that this system produces in a set of periodic tasks which are implemented according to the following outline:

```
task body Periodic is
begin
  loop
    Application_Code;
    Next_Activation := Next_Activation + Period;
    delay until Next_Activation;
  end loop;
end Periodic;
```

When adapting the results of the works described in [6] and [4] to our system, the terms of equation 1 can be calculated in the following way:

$$C_i = (C_i^{user} + JST_i) \tag{2}$$

$$B_i = B_i^{activation} + B_i^{critical\_section} \tag{3}$$

$$B_i^{critical\_section} = max((SB_k : k \in lp(i)) \cup B_i^{user}) \tag{4}$$

$$B_i^{activation} = \sum_{\forall k \in lp(i)} \left\lceil \frac{R_i}{T_k} \right\rceil DUW_k \tag{5}$$

The meanings of the terms of these equations are the following:

- $C_i^{user}$: worst-case execution time of $\pi_i$, only taking into account the code at application level.
- $B_i^{user}$: blocking which takes place from critical sections at application level.
- $Job\_System\_Time_i$ ($JST_i$): System execution time of the task $\pi_i$.
  This execution includes the execution of the statement `delay until` and other system executions requested by the task.
- $System\_Blocking_k$ ($SB_k$): System-level blocking that can produce the task $\pi_k$.
  Some parts of the execution of the GNAT run-time system are critical sections protected by POSIX mutexes . The ceiling of these mutexes is the maximum priority of the system, so the duration of these critical sections must be considered as possible blocking for any task whose base priority is greater than $\pi_k$.
- $Delay\_Until\_Wakeup_i$ ($DUW_k$): System execution time which is necessary to wake up a task $\pi_k$ that is suspended in a statement `delay until`.

Programmers need to know the following metrics in order to know the overload that the system has in the real-time application: $Job\_System\_Time$, $System\_Blocking$ and $Delay\_Until\_Wakeup$. Next, we describe how a metric can be implemented, using the metric $Delay\_Until\_Wakeup$ as an example.

In our system, when a task executes the statement `delay until`, a call takes place to the GNAT run-time system, which registers the new state of the task and calls the POSIX function `pthread_cond_timedwait` so that the task is suspended in a POSIX condition. MaRTE OS services this call by scheduling a temporal event and selects a new task for execution. To measure the time that the system needs to wake up a task that is suspended in a statement `delay until`, we can build the automaton shown in Fig. 3.

This automaton (in state 0) waits until the hardware timer interrupts. When this interrupt takes place, the MaRTE OS timer interrupt handler is executed and a measurement begins (state 1). State 4 is reached after the first phase of the timer interrupt handler and MaRTE OS begins to extract the temporal events that have expired at this instant. If one of these events implies waking up the task $i$, the automaton changes to state 7. Otherwise, the measurement is canceled (state 6) and the automaton returns to its initial state.

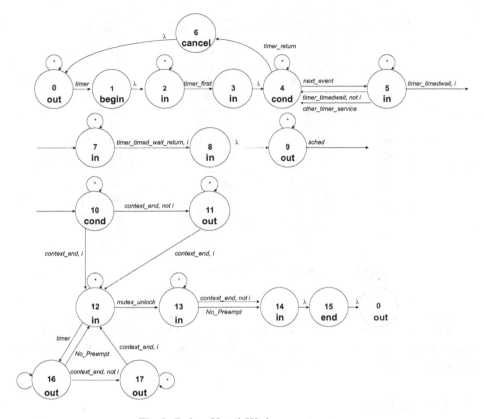

**Fig. 3.** $Delay\_Until\_Wakeup_i$ automaton

If the measurement continues, the automaton waits until the service for the function pthread_cond_timedwait for the task $i$ completes (state 8). In state 9, the system activity that is related to the extraction of other temporal events is discarded. MaRTE OS next calls the scheduler to select a task to execute. Once the task $i$ continues its execution (state 12), the automaton can wait until the GNAT run-time system releases the mutex that was acquired at the beginning of the execution of the statement delay until (state 13). The measurement then completes when MaRTE OS decides what task will continue being executed (state 14). If new interrupts of the hardware timer occur when the automaton in in state 12, states 16 and 17 discard the execution that may occur due to this interrupt.

To obtain the three metrics from MaRTE OS, we prepared a test made up of a group of 15 periodic tasks. The periods of these tasks ranged between 25 and 200 ms and their computation times ranged between 2 and 20 ms. This task set was executed on a 200 Mhz Intel Pentium processor. Storing a trace event in a trace stream takes around 1 microsecond on this processor. Figure 4 shows a summary of the results obtained for the three metrics in which the interference caused by the tracing system has been subtracted. These values correspond to the largest measurement obtained for each task.

For example, when analyzing the results for task 1, we observe the following. The time that the system dedicates to each job of this task is at most 84.838 $\mu s$. The exe-

| Task | Job_System_Time | System_Blocking | Delay_Until_Wakeup |
|------|-----------------|-----------------|--------------------|
| 1 | 84.838 | 27.837 | 44.235 |
| 2 | 81.521 | 27.972 | 43.680 |
| 3 | 73.326 | 27.131 | 36.207 |
| 4 | 74.768 | 28.257 | 35.291 |
| 5 | 83.278 | 28.632 | 43.380 |
| 6 | 77.258 | 28.302 | 37.602 |
| 7 | 80.952 | 27.807 | 43.726 |
| 8 | 80.472 | 26.921 | 42.435 |
| 9 | 74.753 | 26.786 | 37.392 |
| 10 | 81.327 | 27.687 | 42.241 |
| 11 | 77.814 | 28.377 | 37.617 |
| 12 | 76.778 | 28.182 | 36.762 |
| 13 | 76.540 | 28.572 | 35.936 |
| 14 | 76.209 | 28.047 | 36.041 |
| 15 | 75.488 | 27.282 | 36.552 |

**Fig. 4.** Results of the metrics (time in microseconds)

cution of the task 1 may be delayed up to 28.632 $\mu s$ by system blocking coming from tasks of lower priority. In the worst case, if tasks 2 to 15 are activated while task 1 is running, the execution of task 1 will be delayed 544.862 $\mu s$, which is the sum of the Delay_Until_Wakeup metric for tasks 2 to 15.

## 5   Conclusions

In this paper, we have presented an interface that allows programmers to obtain system metrics from POSIX trace streams. The approach that we have used offers several advantages. Programmers can obtain system metrics from trace streams which are generated by a system whose implementation is unknown. The proposed interface is easily portable since it is implemented on top of POSIX standard services. The use of automata facilitates the creation of new metrics as well as the adaptation of this interface to other systems where the events generated could possibly be different. We are currently using this interface for the analysis of Ada real-time applications in the operating system MaRTE OS, for which we have defined a wide set of system metrics.

## References

1. 1003.1q-2000 IEEE Standard for Information technology-Portable Operating Systems Interface (POSIX®)-Part 1: System Application Program Interface (API)-Amendment 7: Tracing [C Language]
2. 1003.13-1998 IEEE Standard for Information Technology–Standardized Application Environment Profile (AEP)–POSIX® Realtime Application Support. [0-7381-0178-8]
3. N. Audsley, A. Burns, M. Richardson, K. Tindell and A. Wellings. Applying new scheduling threory to static priority pre-emptive scheduling. Software Engineering Journal, 8, (5), 284–292. (1993)

4.  A. Burns, A.J. Wellings. Engineering a hard real-time system: from theory to practice. Software Practice and Experience, vol 27(7), 705–726
5.  A. Espinosa, A. García-Fornes, A. Crespo. An Ada Binding to the IEEE 1003.1q (POSIX Tracing) Standard. 7th Ada-Europe International Conference on Reliable Software Technologies, Vienna. LNCS 2361, p. 321.
6.  D. Katcher, H. Arakawa, J. Strosnider. Engineering and analysis of fixed priority schedulers. IEEE Transactions on Software Engineering, vol. 19, 920–934.
7.  M. Aldea, M. Gonzalez. MaRTE OS: An Ada Kernel for Real-Time Embedded Applications. 6th Ada-Europe International Conference on Reliable Software Technologies, Leuven. LNCS 2043, p. 305

# Some Architectural Features
# of Ada Systems Affecting Defects

William M. Evanco and June Verner

Drexel University, 3141 Chestnut Street
Philadelphia, PA 19104
{William.Evanco,June.Verner@cis.drexel.edu}

**Abstract.** In this study, we discuss some software architecture tradeoffs for Ada83 systems. External and internal complexities are specified for Ada packages and these complexities are related to the numbers of defects uncovered during testing using a non-linear model based on the negative binomial probability distribution. The non-linear associations among the software measures that we investigate are exploited to identify optimal relationships between context coupling and the number of visible declarations. The data we used to develop our model consists of Ada package characteristics from four software systems. Some of the packages are reused from previous projects and non-defect changes may have been made to these packages during the software development process. Therefore, in our defect model, we controlled for these two factors. For the 262 packages that we considered, our research shows that with respect to software defects there is too much context coupling, indicating that the architecture is far from optimum. If less of the workload were delegated to other packages and instead implemented within the package through hidden declarations, then context coupling might be reduced.

## 1 Introduction

Interest in improving the quality of software systems has led to numerous studies analyzing the relationships between software quality outcomes and the complexity characteristics of software [1,2,3]. One indicator of software system quality is its reliability as measured by the number of defects in its constituent components. By relating the number of defects to software complexity characteristics such as module cohesion and coupling, we can get an early indication of the software's quality.

A software designer must make decisions regarding system partitioning and this partitioning impacts both component composition and coupling. Software quality attributes such as reliability and maintainability are influenced by the precise arrangement of components and their coupling [4]. Designers may choose to build systems with large components having relatively little coupling, small components that are extensively coupled to each other or something between these two extremes.

Although we have been told that modules should be constructed to exhibit high cohesion and low coupling [5], such qualitative statements provide no real guidance as to how low the coupling should be, nor what level of cohesion we should try to achieve. There is substantial evidence that highly coupled systems exhibit more defects [6]. On the other hand, the concept of cohesion is problematic. While high cohesion appears, at face value, to be a desirable trait for modules, there are both theo-

J.-P. Rosen and A. Strohmeier (Eds.): Ada-Europe 2003, LNCS 2655, pp. 232–245, 2003.

retical and empirical problems with the concept of cohesion. Cohesion tends to be strongly related to the semantics of the problem domain and, for the most part, seems to defy easy definition and measurement [7]. Moreover, cohesion measures that have been developed appear thus far not to have a consistently significant relationship to defects in software [6]. Consequently, the idea of cohesion may be a less useful concept in terms of providing guidance for making software architecture decisions.

Much effort has been devoted to identifying and quantifying the characteristics of software components that may contribute to low system reliability as measured by defect counts. These characteristics are then related to software defects through models that are estimated by various means. Because the mechanisms underlying the generation of software defects are not well understood, exact models cannot be specified on the basis of theory. We rely on empirically estimated defect models that incorporate these characteristics as explanatory variables and provide an approximation to the exact models over a range of the explanatory variables. The resulting models generally focus on the quality attributes of reliability, as measured by faults or defect counts [1–3], and maintainability, as measured by the efforts to repair or change software [8,9].

Deciding upon the "best" partitioning of a software system requires models relating software quality outcomes to measurable characteristics of a software system that can be obtained during the design stage of software development. With appropriate models, one can examine tradeoffs among the measurable characteristics and select those characteristics to optimize one or more quality outcomes. Many models have been proposed relating software characteristics to quality outcomes [2,3,10]. Some of these models are based on statistical techniques such as multivariate regression analysis. Multivariate regression techniques are then used to determine the values of the parameters associated with each of the linear terms. Once we have such a model, it can be used to predict a quality factor associated with a new project based on its software characteristics. The advantage of such models is that the software characteristics are generally known long before the software quality factors.

However, as a general rule, these linear models do not lend themselves to an examination of optimal software components. Linear models are monotonic in their behaviors with respect to the explanatory variables and these variables are not linked in any way that would allow for an examination of tradeoffs. A greater value of some software characteristic, such as cyclomatic complexity for example, leads to a greater expected number of defects. In linear models, cyclomatic complexity is not coupled to nesting depth or to the call structure, for example, so that tradeoffs cannot be examined among these independent variables with regard to their impacts on defects.

Other modeling approaches use pattern recognition [2], neural networks [11], genetic algorithms [12], and Bayesian Belief Networks [13]. These approaches, though oriented toward the identification of defect-prone modules, are nonparametric and obscure explicit functional relationships between characteristics and outcomes. For example, a major drawback of neural network approaches is that analysis results, expressed in terms of neuron weights, cannot be interpreted in simple terms that are related to the initial software measures. Analyses using genetic algorithms focus on the use of software characteristic variables to predict defects, but there has been little concern about the statistical significance of these variables. In addition, the emphasis of these approaches on "defect-proneness" (a binary variable) as an outcome variable, rather than a more disaggregated measure such as the actual number of defects, leads to a relative insensitivity of the outcome variable to changes in the software characteristics.

In this study, we attempt to extend our previous work [14] to object-based programming languages such as Ada83, which support the specification of objects but do not support inheritance relationships. In the next section, we discuss the problem statement. In Sect. 3, the methodology used to attack the problem, based on nonlinear regression modeling, is presented. In Sect. 4, the model variables and empirical data are discussed. In Sect. 5 we present the empirical results, while Section 6 includes the discussion and conclusions.

## 2    Problem Statement

Ada packages encapsulate logically related entities and computational resources. Entities may be objects or types while computational resources are subprograms or tasks. The package consists of a specification, identifying the visible declarations of entities and computational resources, and a body, implementing the details of the package[1]. The implementation details are hidden from the package user. Context clausing (i.e., a "with" statement) allows the visible declarations to be exported to (imported by) other program units such as subprograms, packages, tasks, or generic units.

Some of the details of the visible declarations in the package specification may be implemented within the corresponding package body, while other implementation details are deferred by the use of context clausing. Thus, for example, using algorithmic and control statements, parts of a visible procedure may be elaborated in the procedure body, which is encapsulated in the package body. The implementation of other parts of the procedure may be deferred by calling subprograms that are imported from other packages. In this study, we define and measure two top-level features of system architecture, namely, the external complexity of packages as indicated by their coupling to other packages and their internal complexity as indicated by their "workload." We then consider the tradeoffs that may exist between internal complexity and external complexity. As noted earlier we may reduce the internal complexity of a package by increasing its external complexity and, conversely, we might reduce external complexity, but at the expense of increasing internal complexity.

This situation is similar to that expressed in Card and Agresti [15] and Evanco and Verner [14] for procedural languages in which one can defer the implementation details of a subroutine or function through call statements. By establishing the impact of both internal and external complexity on software defects, one can identify the optimal call structure given that a subroutine has a certain level of input/output data [15] or unique operands [14]. Similarly, for the case of Ada packages, we can pose the question: Given the visible declarations of a package, is there an optimal relationship between internal and external complexity such that the number of defects will be minimized?

---

[1] Subunits may also be specified to implement details outside the package body.

A long-standing belief in the software metrics literature – one supported by empirical evidence – is that software defects uncovered during the testing process occur more frequently for software components that are more complex [10]. Software complexity begins to reveal itself during the design phase and the subsequent coding phase [1]. Thus, if models for defect counts are developed during the design or implementation phases prior to testing, one may be able to analyze tradeoffs among design parameters so as to optimize some quality factor.

Software complexity has been defined differently by different authors. What has come to be commonly believed is that software complexity cannot be specified by a single software characteristic measure [1,16]. Rather, software complexity is a multi-dimensional concept and different facets of complexity are measured by different complexity metrics. We focus, in this study, on two complexity factors, related to internal and external complexity, which can easily be collected during top-level design.

In addition to software complexity, other features of software or its development environment may affect software defects [1]. We can view these features as contributing to a more general complexity that is related to defects. For example, a software component that is reused verbatim from a reuse library or from another project may be viewed as less complex than one that is newly created. The component has been tested and used previously and should have far fewer defects than a newly constructed component. In addition, the reuse of a component may only require the staff to understand its required inputs and the delivered outputs, rather than its internals.

Another factor that may contribute to complexity is the number of non-defect changes made to a component either to adapt a reused component for a new system or to add new features or requirements during the testing phase. Such changes generally are unanticipated and may result in greater software complexity. An initial design may not easily accommodate these changes, leading to a convoluted design or code. In addition, the number of non-defect changes may be viewed as indicative of a more complex development environment in which requirements not well-specified initially force adjustments later in the project.

## 3  Methodology

In this study, we restrict ourselves to the analysis of Ada packages consisting of a specification, body, and possible subunits. Thus, the Ada packages will all involve subprograms and those Ada packages consisting only of named collections of declarations will not be considered.

A small software component such as a package may consist of a relatively small number of source lines of code. Some packages have zero defects while others have only small numbers of defects. Thus, in any sort of analysis that we might perform, the defect count should be treated as a discrete countable variable and appropriate analytical methods must be identified for our model building exercise. Previous efforts in this area have utilized Poisson regression models [17]. However, a drawback of Poisson models is that the standard deviation is required to be equal to the mean. Consequently, Poisson models may be inappropriate for data exhibiting over-dispersion—when the variance of the defect counts is greater than its mean [18]. In such a case, the standard errors associated with the parameter estimates may be overly optimistic (too small) and the chance of including insignificant variables will increase. In the case of under-dispersion, we would expect the standard errors asso-

ciated with the parameter estimates to be larger than their true values yielding pessimistic estimates.

In order to deal with this problem, we propose a count model based on a negative binomial statistical distribution, which treats the variance as a quadratic function
of the mean, with a dispersion parameter that is estimated on the basis of the empirical data. Negative binomial count models have been discussed extensively in the statistical literature [18] and applied to the analysis of software defects [19]. The methodology allows us to determine the expected number of defects in a package. A major difference between this study and another by Evanco [19] is that in this study we
focus on non-linear models that can be used for prescriptive purposes.

The negative binomial regression model specifies that the number of defects, $n_i$, for
subprogram i is drawn from a negative binomial distribution with parameters $\lambda_i$ and $\theta$.

$$f(n_i \mid X_i) = \frac{\Gamma(\theta + n_i)}{\Gamma(n_i + 1)\Gamma(\theta)} s_i^{n_i} (1 - s_i)^{\theta} \tag{1}$$

where $\Gamma(.)$ is the gamma function and

$$s_i = \frac{\lambda_i}{\lambda_i + \theta} \tag{2}$$

The mean of the negative binomial distribution is given by $\lambda_i$ and its variance by
$\lambda_i(1 + \lambda_i/\theta)$ where $\theta$ is a parameter measuring the extent of dispersion[2]. The parameter, $\lambda_i$, is interpreted as the expected number of defects in subprogram i and is related
to the vector of characteristics, $X_i$, that influence the number of defects by

$$\ln(\lambda_i) = \beta' X_i \tag{3}$$

where "ln" represents the natural logarithm, $\beta$ is a column vector of parameters to be
estimated (the prime represents its transpose), and $X_i$ is a column vector of explanatory variables. The column vector $X_i$ may include basic characteristics of subprograms or higher order terms in these characteristics (e.g., quadratic terms). In this
study, we keep up to quadratic terms, specifying the natural logarithm of the expected number of defects by

$$\ln(\lambda_i) = a_0 + \sum_{i=1}^{n} a_i x_i + \frac{1}{2} \sum_{i=1}^{n} \sum_{j=1}^{n} \gamma_{ij} x_i x_j \tag{4}$$

where the $a_i$ (i=0, 1, ..., n) and the $\gamma_{ij}$ (i,j=0, 1, ...,n) with $\gamma_{ij} = \gamma_{ji}$ are the parameters to
be estimated. This functional form is better able to model more of the non-linearity
in the explanatory variables and may provide a better fit to the data. As is usual with
any regression analysis, the validity of the model is constrained to within the range
of values of the explanatory variables used to estimate its parameters. For example, if
our collection of packages used to calibrate the model ranged between 20 and 200

---

[2]    For very large $\theta$, the mean $\lambda$ is approximately equal to the variance and a Poisson model
would apply.

source lines of code, we would not expect the model to be applicable for predicting the defects of a package with 500 source lines of code.

The parameters, $a_i$, i=1,...,n, and the $\gamma_{ij}$, i,j=1, 2,...,n are estimated based on maximum likelihood estimates using the GENMOD procedure of the Statistical Analysis System (SAS). The GENMOD procedure allows the use of both Poisson and negative binomial probability distributions when one has a dependent variable, such as the number of defects, which is based on count data.

# 4    Model Variables and Empirical Data

As noted earlier, in this study, we focus on the analysis of defects in Ada packages having a specification, a body, and possible subunits; we will regard this component group as a "package."

## 4.1    Model Variables

Software architectural decisions occurring during top-level architectural design affect the complexity of the resulting software. We can represent the various facets of this complexity by measures. Moreover, these measures can be collected by automated means using a software analyzer if the top-level design representation is available in compilable form. Then, if we have a model that relates these measures to software development outcomes such as the number of defects in packages, we can predict these numbers based on early architectural characteristics. Similarly, if we know the relationships between the different facets of software complexity and the numbers of expected defects, then we may be able to identify architectural characteristics of the software that may minimize the numbers of defects.

The number and types of visible declarations contained in a package are determined by design considerations. For example, a problem domain may be partitioned into objects that represent the conceptual view of that domain. Those objects are defined in terms of their internal content and their relationships to the rest of the problem domain. There is some flexibility, however, in the level of disaggregation that a designer chooses when objects are specified. For example, a house may be defined as an object and implemented at that level, or the individual rooms within a house may be defined as objects. Thus, a designer may be able to influence the quality of software as measured by the numbers of defects by paying attention to the granularity of the objects that are defined during the partitioning process.

During top level design, the visible declarations in a package specification are identified. These declarations can be exported to and used by other packages. The declarations can be either data or program unit related. The details of the visible declarations may be implemented, in part, within the body of the package and are not visible to other packages. These details may include hidden subprograms, tasks, and data declarations. Thus, the *number of visible declarations*, denoted by $X$, implies a workload for the package that is reflected in the package body.

While some of the details associated with the visible declarations may be implemented in the corresponding body, other implementation details may be delegated or deferred to other packages by the use of context clauses ("with" statements). A context clause allows a package to import the visible declarations of another package and to use these declarations in the implementation of its body. Thus, part of the workload implied by the visible declarations of a library package can be implemented within its body while another part can be delegated to the resources of other packages. The *number of unique "with" statements for a package*[3], denoted by $W$, is a measure of the external complexity of the package.

This external complexity measure was first proposed by Agresti and Evanco [1] and is comparable to the coupling between object (CBO) metric proposed by Chidamber and Kemerer [20] and the coupling factor (COF) proposed by Abreu et al. [21]. The CBO metric makes no distinction between import and export coupling in the sense that two classes are coupled if one uses the other or vice versa. The COF metric does distinguish between the two cases. The context coupling measure used in this study is concerned only with import coupling where a package uses the resources of another package. Like the COF and CBO measures, the context coupling measure does not distinguish among actual method invocations and attribute references in a package.

We expect the internal complexity of a package to increase as the number of visible declarations increases if the workload associated with the visible declarations is implemented within the package body. However, insofar as the workload associated with the visible declarations is deferred to other packages through object coupling, the internal complexity of a package is reduced. This reduction of internal complexity is made at the cost of increasing the external complexity through object coupling. We take as a measure of internal complexity the ratio $X/W$–reflecting the fact that external complexity is increased as the number of visible declarations is increased and is decreased as the object coupling increases, deferring the workload to other objects. To avoid a singularity, when $W=0$, we set $X/W=0$ in our data.

This approach to external and internal complexity is similar to that of Card and Agresti [15]. They defined these complexities for procedural languages such as FORTRAN. The fan-out of a module measures the number of calls that the module makes to other modules and is regarded as a measure of external complexity. The number of I/O variables divided by the fan-out is a measure of the internal complexity. Similar to our situation, the I/O variables represent a workload for a module. This workload can either be implemented within the module or deferred to other modules by means of call statements.

Our data for the Ada packages is heterogeneous. The constituents of packages may have been reused from previous projects. For example, an Ada package may contain both subprograms that are newly developed and ones which are reused. In addition, packages experience different levels of non-defect changes throughout the development process, reflecting requirements changes. These two factors may be expected to influence the numbers of defects and may interact in non-linear ways with the architectural characteristic variables discussed above. Therefore, we include them in our analysis as discussed below.

Each component of a package (i.e., specification, body, subunits) is characterized by its origin. A component may be: newly developed, reused with extensive

---

[3]    A package may be "withed" into another package at the specification or body level. We make no distinction between the two cases in the count of "with" statements. However, if a package is "withed into both the specification and body of another package, then this increases the number of unique "with" statements by unity.

modifications ($\geq$25% of source lines of code changed), reused with slight modification (<25% of source lines of code changed), or reused verbatim (i.e., no changes)[4]. We compute the fraction of source lines of code in a package that are newly developed or reused with extensive modifications, denoted by FNEM. Components that are newly developed or reused with extensive modifications have been shown to exhibit substantially more defects than components that are reused with slight modifications or reused verbatim [10]. Thus, FNEM can be viewed as a measure of complexity contributing to the presence of defects.

The number of non-defect changes made to a package, denoted by CHG, also contributes to the composite complexity and to the number of defects. The changes made to adapt reused components for use in a new project *are not* included in the CHG number. The expected value of the number of defects, l, is thus regarded as a function of W, X/W, FNEM, and CHG given by f(W, X/W, FNEM, CHG).

The functional form of the expected defect number is not known on *a priori* theoretical grounds. Typically, linear regression models are proposed which effectively are Taylor expansions of some unknown functional form where only the lowest order (linear) terms are kept. The unknown parameters associated with the linear terms are then estimated by, for example, a least squares approach. However, in order to explore potential non-linear behaviors (and potential tradeoffs among the architectural variables), we Taylor expand the composite complexity, keeping up to quadratic terms, as follows:

$$\begin{aligned}
f(W, X/W, FNEM, CHG) = {}& a_0 + a_1{*}W + a_2{*}(X/W) + a_3{*}FNEM + a_4{*}CHG \\
& + b_{11}{*}W^2 + b_{22}{*}(X/W)^2 + b_{33}{*}FNEM^2 + b_{44}{*}CHG^2 \\
& + b_{12}{*}W{*}(X/W) + b_{13}{*}W{*}FNEM + b_{14}{*}W{*}CHG \quad\quad (5) \\
& + b_{23}{*}(X/W){*}FNEM + b_{24}{*}(X/W){*}CHG \\
& + b_{34}{*}FNEM{*}CHG
\end{aligned}$$

## 4.2  Empirical Data

Data from four Ada projects developed at the Software Engineering Laboratory of the NASA Goddard Space Flight Center are used in this study. The application domains were flight telemetry and dynamic simulations. The four projects yielded a total of 262 packages, which are the units of observation in the statistical analysis. The software data were obtained using the Ada Source Analyzer Program (ASAP) [22] to extract raw data from the code from which the measures were then derived. The means and standard deviations (in parentheses) of relevant variables characterizing the Ada packages are shown in Table 1 for each of the four projects.

Across all projects, the mean number of source lines of code (SLOC) per Ada package was 575 ranging from about 400 to 761 across the four projects. These SLOC supported an average of about 34 visible declarations for all projects, ranging from about 24 to 43 visible declarations per package across the projects. To support the implementation of a package, there was an average of almost 8 context clauses (unique "with" statements) across all projects, ranging from 6.4 to 9.1 context clauses within individual projects. The exports per "with" statement averaged 9.2

---

[4]  This particular classification was used because the component origin data was provided to us in this form only.

across all of the projects and ranged from 7.5 to 10 within the projects. The fraction of new and extensively modified code had a mean of .46 for all of the projects and ranged between .04 and .66 across the projects. There was an average of 3.9 changes per package ranging from .19 changes to 7.2 changes per package across the projects. The low of .19 changes per package for Project D was due to the heavy reuse of components in this project. The mean number of defects per package found during unit, system, and acceptance testing was 2.4 ranging from .08 to 4.2 among the projects. The defects are highly correlated with the value of FNEM.

**Table 1.** Statistical characteristics of packages

| Variable | Project A Mean (S.D.) | | Project B Mean (S.D.) | | Project C Mean (S.D.) | | Project D Mean (S.D.) | |
|---|---|---|---|---|---|---|---|---|
| SLOC | 761. | (934.) | 652. | (1039.) | 397. | (344.) | 413. | (407.) |
| X | 28.8 | (42.5) | 24.1 | (25.7) | 43.5 | (33.1) | 41.7 | (33.6) |
| W | 9.1 | (9.1) | 8.0 | (8.9) | 7.0 | (10.5) | 6.4 | (9.5) |
| X/W | 9.3 | (31.9) | 7.5 | (13.9) | 10.0 | (19.8) | 9.6 | (19.3) |
| FNEM | .66 | (.44) | .49 | (.46) | .58 | (.48) | .04 | (.18) |
| CHG | 7.2 | (9.6) | 4.5 | (13.0) | 2.4 | (5.2) | .19 | (1.1) |
| Defects | 4.2 | (6.3) | 2.7 | (5.4) | 2.1 | (2.7) | .08 | (.33) |
| Pkgs | 86 | | 56 | | 58 | | 62 | |

## 5 Analysis Results

In the statistical analysis of this section, we estimate the various coefficients of the model presented in (5). Not all of the coefficients may be statistically significant, so in such cases the associated terms are dropped from the analysis. The next subsection discusses the statistical results and in Sect. 5.2, we use the resulting equations for prescriptive analyses.

### 5.1 Statistical Analysis

The maximum likelihood estimates of the parameters, obtained from the GENMOD procedure of the SAS statistical analysis system, are shown in Table 2. The model keeps only the linear and quadratic terms as shown in the expansion (5) that are statistically significant. The numbers in parentheses are the standard errors associated with the coefficient estimates. The coefficient estimates are all statistically significant at the 1% level with the exception of XPW, which is significant at the 5% level. The coefficients of the expected number of defects are all positive implying that as internal complexity (X/W) and external complexity (W) increase so does the expected number of defects. Similarly, the larger the fraction of new and extensively modified reused code in a package and the more changes made to the package, the greater the expected number of defects.

The coefficient of determination, $R^2$, indicates that 71% of the variation is explained by the four variables introduced into the analysis. For the kind of micro-data being considered (i.e., Ada packages), this is a relatively large value. Finally, a significant value of the dispersion parameter of .34 indicates that our data exhibits substantial over-dispersion and that we were justified in using a count model based on the negative binomial probability distribution rather than the Poisson distribution.

**Table 2.** Negative binomial model results

| Coefficient | Parameter Estimates |
|:---:|:---:|
| Constant | -1.82 (.28) |
| X/W | .03 (.009) |
| W | .07 (.02) |
| $(X/W)^2$ | -.0001 (.00004) |
| $W^2$ | -.0014 (.0004) |
| FNEM | 5.99 (1.05) |
| $FNEM^2$ | -4.71 (.96) |
| CHG | .08 (.01) |
| $CHG^2$ | -.0008 (.0002) |
| FNEM*W | .04 (.02) |
| Dispersion | .34 (.11) |
| $R^2$ | .71 |

## 5.2    Prescriptive Analyses

Using the results of these analyses, we can now pose prescriptive questions regarding tradeoffs between internal and external complexity and the impact of these tradeoffs on package defects. Given that the number of visible declarations in a package is governed by object-based principles for decomposing a problem space into objects, for analysis purposes we regard this number

$$\frac{d\lambda}{dW} = \lambda * \left[ .077 - .034 * \frac{X}{W^2} - .0028 * W + .0002 * \frac{X^2}{W^3} + .039 * FNEM \right] \tag{6}$$

as a parameter.

The first derivative of the expected number of defects, $\lambda$, is given b.

Setting this derivative equal to zero and solving numerically for W given a specific X value provides us with a critical point of the function for $\lambda$. The second derivative of function, $\lambda$, at the critical point is given by:

$$\frac{d^2\lambda}{dW^2} = -.0028.068*\frac{X}{W^3} - .0006*\frac{X^2}{W^4} \tag{7}$$

The second derivative is positive for values of X and W, so that the function, $\lambda$, is a minimum for these values. Note that the value of FNEM affects our results, but the number of non-defect changes represented by C does not because C enters into equation (5) only linearly.

The plot of W vs. X giving minimum values of expected defects is shown in Fig. 1 for FNEM=1. Since W is an integer value, when dealing with specific packages, we must round to the nearest integer when using this plot to estimate the value of W, yielding the minimum expected defects for a package with X visible declarations. The integer-rounded W values are shown in Figure 1 for different ranges of X. For example, for a package with 15 visible declarations, the optimum W=2. The optimum number of "with" statements rises with the number of visible declarations. For an average package having about 34 visible declarations, the optimum number of "with" statements is three, while for ninety visible declarations, the optimal value is five "with" statements.

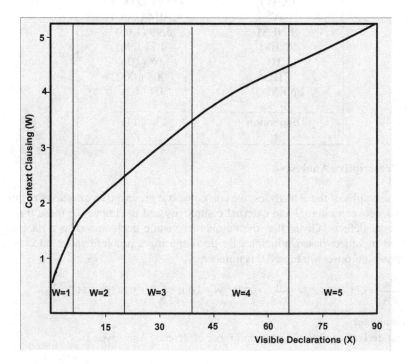

**Fig. 1.** Optimal context clausing vs. visibile declarations, FNEM=1

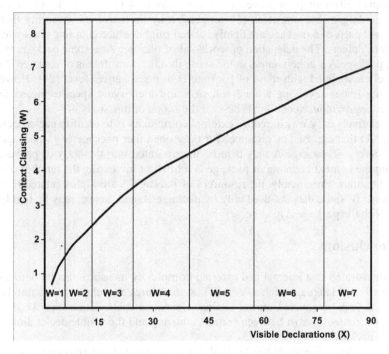

**Fig. 2.** Optimal context clausing vs. visible declarations, FNEM=0

Figure 2 shows the results of equivalent calculations for FNEM=0 (i.e., the package contains new and extensively modified reused code. Context clausing in this case rises a bit more steeply than that indicated in Fig. 1. However for the average package containing 34 visible declarations, the optimum context clausing still remains three. Reused verbatim and slightly modified code is more tolerant of context clausing, rising faster as the number of visible declarations increases. For the average package of 34 visible declarations, the optimum context clausing is four.

We use the model to calculate the relationship between the variables characterizing a package – namely X and W – that minimizes the numbers of defects in that package. Given that the number of visible declarations in packages are determined by object-based principles for decomposing a problem space into objects [22], in the analysis that follows, we will regard this number as a parameter.

From Table 1, we note that for the four projects in our data, the mean number of visible package declarations ranges from about 29 to 44, while the mean context clausing ranges from 6.4 to 9.1. Thus, it appears that, on the average, the packages of these projects make too much use of context clausing, consequently delegating too much of the workload implied by the mean number of declarations to other packages. A typical example of this delegation may be a call statement in a subprogram within a package to another subprogram of a package that is context coupled to the package. Consequently, if less of the workload were delegated and, instead implemented within the package through hidden declarations contained in the body, the context coupling might be reduced.

From a different perspective, if the elements in a package lack cohesiveness, then according to Stevens et al. [5] coupling may be increased. Assuring that the constituent parts of a package are tightly related might reduce coupling and lower the number of defects. The delegation of workload of package A to some package B may reduce package A's cohesiveness in the sense that the constituents of package B may be closely associated with those of package B at a semantical level [24]. However, examining issues revolving around cohesion and identifying specific means to increase cohesion/reduce coupling is beyond the scope of this study.

A relatively easy way to reduce external complexity is to identify packages (e.g., package A) that are used as resources for only one other package (e.g., package B). The resources of package A may then be incorporated into the body of package B, reducing the context coupling of package B while not increasing the number of visible declarations. Presumably, the resources of package A are highly related to those of package B, since they are used only by package B, and, hence, may be highly cohesive with the package B resources.

## 6    Conclusions

We demonstrated that internal and external complexity measures can be defined for an object-based language such as Ada. These measures are related to the number of package defects uncovered during testing. Based on this relationship, we demonstrated that tradeoffs exist between context clausing and the visible declaration numbers in a package.

The results of this study can be used in a prescriptive fashion as one input in deciding among different designs, in guiding the evolution of an initial design, or in modifying an existing system. A software designer may not be able to strictly adhere to these prescriptive guidelines for all packages in the system. However, these guidelines can be used to identifypotential problem areas that may be either modified or subjected to additional testing. While the analysis was conducted for Ada software systems, the results may be applicable to other object-oriented/based languages. The extent of this applicability is currently being explored.

The model presented in this paper is relatively simple, focusing only on some of the architectural features that manifest themselves during top-level design. In future work, we plan to introduce software features into our models that reflect some of the lower level architectural decisions. For example, the cyclomatic complexity and the numbers of parameters involved in the subprograms of the packages may affect defect numbers. Such models will be useful at later design and implementation phases. We might expect these models to be substantially more complex than the ones discussed here and to exhibit additional tradeoffs among the software characteristics that emerge throughout design and implementation.

## References

1.    Agresti, W. and Evanco, W. (1992), Projecting Software Defects from Analyzing Ada Designs, *IEEE Transactions on Software Engineering*, 18 (11), pp. 988–997.
2.    Briand, L.; Basili, V.; Thomas, W. (1992), A Pattern Recognition Approach for Software Engineering Data Analysis, IEEE Transactions on Software Engineering, 18 (11), pp. 931–942.

3.  Munson, J. and T. Khoshgoftaar (1992), The Detection of Fault-Prone Programs, *IEEE Transactions on Software Engineering*, 18 (5), pp. 423–433.
4.  Kazman, R.; Klein, M.; Barbacci, M.; Longstaff, T.; Lipson, H. Carriere, J. (1998), The Architecture Tradeoff Analysis Method, *4th International Conference on Engineering of Complex Computer Systems*.
5.  Stevens, W.; Myers, G.; Constantine, L. (1974), Structured Design, *IBM Systems Journal*, 13 (2), pp. 115–139.
6.  Briand, L. and Wust, J. (2002), Empirical Studies of Quality Models in Object Oriented Systems, *Advances in Computers*, 56, pp. 97–166.
7.  Mišic, V. (2001), Cohesion is Structural, Coherence is Functional: Different Views, Different Measures, *7th International Software Metrics Symposium*, pp. 135–144, London, England.
8.  Evanco, W. (1995), Modeling the Effort to Correct Faults, *Journal of Systems and Software*, Volume 29, pp. 75–84.
9.  Rombach, D. (1987), A Controlled Experiment on the Impact of Software Structure on Maintainability, IEEE Transaction on Software Engineering, 13, pp. 344–354.
10. Evanco, W. and W. Agresti (1994), A Composite Complexity Approach for Software Defect Modelling, *Software Quality Journal*, Volume 3, Number 1, pp. 27–44.
11. Khoshgoftaar, T. and Szabo, R. (1996), Using Neural Networks to Predict Software Faults During Testing, *IEEE Transactions on Reliability*, 45 (3), pp. 456–462.
12. Baisch, E. and Liedtke, T. (1997), Comparison of Conventional Approaches and Soft-computing Approaches for Software Quality Prediction, *IEEE International Conference on Systems, Man, and Cybernetics: Computational Cybernetics and Simulation*, Volume 2, pp. 1045–1049.
13. Fenton, N. and Neil, M. (1999), A Critique of Software Defect Prediction Models, *IEEE Transactions on Software Engineering*, 25 (5), pp. 675–689.
14. Evanco, W. and Verner, J. (2001), Revisiting Optimal Software Components, *Proceedings of the 12th European Software Control and Metrics Conference*, pp. 117–124.
15. Card, D. N. and W. W. Agresti (1988), Measuring Software Design Complexity, *Journal of Systems and Software*, Volume 8, Number 3, pp. 185–197.
16. Munson, J. and T. Khoshgoftaar (1988), The Dimensionality of Program Complexity, *Proceedings of the 11th International Conference on Software Engineering*, pp. 245–253.
17. Evanco, W. (1997), Poisson Analyses of Defects for Small Software Components, *Journal of Systems and Software*, 38 (1), pp. 27–35
18. Cameron, A. and Trivedi, P. (1986), Econometric Models Based on Count Data: Comparison and Application of Some Estimators and Tests, *Journal of Applied Econometrics*, 1, pp. 29–54.
19. Evanco, W.M. (2000), Subprogram Defect Predictions Based on Negative Binomial Models, *Proceeding of the 13th International Conference on Software and Systems Engineering and Applications*, Volume 2, Paris, France.
20. Chidamber, D. and C. Kemerer (1994), A Metrics Suite for Object Oriented Design, *IEEE Transactions on Software Engineering*, 20 (6), pp. 476–493.
21. Abreu, F.; M. Goulao; R. Esteves (1995), Toward the Design Quality Evaluation of Object-Oriented Software Systems, *Proceedings of the 5th International Conference on Software Quality*, Austin, TX.
22. Doubleday, D. L. (1987), *ASAP: An Ada Static Source Code Analyzer Program*, TR-1895, Department of Computer Science, University of Maryland
23. Booch, G. (1983), *Software Engineering with Ada*, Benjamin Cummings, Menlo Park, CA.
24. Henderson-Sellers, B. (1996), *Object-Oriented Metrics: Measures of Complexity*, Prentice Hall, Upper Saddle River, NJ.

# Evidential Volume Approach for Certification

Silke Kuball and Gordon Hughes

Safety Systems Research Centre, Department of Computer Science
University of Bristol, Merchant Venturers Building, Woodland Road
Bristol BS8 1UB, UK
silke@cs.bris.ac.uk

**Abstract.** In this paper we describe an approach to capture the degree of compliance of a product with the international standard for functional safety of E/E/PE systems, IEC 61508. We call this the *evidential volume* of an assessment scenario. It is based on compiling observed evidence according to assigned weighting factors, which describe the relative importance of each piece of evidence. The evidential volume can by itself be used as an indicator to compare different assessment scenarios. This could form the basis for improved consistency in assessment. We suggest a model to relate the evidential volume to the probability of having achieved a product of required safety integrity. Developing such a relationship can lead to a decision-aid on acceptance or rejection or can be used to decide whether additional evidence, such as statistical testing could be used to achieve target safety integrity. The model we suggest is based on the Success Likelihood Index Model (SLIM) and it poses an initial step towards decision-support for assessment.

## 1 Introduction

Systems comprising software and/or programmable electronic systems are increasingly used within safety-critical applications such as nuclear protection systems or avionics. From this stems an increased need for guidance on the development and assessment of safe and reliable systems. One of the routes that support the development and maintenance of safe and reliable systems is the use of industry standards such as DO-178B (avionics), IEEE/EIA 12207 (defense), IEC 61508 (all industry sectors) PES Guidelines (nuclear). Even though the use of standards and guidelines itself does not garantuee reliability, it helps to implement techniques that are considered to be the basis of safe and reliable products. Thus, the use of standards can be seen as a technique complementary to more quantitative methods such as statistical testing (ST), which estimate a reliability figure but do not directly scrutinize the process of design and development. Our aim is to work towards combining both routes of assessment: standards/certification and ST. Therefore, we set out to consider how the wide range of tasks performed when following a standard can be transformed into a quantitative figure measuring something like the foundation on which ST methods then start to operate. On the long-term, the following questions should be answered: "Can we combine the evidence seen during certification and during

J.-P. Rosen and A. Strohmeier (Eds.): Ada-Europe 2003, LNCS 2655, pp. 246–257, 2003.
© Springer-Verlag Berlin Heidelberg 2003

ST into one reliability measure?" "Does the groundwork performed in design and development for any particular component look sound enough to justify launching into the effort of ST?" Due to the interests of the industry we work with, we focus our considerations here on two standards used by this industry: IEC 61508 [1] and the PES Guidelines [2]. However, it should be noted that any other standard could be inserted throughout the paper, the methodology proposed would remain the same.

The international functional safety standard IEC 61508 (Functional safety of E/E/PE safety-related systems) aims at providing guidance on the development and assessment of safe and reliable systems consisting of hardware, software or a combination of both. It describes the overall safety lifecycle of a system or component as consisting of 16 phases and provides generic guidance for all safety lifecycle activities. The achievement of functional safety is supported through a wide rande of measures, techniques and principles relating both to the process and the product. IEC 61508 uses the concept of Safety Integrity Levels (SILs), which are linked to upper bounds on the tolerated system probability of failure on demand (pfd). Examples for claims on the pfd for different SILs are:

$$SIL2: \quad pfd \in [10^{-3}, 10^{-2}[, \quad SIL3: \quad pfd \in [10^{-4}, 10^{-3}[. \tag{1}$$

Starting with a Hazard and Risk Analysis, the safety requirements are identified for each safety function and then later translated into SILs for components implementing the function. For each SIL, the set of measures, techniques and principles recommended in [1] changes, becoming more stringent for higher SILs. Throughout this paper, we refer to the techniques/principles listed in [1] as *requirements*. Implementation of the applicable requirements does not guarantee having developed a component of tolerable failure probability, but it shows that all reasonable activities to prevent intolerably high failure probabilities have been carried out. Thus we are reasonably confident in having achieved a component of required safety integrity. This is based on the fact that the standard is a compilation of the best knowledge and understanding currently available from experts in the field. Thus the assumptions that following a standard supports a reliability claim is an expression of expert judgement. When certifying a system, expert judgement comes in again: an assessor makes the decision whether or not to accept a product based on the evidence he/she has seen. This decision is mostly seen as a binary process: accept or reject. In the case where each requirement has been demonstrably met, the decision for acceptance seems obvious. However, in the case where not all expected evidence is available, maybe because it was replaced by alternative evidence or a technique wasn't documented or simply not applied, the decision will be less clear. Depending on the type of information missing, the quality of alternative evidence provided, the overall picture of evidence, the assessor will decide in favour of or against acceptance. It is such cases that seem to potentially benefit from decision-support because decisions here are based on subjective indiviudal judgement and individual experiences that cannot easily be replicated or compared between scenarios, thus allowing for inconsistencies to arise. It is the aim of this paper to explore the possibility of a quantitative measure expressing the "overall picture" of evidence available with respect to a standard, a measure for the total volume of related evidence. This measure should be repeatable and provide a high degree of consistency between assessment scenarios. It could be used alongside expert judgement and aid in deciding on acceptance or rejection of a system. In this paper we explore a possible way of achieving such a quantitative decision-aid.

## 2   Background: Structure of Evidence

In Fig. 1, we have sketched part of the structure of evidence contained in IEC 61508. The output generated during one phase serves in most cases as input to succeeding phases. The structure in Fig. 1 is laid out over 4 levels. At level 4 the requirements are listed. The remaining levels are used to indicate that phases can consist of sub-phases, which again can consist of sub-phases. For example, the Level 1 phase "Software Realisation" contains among others the sub-phases "Safety Requirements Allocation" and "Design and Development". Level 2 phase "Design and Development" contains a series of sub-phases building on each other such as "Software Architecture", "Support Tools & Programming Languages (PLs)", "Code Implementation" etc. At the final level a set of require-ments is listed for each phase in the form of clauses and tables. For example: requirements listed under "Support tools & PLs" and "Code Implementation" are: "Programming language unambiguously defined", "No dynamic objects", "Limited use of pointers". Please note that Fig. 1 does not attempt to replicate the safety lifecycle as given in [1], it only shows an excerpt for illustrative pur-

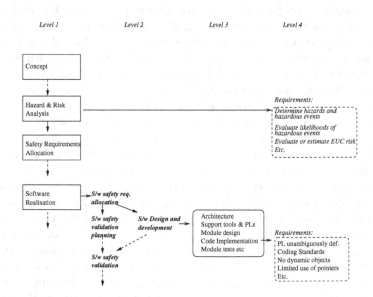

**Fig. 1.** Excerpt from safety lifecycle as described in IEC 61508.

poses. The PES Guidelines contain a similar structure of phases and sub-phases but are a smaller and more accessible document tailored to the needs of PES in Nuclear Safety Applications. Examples of requirements listed here in sub-phase "Design for Safety" are: "Avoid use of interrupts, pointers, tasking. Safe language subsets (e.g. SPARK Ada) help to reinforce these restrictions", "Semantics of language should support Formal Proof", "Use safe language subsets: no dangerous constructs that may be ambiguous".

Using a language that meets the requirements listed above thus seems to support the creation of a safe and reliable system/component. If all language-sensitive objectives and guidance are satisfied then the language is seen as suitable for the purpose of a safety-critical system of given target integrity. But, how does the particular evidence pertaining to coding and programming-languages contribute to the overall system quality? How does the evidence collected during assessment on both process and product add up to build confidence in system safety? In some way, the degree to which a sub-phase complies with the requirements listed will be indicative of the quality achieved for its parent phase, i.e. the next level down in Fig. 1, and finally this will contribute to the overall compliance of the product with the safety claim. If important evidence is missing, this will affect the trust we have in the product reliability which is expressed through the probability of failure on demand (pfd). For example, the use of C as a programming language may affect the degree to which conformance has been achieved within the phases "Code Implementation (IEC 61508) or "Design for Safety" (PES Guidelines). The use of SPARK Ada on the other hand lends itself well to achieving conformance within these phases and to achieve this rather efficiently. Depending on the relative importance of these phases in the context of the overall safety lifecycle, this will more or less affect our belief in having achieved an end-product of sufficient reliability. Due to the hierarchical structure of evidence in the safety lifecycle, a flaw in one of the phases, e.g. lacking rigour in chosing programming techniques will filter down and potentially undermine success of the project, even if all the subsequent phases were carried out to high standard and in compliance with the requirements. Even though all requirements are important, there will most likely be some, the absence of which would influence an assessor's confidence in the product more severely than others. Hereby absence could mean that a recommended technique was not documented or not implemented and not repaced by a suitable alternative technique. Also, if a large number of pieces of evidence are missing that would by themselves not affect our belief in the end-product, then the total mass of missing evidence may create again a considerable plunge in confidence leading to rejection of a product. It seems beneficial to formalize these concepts through the introduction of a quantitative measure to aid different assessors to come to the same conclusion (accept or reject) for the same set of evidence. In order to make a step towards such a formalization, we introduce the following approach.

# 3    Evidential Volume Approach

Let $P_j$ be used to indicate a phase in the overall safety-lifecycle. We assume that for $P_j$ a set of requirements $R_{1j}, ..., R_{nj}$ is listed. An example could be the "Code Implementation" sub-phase in Fig. 1 and the set of requirements listed for that sub-phase. An assessor observes whether a requirements has been met or not met. Let $I_{1j}, ..., I_{nj}$ be an indicator taking on the value 1 if the assessor observes that $R_{ij}$ is met and 0 otherwise. One could also allow for a value between 0 and 1 being assigned to express the quality by which a requirement has been met, however, at this stage we model the observation of an assessor as a binary value. The set $I_{ij}$, $i = 1, ..., n$ describes the whole set of evidence present at the end of the assessment. However, when the assessor makes his/her decision on whether to accept or reject a system or component, not only the number of requirements met will be taken into account, but the relative importance of the evidence present and of the evidence missing has to be accounted for as well. We model this by introducing a set of weights, $w_{1j}, ..., w_{nj}$ attached to requirements $R_{1j}, ..., R_{nj}$. These weights describe a ranking of the importance of different requirements in the whole set of listed requirements with respect to achieving success in phase $P_j$. Success could mean for example: "code implementation has been performed in accordance with the recommendations in the standard". We assume the weighting factors to be normalized such that $\sum_{i=1}^{n} w_{ij} = 1$. The technique that is considered most important would receive the highest weight, if no preference can be given to any technique, all weights are chosen to be equal. The weighting factors should be elicited through discussions with a panel of experts. This process would have to be undergone once for each SIL in order to fix weights for each requirement depending on what the total set of requirements for that SIL is and depending on the level of recommendation the requirement receives in IEC 61508 for that particular SIL. The weights thus identified could then be used for every assessment scenario based on IEC 61508. From the assessor's observations $I_{1j}, ..., I_{nj}$ and the generic set of weights $w_{1j}, ..., w_{nj}$ we compile a measure for the degree of compliance of the product under assessment with the principles of IEC 61508:

**Definition 1.** *Let $P_j$ be a sub-phase of IEC 61508 for which requirements $(R_{1j}, ..., R_{nj})$ are listed. Let $(I_{1j}, ..., I_{nj})$ be a set of indicator values assigned by the assessor with $I_{ij} = 1$ if $R_{ij}$ is met and 0 otherwise, $i = 1, ..., n$. Let $(w_{1j}, ..., w_{nj})$ be weights describing the relative importance of $R_{1j}, ..., R_{nj}$ in achieving success in phase $Pj$. Then,*

$$EV_j := f_j(I_{ij}, w_{ij}, i = 1, ..., n) = \sum_{i=1}^{n} w_{ij} \cdot I_{ij}. \tag{2}$$

*is called evidential volume for $P_j$.*

Generally $f_j$ is an aggregation function, which can adopt different forms. At the level of aggregating single requirements, we have chosen the additive form. In the next section we will brainstorm some possibilities of assigning weighting factors to each evidence unit and again comment on the choice of aggregation functions.

## 3.1    Assigning Weighting Factors

Weighting factors represent a ranking between requirements. They should be assigned by a panel of international experts for each SIL or equivalent integrity target. There are different ways of assigning such weights. If no preference can be given to any phase or requirement, then the weights pertaining to one level could be all chosen equal. Weights can be identified through pairwise comparisons. Within the phase "Code Implementation" "Use of coding standard" could be considered as twice as important as "Limited Use of Pointers" due to a level of recommendation identified in the standard. Another requirement "No Dynamic Objects" could be for example equally weighted as "Use of coding standard" and if they represented a full set of recommended techniques, the assumption that they should add up to 1 would be enough to calculate the weights. Weights could also be thought of initially as qualitative statements. Discussion of this work with our industrial sponsors led to identification of three preliminary ranking-levels for requirements: "Important", "Less Important" and "Very Important" depending on the safety integrity target(s) for which the requirements were listed. Defining relations between the rankings (e.g. "Important=2*Less Important", "Very Important=3*Less Important") can then lead to an identification of weights. These are however initial ideas that have not been validated yet and require more input from experts in order to devise a suitable way of allocating weights.

A further possiblility to allocate weights is to use an approach introduced by Li et al (2000) [3]. Here Multi Attribute Utility Theory is used for the ranking of software engineering measures with respect to their ability to contribute to reliability. This approach seems very suitable to be used in the context of ranking requirements within IEC 61508 for their importance when it comes to generating a reliable product. In [3] for a set of software engineering measures a score is elicited from experts for a set of attributes. Example measures are error distribution and failure rate. Examples of chosen attributes that were chosen are credibility, experience, repeatability, validation, cost, benefit and relevance to reliability. Measures are rated with respect to an attribute on a scale from 0 to 5 or 6, the high scores representing the highest relevance. The scores given by the experts are compiled into a single number using a linear additive aggregation equation with equal weights. In the context of IEC 61508 the same or similar attributes could be used to assess the requirements listed. Each requirement could be rated according to the set of attributes and the total score achieved from a panel of experts would then point towards an appropriate weighting factor.

## 3.2    Aggregation Functions

The choice of aggregation functions depends on the structure of phases and the relationship between phases. There are several levels of aggregation needed. First of all on the sub-phase level we need to aggregate observations on the single requirements. Requirements listed for one sub-phase consist of different techniques that are all perceived to contribute to successful implementation of the same aspect of the safety lifecycle. For example in the sub-phase code implementation, the techniques such as no dynamic objects, use of coding standards, limited use of pointers etc. provide a different component of a reasonable programming regime. Thus these techniques contribute all to some degree to the successful implementation of that particular sub-phase. If one is missing, still some good practice has been followed and thus a certain quality of programming achieved. For this level of aggregation, we have chosen the linear aggregation function of Eq. (1). We choose the linear function here because the set of requirements can be basically seen as separate evidence contributing to some degree to the same measure, e.g. quality of code implementation. After having established the evidential volumes for each phase, we need to aggregate the evidential volumes of the phases into one overall evidential volume. In the overall safety lifecycle, phases heavily depend on each other: the input from one phase serves as output to subsequent phases. If one phase is completely missing, and is not replaced or compensated for by an alternative technique[1] the entire process is invalid, if one phase is flawed this cannot be compensated for by other phases. The same relationship holds for example between the sub-phases of "Software Realisation". This should be captured through the aggregation function used to combine the EVs of different phases into one EV. Aggregation functions should be chosen to comply with experts' belief. To achieve this, a set of axioms could be formulated reflecting the experts' beliefs. Possible axioms that seem reasonable are given below, see also [4]

i)   The calculation of EV is uniquely determined.
ii)  EV shall be not larger than the largest value $EV_j$ and not smaller than the smallest value $EV_j$.
iii) Small improvements in $EV_j$, $j \in \{1, ..., m\}$ do not make EV smaller.
iv)  EV shall take on its highest possible value (1) only when all $EV_j$ take on the highest possible value.
va)  EV shall take on the smallest possible value (e.g. 0) only if all $EV_j$ take on the smallest possible value OR
vb)  EV shall take on the smallest possible value if at least one $EV_j$ takes on the smallest possible value.

Axiom vb) constitutes a "K.O."-criterion, which means that the total EV is 0 already when one $I_{ij} = 0$, i.e. when one $EV_j$ is missing. This means that if evidence for one phase is completely missing, this cannot be compensated for by another phase having been performed exceptionally well. Such a criterion could apply to certain levels of aggregation, as an example, in Fig. 1, it could apply to levels 1 and 3 of the depicted evidence structure. A possibility of modelling such a criterion is via a multiplicative aggregation function.

$$EV = \prod_{j=1}^{m} EV_j^{\alpha_j}. \tag{3}$$

---

[1] Hereby, replacement of a technique with an alternative method should be performed according to the principles acceptable in the context of the standard.

Hereby $\alpha_j$, $j = 1, ..., m$ are again weighting factors chosen to fulfil a condition such as $\sum_{j=1}^{m} \alpha_j = 1$. They express the importance of phase $P_j$ in the entire process, or the severity of loss in confidence when putting no demonstrable effort into that phase. When using the function in Eq. (3), a change in the factor with the largest $\alpha_j$ will be noticed more than a change in a factor with a smaller $\alpha_j$. For the factor with largest $\alpha_j$, the larger the change, the higher the impact on the overall evidential volume. Other aggregation functions are possible such as

$$EV = \min_{j=1,...,m} (EV_j)^{\alpha_j}.$$

Note that the exact choice of aggregation function models our prior belief about how the quality of the overall product is influenced by the quality of sub-phases. Therefore, the aggregation functions too need to be checked with and agreed upon by experts. Depending on the "score" obtained in a specific sub-phase, the place the sub-phase takes in the overall safety lifecycle and the weight assigned to it, the outcome of a sub-phase will influence the overall evidential volume. This can help decide what the benefit of a specific technique (e.g. SPARK Ada) is in the context of the overall evidence gathered. It can also help to compare techniques by investigating the effort needed to obtain a certain EV with different techniques, i.e. how efficient are they in achieving a high EV. For example, the application of a technique such as SPARK Ada lends itself well to fulfillment of the requirements of sub-phase "Code Implementation". At the same time, it supports static analysis [5] and thus the sub-phase "Software Verification", thus may contribute to obtaining a high EV in two phases.

## 4   Use of Evidential Volume in Decision-Making

The weighting factors in the evidential volume approach should be generically specified through consensus of a suitable team of experts. For a given SIL, they should be a fixed set of numbers. Thus, once they are specified, the outcome of an assessment as measured by the evidential volume is repeatable and can be compared to the outcome of other assessment scenarios. For all products for which the same EV is achieved, the same decision outcome should be made: accept or reject.

Potentially, the evidential volume can be used to formulate a prior belief on the reliability of the product under assessment, which could form the basis of deciding whether or not to perform further more quantitative assessment such as statistical testing. Such a prior belief could be for example expressed in the form of a probability of success or a confidence in having achieved the required integrity target. A possible approach to link the EV to a success probability, is based on the success likelihood index model (SLIM) [6]. In SLIM, a success likelihood index $S$ is calculated via $S = \sum_{i=1}^{n} I[PSF_i] \cdot w_i$. Hereby, $PSF_i$ are "performance shaping factors" contributing to the task at hand. $w_i$ describe the relative importance of $PSF_i$ for the task, hereby $\sum_{i=1}^{n} w_i = 1$. $I[PSF_i]$ rate the quality of $PSF_i$ in the current context. The EV is a similar compilation of weights and ratings for a set of factors, all influencing the quality of the final (human) task of producing a system of suitable reliability. Thus, we interpret the EV as a success likelihood index. The use of SLIM in the context of reliability estimation is also mentioned in [7]. Using SLIM, the following relation can be established between $EV$ and a probability of success $P^{success}$ [6].

$$Log(P^{success}) = a \cdot EV + b. \tag{4}$$

$P^{success}$ can be defined as the probability of having successfully met the claimed integrity target, i.e. $P^{success} := Pr(pfd < \text{Integrity Target})$. The parameters $a$ and $b$ in (4) are determined through the boundary conditions

$$EV = 0 \Rightarrow P^{success} = P_{Lower}, \quad EV = 1 \Rightarrow P^{success} = P_{Upper},$$
$$Log(P_{Lower}) = a \cdot 0 + b, \qquad Log(P_{Upper}) = a \cdot 1 + b. \tag{5}$$

SLIM is originally used to calculate human error probabilities. The task of developing a software product from a given specification to a certain level of reliability is a human task, thus use of this model seems suitable. It has to be noted that even though SLIM is based on experimental data and theoretical considerations [6], it is to be seen as a *model* compiling expert judgement and thus prior belief rather being derived from statistical inference. Thus it is important to check the message the model conveys against expert beliefs for example for a series of example scenarios. In (5), for two extreme cases such as EV=0 or 1 one formulates a prior belief about $P^{success}$. This could take on the following form. We consider the example of aiming at a SIL 2 system, i.e. $pfd < 0.001$, see (1). In the case of EV=1, we have encountered the "ideal" assessment scenario were all evidence is demonstrably met. In this case, because the standard comprises the consensus on what is considered to be sound development practice, and because compliance with the standard is associated with complying with a claimed integrity level, one could model $P_{Upper}$ as for example 0.99. This would be the same level of confidence that one would aim at with statistical testing. Other values for $P_{Upper}$ are possible, the importance is on checking it with expert belief. The value $P_{Lower}$ could be modelled as follows. In the absence of all evidence, we could assume that we do not know anything about the system's pfd. This can be modelled through a uniform prior distribution over the pfd, which results in $Pr(pfd < 0.001) = 10^{-3}$. Thus a possible choice is $P_{Lower} = 10^{-3}$. The assumption $P_{Lower} = 0$ is not suitable due to the $Log(P_{Lower})$ in Eq. (5).

Using (4) and (5), one can then calculate $P^{success}$ for EV $\in ]0, 1[$. From this relationship one can potentially determine the utility or risk of accepting a system, and a cut-off point, below which a system should be rejected. In the following we shall perform some example calculations. We use the examples of having a a) SIL2 target and b) SIL3 target. We define boundary conditions as described above, and calculate success probabilities for different EVs.

**Example:** We define $P^{success} := Pr(pfd < T^{SIL})$. Hereby $T^{SIL}$ is the safety integrity target for the given SIL, which can be interpreted as the lower bound of the pfd interval in (1), see also [1], Part 1, Table 2. We assume the following boundary conditions:

1) SIL 2. $P^{success} := Pr(pfd < 0.001)$. $P_{Lower} = 0.001$, $P_{Upper} = 0.99$.
2) SIL 3. $P^{success} := Pr(pfd < 0.0001)$. $P_{Lower} = 0.0001$, $P_{Upper} = 0.99$.

We obtain the following results:

SIL 2:

| EV | $P^{success}$ |
|---|---|
| 0.95 | 0.7 |
| 0.9 | 0.497 |
| 0.8 | 0.249 |

SIL 3:

| EV | $P^{success}$ |
|---|---|
| 0.95 | 0.624 |
| 0.9 | 0.394 |
| 0.8 | 0.15 |

Using SLIM with the above boundary conditions yields for SILs 2 and 3 a relatively small estimated success probability already when $5 - 10\%$ of the total evidential volume is missing. This may be considered as too pessimistic by experts, in which case the model needs to be revised to capture these beliefs. However, 5% of the total evidential volume missing may be considered as a case where a considerable decrease in confidence in the product from assessment data alone is justified. In this case, it could be investigated whether a higher level of confidence can be achieved by adding evidence such as statistical testing. In order to investigate what can cause a loss of $5 - 10\%$ of evidential volume, we have considered an example scenario. We assume a series of 14 phases $P_j$, each containing a list of $n_j$ requirements. The evidential volume of $P_j$ is calculated as $EV_j = \sum_{s=1}^{n_j} w_s \cdot I_s$. In this example, we assume that the weights $w_s$, $s = 1, ..., n_j$ within each phase are equal. The total evidential volume is calculated as $EV = \prod_{j=1}^{14} EV_j^{\alpha_j}$. For the weights $\alpha_1, ..., \alpha_{14}$, we establish a set of comparative equations.

$$\alpha_1 = 1 - \sum_{j=1}^{14} \alpha_j; \alpha_1 = 3 \cdot \alpha_3; \alpha_3 = 3 \cdot \alpha_2; \alpha_4 = 0.5 \cdot \alpha_3;$$
$$\alpha_5 = \alpha_6 = \alpha_7 = ... = \alpha_{10} = \alpha_4; \alpha_{11} = 2 \cdot \alpha_4; \alpha_{12} = ... = \alpha_{14} = \alpha_{11}.$$

From these conditions, the following set of weights are calculated:

$$\alpha_1 = \frac{18}{71}, \alpha_2 = \frac{2}{71}, \alpha_3 = \frac{6}{71}, \alpha_4 = \frac{3}{71} = ... = \alpha_{10}, \alpha_{11} = \frac{6}{71} = ... = \alpha_{14}.$$

For the numbers $n_j$ of requirements within $P_1, ..., P_{14}$, we have assumed:

$$n_1 = 11, n_2 = 5, n_3 = 12, n_4 = 3, n_5 = 6, n_6 = 5, n_7 = 2, n_8 = 4, n_9 = 5,$$
$$n_{10} = 8, n_{11} = 10, n_{12} = 8, n_{13} = 13, n_{14} = 3.$$

This gives us a total of 95 requirements. We have now assumed the following scenarios:

1) 3 requirements are missing in $P_1$ (large weight) and 3 requirements are missing in $P_{13}$. Both $P_1$ and $P_{13}$ possess large weights. The resulting total evidential volume is $EV = 0.91$, this yields a success probability of ca. 40% if we are performing SIL 3 assessment, and 50% if we were assessing for SIL 2. Approximately 6% of all requirements are missing, mostly in important phases, and the resulting $EV$ has decreased by ca. 10%.

2) A total of 12 requirements is missing, but $P_1$ is complete. The missing requirements are distributed over 9 phases. This lead to a total evidential volume of 89%, thus not a large change in comparison to case 1), however double the amount of requirements are missing.

Thus, in the compilation of the evidential volume, the number of missing requirements, their weights and the weights of the phases they are associated with within the overall lifecycle is of importance. The model introduced here describes the belief that missing a small amount of important evidence may already lead to a considerable decrease in confidence in the product, which then has to be dealt with by revising the development process, generating additional evidence, such as statistical testing, or else rejecting the product. Linking the evidential volume to a success probability is useful because it can serve as the basis of calculating the utility of a decision, for example by multiplying the risk involved when accepting a product that is not safe enough with $(1-P^{success})$ and similarly multiplying the risk (or cost) of rejecting a safe product unnecessarily with $P^{success}$ and comparing the figures. The success probability can also form the basis of building a prior belief function for statistical testing. This could be used to predict the probability of the system under test to pass statistical reliability testing.

## 5   Outlook

The intention of this paper is to introduce the EVA to the reliability community for discussion and feedback. The model introduced here, needs to be discussed with experts in order to be validated as a suitable model for expert belief. This requires iterative calculations of example scenarios and this should be helped by tool-support. We have started to develop a simple support-tool that contains the requirements listed in the PES Guidelines and a set of preliminary weights. The tool takes as inputs the decision whether any particular requirement is "met", "not applicable" or "missing". If the requirement is rated as "met", a set of security questions pops up to determine whether the evidence has been directly met or through some allowed alternative route. If not all relevant questions can be answered with "yes" but still "requirement met" is ticked, a warning message will be issued. The tool calculates the EV for all main phases as well as the overall EV. The purpose of this simple prototype tool is to distribute it to a set of experts as a basis of discussion of the approach and to demonstrate how the approach works. This will help identify the strengths and limitations and provide useful feedback on how to take this work further. Working with a tool can also be useful when discussing the weights in the approach. Example EV's for known assessment scenarios can be calculated and the results compared with the actual assessment decision taken. This can shed some light on suitable choices of weights. On the longer term, a tool for the calculation of an EV - based on a set of weights assigned by a panel of experts from the relevant industry, academia and government institutions- might also be beneficial to developers of safety-critical components to assess the degree of potential compliance achieved with their product at any stage of the development cycle in a user-friendly way. To propose how to setup a panel of experts is beyond the scope of the project performed here. This could potentially be part of a future project. No application to real cases has been performed yet, but hopefully at some stage we will be able to perform example calculations on real assessment scenarios and examine how well the results obtained with the EVA fit the original assessment decisions. The first step still to be performed is however to arrive at a useable set of weights.

## 6    Conclusions

In this paper we have described a model to capture the degree of compliance of a product with an industrial standard such as IEC 61508. This figure, the *evidential volume* can by itself be used as an indicator to compare different assessment scenarios. This could form the basis for improved consistency in assessment. A model was suggested for relating the evidential volume to the probability of having achieved a product of required safety integrity. Developing such a relationship can lead to a decision-aid on acceptance or rejection or can be used to determine the amount of additional evidence, such as statistical testing, required to help with demonstrating successful product development. The model described here is based on SLIM, it poses an initial step towards decision-support for assessment and poses the basis on which experts can start formulating their prior belief arising from assessment quality.

## References

[1] "Functional safety of electrical/electronic/programmable electronic safety-related systems, International Electrotechnical Commission," 1998.
[2] L. Winsborrow and A. Lawrence, "Guidelines for using programmable electronic systems in nuclear safety and nuclear safety-related applications," *British Energy Generations Ltd.*, 2001.
[3] M. Li, C. Smidts, and R. W. Brill, "Ranking software engineering measures related to reliability using expert opinion," *Proceedings of ISSRE 2000*, 2000.
[4] H. Dehlinger, "Deontische Fragen, Urteilsbildung, Bewertungssysteme," *Arbeitsbericht aus dem Fachgebiet Design Theorien and Methoden*, vol. 7/94.
[5] R. Chapman, "Industrial experience with SPARK," *Proceedings of ACM SigAda 2000*, pp. 64–68.
[6] G. Salvendy, *Handbook of Human Factors and Ergonomics*. Wiley Interscience, 1997.
[7] C. Smidts and D. Sova, "An architectural model for software reliability quantification: sources of data," *Reliability Engineering and System Safety*, vol. 64, pp. 279–290, 1999.

# A Survey of Physical Unit Handling Techniques in Ada

Christoph Grein[1], Dmitry A. Kazakov[2], and Fraser Wilson[3]

[1] ESG-Elektroniksystem- und Logistik-GmbH, D-81675 München
[2] cbb software GmbH, D-23560 Lübeck
[3] Anago b.v., NL-3995AA Houten

**Abstract.** There has always been a demand to be able to compute with physical items where dimensional correctness is checked. A survey of methods is presented how to do this with the features of Ada. Compile-time methods use the type checking mechanism whereas run-time methods use additional components to represent dimensions.

## 1 Introduction

When dealing with physical equations, physicists are used to checking their results for dimensional correctness. So they feel a certain loss when they transfer their equations to computers since none of the commonly used programming languages provide features to deal with physical dimensions. Programmers resort instead to work-arounds within the language-provided features to keep track of dimensions or do completely without them.

In the following chapters, a few methods are presented how the features of Ada can be used to compute with dimensions.

We would however like to state that it is not the wrong equations that introduce the actual problems in big software systems (these can be spotted by code inspections); their causes normally are buried much earlier in wrong system or software designs, as has been shown by some recent spectacular failures.

*Mars Lander*'s problem was the mixing of metric and British units in the communication between two CSCIs, which none of these methods would have been able to avoid since they apply only within a CSCI (Computer Software Configuration Item – a separate program communicating with other programs via e.g. global memory).

The ideas to import dimensions into Ada (or other programming languages) are not new. Do-While Jones [1, 2] published such a method using private types, which however becomes unwieldy when dimensions are mixed; it also does not include mathematical functions.

Gehani [3] compared the appearance of expressions in Ada with a hypothetical language which includes dimensions as data attributes and came to the conclusion that only the latter method can solve all problems elegantly and satisfactorily. Since at that time, he considered an Ada language change as most improbable (the Ada9X process has proven him correct), he proposed a solution how correctness of physical expressions can be checked during run-time (by using discriminants).

Gehani's method suffers however from its notation of declarations and statements that differs considerably from the common technical and scientific one. The nota-

J.-P. Rosen and A. Strohmeier (Eds.): Ada-Europe 2003, LNCS 2655, pp.258–270, 2003.

tional problems of both Gehani's and Jones' methods can be cured if Ada's features are used cleverly as Hilfinger [4] has shown, who based his ideas on Gehani's work.

In passing, we mention as a third method beside the two above: the use of pre-processors. Schneider [5] presents such a pre-processor for Pascal, implemented in C. He also provides an overview over attempts to include dimensions into programming languages. For further references, see his paper.

A universal unit support faces contradictory requirements which are difficult if not impossible to satisfy:

1. **Compile-Time Checks.** The dimension errors should be discovered as early as possible. For example if dimensions are known at compile time, then all unit errors shall be detected at this stage.
2. **No Memory / Speed Overhead at Run-Time.** If dimension information is not used at run-time (because of 1), then at run-time neither memory should be allocated to keep it, nor checks should be made.
3. **Support for Derived Types.** Dimensioned operations should involve not only the isles of scalar types. There should be a way to declare other dimensioned types, for example, an array of dimensioned scalars; a dimensioned record type; a dimensioned private type, with all necessary cross operations.
4. **Generic Programming.** Here generic is used in a wider sense, as an ability to write code dealing with items of types from some type sets. In our case it means items of different dimension. For instance, it should be possible to write a dimension-aware integration program, which would work for all valid combinations of dimensions.
5. **No Precision / Range Loss.** Very often dimensioned values are measured in units different from the base ones. There should be a way to deal with such values without a necessity to convert them to the base units, which is inevitably connected with a precision / range loss. It is hard to expect and wrong to require that an astrophysical program would use meters to calculate distances between galaxies.
6. **Scales.** A more generalized variant of 5 is when data are measured in scales different from standard, or additional constraints are imposed by the nature of the measured values. Examples of scales are logarithmic values, Celsius temperature scale, time versus duration etc. Though values of different scales might have the same units, they cannot be mixed in standard arithmetic operations. For instance in case of time and duration, *time –time* gives *duration*, *time + time* is not defined.

All varieties of possible solutions of the dimension problem can be roughly subdivided into two big classes according to the answer to the question: "Should the dimension information be mapped to different types?"

A positive answer would automatically provide requirements 1 and 2 with little or no efforts, because types are fully checkable at compile-time in Ada. Generics can be used to provide requirement 4 by transferring the dimension as a generic parameter.

In case of a negative answer other questions arise: "Can the method still allow requirements 1 and 2? Should the dimension information be a part of the type?" When it is not a part of the type, then we loose requirement 2 without any hope. In other words: To fully support requirement 4, one should probably have the dimension information as a type parameter.

The conflict between the requirements 1 and 2 on one side and requirement 4 on the other is the source of the problem of dimension handling.

Basically there are two ways to deal with physical dimensions: checking during *compile-time* and checking during *run-time*. The former is of course the preferred one since it prevents errors before they occur, whereas the latter can only detect errors after they have occurred. The following sections will present four methods, two using derived types and two using discriminants. Only the first one has had industrial use.

# 2    Compile-Time Methods

These methods use the Ada type concept in one way or another to discriminate between different dimensions. Mixing of dimensions is accomplished by operator overloading, so errors are detected during compile-time. The naïve method is to try something like the following, which every experienced Ada programmer knows to fail for several reasons (no documentation of the unit used for a given dimension, combinatorial explosion of overloaded operators, general powers and roots cannot be handled):

```
type Length is new Float;
type Time   is new Float;
type Speed  is new Float;
```

However, used with a grain of salt, this naïve method can be made to work.

## 2.1   From the Big Bang to the Universe

C. Grein [6] presents a method which works without dimension checking in expressions and assignments if only all variables belong to a coherent dimension system for which the SI unit system is chosen; but in critical cases where incoherent units are mixed, strong typing is applied. Having no dimension checking within the SI system is not such a big loss as should be clear from the above. The gain on the other hand is obvious documentation which unit is used for a given item's dimension [e.g. for speeds *m/s*, *km/h*, *knots*]. Mathematical functions exist as predefined operations for items in the SI system. This method has successfully been in use for more than a decade in several hard real-time avionics systems.

Starting point is a generic physical dimension package holding a private type and only dimension-preserving operations and which will be used solely for items not measured in the SI system, and a mathematical library. In a kind of creative act, a Big Bang, basic types are produced possessing all derivable operations that are desired.

```
with Math_Lib, Dimension;
package Big_Bang is
   -- A type with all mathematical operations:
   type Primeval_Float is digits project_defined;
   package Math is new Math_Lib (Primeval_Float);
   -- Take subprograms out of the instantiation to make
   -- them primitive operations on type Primeval_Float:
   function Sqrt (X: Primeval_Float)
      return Primeval_Float renames Math.Sqrt;
   -- A type without mathematical operations:
```

```
      type no_Math is digits project_defined;
      -- A stand-alone dimensioned type:
      package Primeval_Dim is new Dimension (no_Math);
      type Primeval_Unit is new Primeval_Dim.Unit;
   end Big_Bang;
```

Out of the Big Bang evolves the Universe containing all types we need. `Big_Bang` itself is no longer visible, i.e. it is strictly forbidden to use it in any context clause. We agree on the following convention: Within our application, all computations are done in only one coherent system of physical dimensions, the SI unit system. Any other unit will be converted to the corresponding SI unit.

```
   with Big_Bang;   -- This package is taboo for all times!
   package Universe is
      -- Type with mathematics:
      type SI_Unit is new Big_Bang.Primeval_Float;
      subtype Meter is SI_Unit;
      subtype Joule is SI_Unit;   -- further SI units
      -- Types without mathematics:
      type Kilometer is new Big_Bang.Primeval_Dim;
      function km_To_m (km: Kilometer) return Meter;
      function m_To_km ( m: Meter    ) return Kilometer;
      type Hour is new Big_Bang.Primeval_Dim;
      ...
   end Universe;
```

In this way, we can include semantic information into declarations, but *it is the programmer's own responsibility that formulae are physically correct*, no type checking is done!

```
   g: Meter_Per_Second_2 := 9.81;
   t: Second              := 10.0;
   s: Meter               := 0.5 * g * t**2;
```

And in critical cases, strong type binding prevents illegal mixing of types:

```
   Dist: Kilometer := +5.0;
   t := Sqrt (2.0 * Dist / g);   -- illegal
   t := Sqrt (2.0 * Meter (Dist) / g); -- illg. conversion
   t := Sqrt (2.0 * Kilometer_To_Meter (Dist) / g);   -- OK
```

The subtype `Without_Unit` is meant to be used for dimensionless items, i.e. for items with the dimension 1:

```
   Alpha: constant Without_Unit :=
             e**2 / (4.0 * PI * eps_0 * h_bar * c);
```

There are many units in the SI system, more than is probably sensible to define for a given application. The type name `SI_Unit` shall be used in these cases. To indicate the proper unit, we require that in the declaration, the unit shall then be given as a comment:

```
   Magnetic_Field: constant SI_Unit := 0.5E-4;   -- Tesla
```

We can summarise the features:

The method is cheap with respect to run-time; full mathematics is included; programmers are forced to use a coherent unit system; the unit is clearly indicated; in case of incoherent units, strong typing prevents mistakes. Of our requirements above, 2 is clearly fulfilled, 1 and 3 only partly.

## 2.2  Macks

*Macks* (an acronym of Meter/Mole, Ampere, Candela, Kilogram/Kelvin, Second) by Fraser Wilson [7] is an attempt to finesse the problem of writing enormous Ada specifications which cover every possible operator by generating them automatically. A specification file defines physical units - fundamental and derived. Each derived unit is defined as a set of products or ratios of other units. If more than one definition is provided, Macks performs dimensional analysis to check consistency. A vector type package may optionally be generated. Here an example of a simple Macks source file:

```
unit Kilogram is fundamental;
unit Meter    is fundamental;
unit Second   is fundamental;
unit Speed    is Meter / Second;
unit Accel    is Speed / Second;
unit Force    is Kilogram * Accel;
unit Speed_2  is Speed * Speed;
unit Energy   is Force * Meter or Kilogram * Speed_2;
```

Each definition results in a new floating-point type. All units have operators for multiplication and division by scalars generated, while automatically defined (by the Ada language) yet incorrect operations are declared abstract. For each derived unit definition, four operators are defined, e.g. for a *Derived* unit, defined as *Base_1 * Base_2*:

```
function "*" (L: Base_1 ; R: Base_2) return Derived;
function "*" (L: Base_2 ; R: Base_1) return Derived;
function "/" (L: Derived; R: Base_2) return Base_1;
function "/" (L: Derived; R: Base_1) return Base_2;
```

If the derived unit is defined in terms of a ratio, similar operators are defined. Of course, in the case of *Speed_2* above, only two operations are necessary.

The **or** lets you give alternative formulae for derived units, so in this case Energy would have a total of eight operators created. However (Kilogram * Speed) * Speed will not compile because there is no operator for it. An alternative would have been to have Macks automatically create these additional operators, so you could say

```
Energy = Mass * Length * Length / Time / Time
```

and get everything you need, but that pushes the limits of what is reasonable with a compile time solution. Macks also does not handle powers at the moment, because of the problems with mapping them to Ada.

Macks may be used for vector operations as well. If a declaration such as

```
vector Position is Meter * 3;
```

appears in a specification file, then vector addition, subtraction, dot product and (in the case of 3D vectors) cross product operations are generated.

Macks parses any text file. Macks' formal syntax in the Ada RM variant of Backus-Naur notation is as follows:

```
macks_file           ::= option_list declaration_list
option_list          ::= { option }
option               ::= option option_name = option_value;
option_value         ::= string_constant | identifier
declaration_list     ::= declaration { declaration }
declaration          ::= unit_declaration
                         | vector_declaration
unit_declaration     ::= unit unit_name is unit_definition;
unit_definition      ::= simple_definition
                         | compound_definition
                         | alias_definition
simple_definition    ::= fundamental
compound_definition  ::= term { or term }
term                 ::= unit_name operator unit_name
operator             ::= * | /
alias_definition     ::= also unit_name
vector_declaration   ::= vector vector_name is
                             vector_definition;
vector_definition    ::= unit_name * integer_constant
name                 ::= identifier
```

The reserved words are **also fundamental is option or unit vector**. The options are just a way of tailoring the output. The alias definition causes a subtype to be generated.

In summary, Macks mimics a hand-generated dimensioned unit package, but without the hard work and potential for error. Furthermore, it is relatively easy to tweak the generator to change the style of the unit handling; for example, the unit types, which are currently exposed as floating points, could be made private, thus avoiding the problems of implicit operations (though at the cost of losing literals).

Of the requirements above, Macks fulfils 1, 2 and 3. Requirement 5 is partially fulfilled: the fundamental units can be anything you like; however there cannot exist two fundamental units with the same dimension but different scale in one Macks source file, and derived units are always measured with the fundamental scale. Straightforward extensions to Macks would overcome this problem, and fulfil requirement 6 as well.

The Macks Ada source code is licensed under the GPL. Code generated by Macks can be used in any way you like.

# 3   Run-Time Methods

These methods add dimensions to numeric values as record components. Errors will be detected only during run-time when the error occurs. This approach has the shortcoming that a lot of numeric attributes are lost. You can't e.g. query 'First, 'Last, 'Small. Of course the method can provide substitutes. You also cannot provide such a type as a generic formal where a floating (or fixed) point type is expected.

Ada offers many opportunities to have a finer type relationship than simple *same* versus *other*. Having req. 4 in mind, one should consider the following possibilities:

> **Subtypes Constrained by Discriminants.** The discriminant may carry the information about the dimension. Generic programming is achieved by using unconstrained types.
> **Tagged Types.** The type tag may reflect the dimension. Generic programming is achieved by using class-wide operations.

The fundamental disadvantage of all such approaches is problems with requirement 1 and 2. Though neither a tag nor a discriminant is required at run-time if statically known, it is not warranted that the compiler would remove them from the object. In fact it does not, but see Hilfinger [4] who proposed compiler implementations that could optimise most of the run-time overhead away.

C++ provides, contrary to Ada, implicit instantiations of templates. If this feature were available in Ada, methods like those presented below could perform the checks already during compile-time. In fact, there was such a proposal for Ada9X by Shen and Cormack [8].

### 3.1   Units of Measurement for Ada

Dmitry A. Kazakov [9] bases his method on the seven SI units and handles dimensions as discriminants. The proposed solution is oriented mainly on the aspects of the man-machine interface. Thus it focuses rather on the requirement 4, than on 1 and 2.

The dimension Unit is a modular type with values corresponding to dimension formulae like $[L^{-3}MT^2]$. Hence the operations defined on Unit are **, *, / and sqrt. Only whole powers are allowed, so roots are restricted to even powers. For performance sake, all powers of the seven base units are packed into one value. Because Unit is a 32-bit number, the maximal powers are in the range -8 .. 7. All seven base units of SI there are defined constants of Unit.

The type Unit is used as a discriminant (named SI) of the unconstrained type Measure in the generic package Measures. Number is the generic formal type parameter, a floating-point type used to carry the values. Its unconstrained base type is used in arithmetical operations to avoid unexpected range problems. The components Gain and Offset implement a simple linear scale where (X.Gain+X.Offset) *X.SI is the value of the physical item X. Usually Offset is zero. It is used for shifted units like degree Celsius. Standard arithmetic operations are defined on Measure, which keep track of the actual dimension.

```
type Measure (SI: Unit := Units.Base.Unitless) is
   record
      Gain   : Number;
      Offset: Number := 0.0;
   end record;
```

Unconstrained variables of type `Measure` can be used for generic programming. Their actual dimensions can change, an undesirable effect in most cases. Therefore constrained subtypes of type `Measure` can be defined:

```
subtype Speed is Measure (Speed_Unit);
Car_Speed : Speed := 10.0*km/h;    -- OK
begin
Car_Speed := 1.0*A;    -- Illegal, Constraint_Error
```

The way the arithmetic operations are defined on `Measure` takes into account the values' scales. Values with zero offset form a field. They can be added and multiplied retaining the offset. Actually the offset should be another discriminant of `Measure`, but this is impossible in Ada. Values with non-zero offset form only a group with respect to addition. Multiplication is illegal and raises `Unit_Error`. An example of a value with non-zero offset is Celsius degree. $1°C+1°C=2°C$, but $1°C·1°C$ is illegal, because it belongs to another scale. Of course $1K·1K$ is legal and gives $1K^2$. Values of different offsets belong to different scales. They cannot be mixed: $1°C+1K$ is ambiguous, because the scale of the result is not defined.

Because of its man-machine interface orientation, the solution pays much attention to input / output and irregular units. The package `Measures_Irregular` provides some of the great variety of irregular units which are still widely used even in the countries which officially accepted SI. Examples are km/h, °C, rpm etc. In the USA, the first requirement will be: height in *feet* (in air), depth in *fathoms* (at see), distance in *yards* if short, in *miles* if long and in *inches* otherwise. For historical reasons, it is impossible to support all irregular units. Some of them have ill-defined conversion factors; another problem is that miles, pounds, gallons seem varying from country to country.

The package `Measures_Edit` provides conversions of the dimensioned values from and to a human-readable format. It supports all irregular units defined in `Measures_Irregular`. A dimensioned value is in most general form an expression involving numbers and units like 34.5*mm, 65km/h, 1yd².

What about the requirement 6? Though it is partially satisfied, the solution is of course not enough flexible. For example, `Meter*Foot` is legal. This might be desired in some cases and undesired in others. There is still no way to build a unit system based on feet instead of meters and provide conversions between them only when required. As a result, all computations are in effect made in the base SI units. This means that requirement 5 is also not satisfied.

The requirement 3 is also out of sight. The package `Units` could be of course reused to give discriminants to the user-defined private types. Alas array types may have no discriminants, so for them, even `Units` alone does not much help.

The source code is released under the GMGPL.

## 3.2   SI Units Checked and Unchecked

Full generality of physical equations requires rational powers of the basic units. The Gaussian CGS system is not tractable without. And although in the SI unit system all physical items have whole dimensions, intermediate results may have fractional ones, as the Schottky-Langmuir equation $j = \frac{4}{9}\varepsilon_0 \sqrt{\frac{2e_0}{m_0}} \frac{U^{\frac{3}{2}}}{d^2}$ shows.

C. Grein [10] presents a method that is very similar to Kazakov's, the major difference being that arbitrary rational powers of items are allowed, so that physical dimensions are handled in complete generality. It is also based on the seven SI units and includes any units derived from them as well as prefixes like kilo and milli. Other non-coherent units like foot or nautical miles can easily be added. Basic mathematical functions like exp and sine are also available.

The method comes in two variants: Dimensions as *simple components* or as *discriminants*. By changing only a few lines of code, dimension checking can be switched off in the final code (however big it may be), thus reducing the overhead of dimension checking to null. The source code is released under the GMGPL.

Discriminants as in Kazakov's method cannot be used because of the language limitation that discriminants must be discrete. Therefore for the moment we will do without discriminants. We call the package Unconstrained_Checked_SI and represent a dimensioned item by a private type composed of a dimension component and its numeric value. Constants denote the seven base units and the unit 1 and also prefixes like milli and Mega. Operators are defined like in Kazakov's package with the crucial difference that exponentiation is overloaded in allowing whole, rational and real exponents. The following is a rough outline of the package declaration (also the basic mathematical function are available, but omitted):

```
type Item is private;
One  : constant Item;   -- Base
Meter: constant Item;   --  Units ...
Mega : constant Real := 1.0E+6;  -- Prefixes ...
function "**"(B: Item; Exp: Whole  ) return Item;
function "**"(B: Item; Exp: Rational) return Item;
private
  type Dimension is record
    m, kg, s, A, K, cd, mol: Rational;
  end record;
  type Item is record
    Unit : Dimension;
    Value: Real;  -- a numeric generic parameter
  end record;
```

With this declaration, we are able to write in a very natural style

```
Dist: Item := 10.0 * Meter;
Time: Item;
g: constant Item := 9.81 * Meter / Second ** 2;
begin
  Time := Sqrt (2.0 * Dist / g);
```

and we can be sure that the *inner* dimensional consistency will be conserved. Erroneous additions like Dist+Time will inevitably lead to an exception. Unfortunately there is however no possibility to prevent erroneous assignments like

```
Length := Sqrt (2.0 * Dist / g);
```

Raising of an exception would only be possible if the declaration of `Length` included the dimension Meter, which we had to omit since dimensions are not discrete.

Beside the seven base units, there are a plethora of derived units like Joule etc. They are put into a child package which, like its parent, has of course to be generic.

```
Newton: constant Item := Kilogram*Meter/Second**2;
```

Another child for text file IO is available, taking into account the dimension. It knows all SI (base and derived) unit names and prefixes with their case-sensitive abbreviations (like mN for Millinewton), so that it can read a value like 1.0*km/s. For output, an additional optional parameter `Dim` defines the dimension string to be used. Further children for temperature scales like Celsius, polynomial operations, linear interpolation, second order curve approximation, and 3-dimensional vectors are also available.

`Item` is a private type, thus no subtypes with range constraints can be defined. Since one of Ada's strengths is just forcing range checks, a substitute is provided. A child package allows defining a new type `Constrained` with a range constraint and provides operations that make it completely compatible with `Item` so that virtually there is no difference to using Ada subtypes.

While this set of packages provides full internal consistency, an object's dimension may change. This is why the package is called *unconstrained*. The only way to prevent this is via discriminants, and thus we have to swallow the bitter pill and, for each dimension, split numerator and denominator and use them separately as discriminants! A dreadful thought. Where is the simplicity of notation?

Now, the situation is not so bad. Let Ada grab into her wizard's bag and she'll turn up with so-called unknown discriminants denoted by `(<>)` – and the ugliness is hidden. We rename our package into `Constrained_Checked_SI` and, in the visible part, only change the declaration of `Item` a bit:

```
type Item (<>) is private;
```

That is all we have to do. With these unknown discriminants, the user is forced to complete every declaration with an initial value; the discriminants are taken from it and are unchangeable thereafter:

```
Length: Item := Meter;
Length := Sqrt (2.0 * Dist / g);
```

now inevitably leads to an exception. (To be honest, we have to admit that

```
Length: Item := Sqrt (2.0 * Dist / g);
```

is still possible, where `Length` takes the dimension `Second` from the initial value.)

The completion of the declaration of `Item` in the private part looks as follows, where `x_N` stands for the numerator, `x_D` for the denominator of dimension x:

```
subtype Pos is Whole range 1 .. Whole'Last;
type Item(m_N, kg_N, ... mol_N: Whole;
          m_D, kg_D, ... mol_D: Pos) is record
   Value: Real;
end record;
```

Outside, nearly everything remains unchanged, and the above example looks like this:

```
Dist: Item := 10.0 * Meter;
Time: Item := Second;
g: constant Item := 9.81 * Meter / Second ** 2;
begin
Time := Sqrt (2.0 * Dist / g);
```

One has to get accustomed to this notation, that is true, but for safety critical applications, initialisation of every variable declaration often is a mandatory requirement, coming from the experience that uninitialised variables lead to mistakes which are difficult to detect – and after all, it is increased safety we are after with all our effort.

Of course we have to pay for the discriminants: They are space consuming since for each item's dimension, we store 14 integers – and rational arithmetics is very run-time consuming. Thus application of this method under hard real-time conditions might seem prohibitive. This is not the case. We start out from the following considerations. A real-time system is generally developed on a so-called host computer before being applied on the target computer. Also the unit tests are virtually always performed on the host. There the correct real-time performance is irrelevant; it cannot work on multi-user time-sharing systems anyway. So during development including unit test, the method presented above is used.

Only when the program is ported to the target are the dimensions switched off. To this end, another package Unchecked_SI is defined, which has the identical visible specification as one of the packages above, but in the private part the dimensions are removed, only the item's numeric value remains; all constants like Meter become the pure value 1, so they can easily be optimised away. In the package body, the dimension arithmetic is completely removed. If on the host 100% unit test coverage is reached, at the same time also dimensional correctness of all statements is proved – the run-time dimension checking can be switched off without detrimental effect.

```
package Unchecked_SI is
   type Item (<>) is private; -- or without discriminant
   ... -- The visible part is unchanged.
private
   type Item is record
      Value: Real;
   end record;
   One  : constant Item := (Value => 1.0);
   Meter: constant Item := One;
end Unchecked_SI;
```

It is remarkable that the completion of Item in the private part does not have a discriminant. Nevertheless outside everything remains unchanged; declarations without initial value are still illegal when a discriminant is specified in the visible part.

How do we proceed when transiting from the host to the target? We will demonstrate it with the following sample program:

```
with Constrained_Checked_SI;
package SI is new Constrained_Checked_SI (Float);

with SI;   use SI;
procedure Test_SI is
   g    : constant Item :=  9.81 * Meter / Second**2;
   Dist:           Item := 10.0  * Meter;
   Time:           Item := Second;
begin
   Time := SQRT (2.0 * Dist / g);
end Test_SI;
```

All we have to do is change the instantiation of the package SI from Unconstrained_Checked_SI respectively Constrained_Checked_SI to Unchecked_SI since the visible part of both packages is the same; the rest of the application remains completely unchanged, however big it may be.

Of our requirements above, although 1 is clearly violated, 2 is fulfilled in a sense; also 3 and 4 are fulfilled.

# 4   Conclusion

In Ada 95, full support of dimensional arithmetics can only be achieved with run-time solutions. Since the Ada0Y process is just reaching its decisive phase, the considerations presented above could perhaps serve WG9 as an impetus to think about ways to ease handling physical items in the next language version.

# References

[1] Do-While Jones, Dimensional Data Types, Dr. Dobb's Journal of Software Tools, 50–62, May 1987
[2] Do-While Jones, Ada in Action, John Wiley & Sons, Inc., 1989
[3] N.H. Gehani, Ada's Derived Types and Units of Measure, Software – Practice and Experience, Vol. 15(6), 555–569, June 1985
[4] P.N. Hilfinger, An Ada Package for Dimensional Analysis, ACM Transactions on Programming Languages and Systems, Vol. 10(2), 189–203, April 1988
[5] H.J. Schneider, Physikalische Maßeinheiten und das Typkonzept moderner Programmiersprachen, Informatik-Spektrum (1988) 11: 256–263
[6] C. Grein, Vom Urknall zum Universum, Ada Aktuell 1.1 (März 1993)
    German: http://home.T-Online.de/home/Christ-Usch.Grein/Ada/Universum.html
    English: http://home.T-Online.de/home/Christ-Usch.Grein/Ada/Universe.html
[7] F. Wilson, Macks, http://www.blancolioni.org/ada/macks

[8]  J. Shen, G.V. Cormack, Automatic Instantiation in Ada, ACM 0-89791-445-7/91/1000-0338 $1.50
[9]  D.A. Kazakov, Units of Measurement for Ada
     http://www.dmitry-kazakov.de/ada/units.htm
[10] C. Grein, Physikalische Dimensionen in Ada, Softwaretechnik-Trends, Band 22 Heft 4, November 2002, http://home.T-Online.de/home/Christ-Usch.Grein/Ada/SI.html

# Charles: A Data Structure Library for Ada95

Matthew Heaney

On2 Technologies, Inc.
mheaney@on2.com
http://home.earthlink.net/~matthewjheaney/index.html

**Abstract.** Charles is a container library for Ada95, modelled closely on
the C++ STL [1,2,3,4]. Sequence containers (vectors, deques, and lists)
store unordered elements, inserted at specified positions. Associative containers (sets and maps) order elements according to a key associated with
each element; both sorted (tree-based) and hashed containers are provided. A separate iterator type [5,6] associated with each container is
used to visit container items and to allow elements to be modified directly. Charles is flexible and efficient, and its design has been guided by
the philosophy that a library should stay out of the programmer's way.
Charles is for getting your work done.

## 1 Introduction

In general, the Charles library is designed using static mechanisms (generic packages and subprograms) rather than dynamic ones (tagged types). "Static polymorphism" is an important programming idea that is often undervalued. A set
of loosely-related but otherwise disparate types each have an identical interface,
and you compile against this "abstract" interface. By programming to the interface instead of to a specific type, then it becomes relatively easy to change the
type behind the interface.

An important goal in the design of libraries is that abstractions should be
safe and easy to use. Therefore, in Charles memory management for (unbounded)
containers is automatic, by implementing container types as a private derivation
of type `Controlled`. The fact that container types are (privately) tagged also
means that equality composes, so it is not possible for predefined equality to ever
reemerge (satisfying this constraint is necessary because the library is deliberately designed to facilitate the composition of one component from another).

However, in library design there is often a tension between flexibility and
safety. In general, where there has been conflict between these goals, in Charles
it has been resolved in favor of flexibility. This does not imply that there should
be no safety at all, but rather that it is best effected by the client himself, via
some mechanism *outside of* the library, at a higher lever of abstraction. Systems
are built from the bottom up.

Another philosophy in the design of Charles is that it is *easy* to do common
things, and *possible* to do less-common things (perhaps not as easily, but possible
nonetheless). For example, creating a container type whose elements are non-
limited and definite (the common case) requires only a single instantiation (the

J.-P. Rosen and A. Strohmeier (Eds.): Ada-Europe 2003, LNCS 2655, pp. 271–282, 2003.
© Springer-Verlag Berlin Heidelberg 2003

common case is easy). Generic formal subprograms always have defaults ("**is** <>"), which simplifies instantiations (especially for generic algorithms) because matching actuals directly visible at the point of instantiation do not need to be passed explicitly. Container types are non-limited, because this makes it easier to compose abstractions (and not because we need assignment, which can be effected via other mechanisms). In general the library is designed so that a container type (or its associated iterator type) can be used directly as the generic actual element type of another container instantiation.

One property of a good user interface is that a user should be able to guess what something does, and his guess should always be right. Therefore type names and coding style closely follow long-established RM conventions. In particular, operation and parameter names were largely inspired by the **Unbounded_String** package. For type names Charles follows the convention of the predefined I/O packages (each of which use **File_Type** as the name of the abstraction) and uses the name **Container_Type** for all of its containers. The symmetry with files is deliberate, because it emphasizes that files are essentially persistent containers.

Note that the actual container types are (publicly) declared as traditional abstract data types, not as tagged types. A theme of the library is that you create a complex abstraction by composing generic components, not by extending a tagged type. It is not necessary for the library to provide dynamic polymorphism for containers, because a client can add that himself using a thin layer (a "type adapter" [5]) of glue code.[1]

A separate type — an *iterator* — is associated with each container. An iterator is a mechanism for manipulating elements in a container. There are operations for navigating among elements, and for selection and modification of the element currently designated by the iterator. An important feature of this schema is that it allows you to view the collection as a just a "sequence of elements," thus abstracting-away the actual container. This is valuable because now a generic algorithm (for sorting, say) can be written in terms of just an iterator, so that you can use the algorithm over any container having an iterator with the requisite operations. Iterators also obviate the need for special container conversion operations, e.g. that return an array object with all the container elements. The iterator mechanism is sufficiently general that you can convert from any container type (including arrays) to any other container type (including arrays).

## 2    Sequence Containers

The elements in a sequence container are unordered, and may be inserted at specific positions. The family of sequence containers comprises vectors and queues, for constant-time random access of elements, and lists, for efficient insertions and deletions in the middle of the sequence.

---

[1]  For this reason I specifically rejected the approach used the by Weller/Wright Ada95 Booch components [7], which implement containers as a family of tagged types.

Charles does not have "stack" or "queue" containers because you can achieve the same effect by simply pushing and popping items from the appropriate end of one of the more general sequence containers. If a container is needed that, say, disallows access to the elements in the middle of the sequence, then the user can implement that himself as a thin layer on top of a primitive component.

## 2.1   Vectors

The canonical container is the vector, which is simply a linear sequence of items. A vector allows random access of elements (with only *unit* time complexity), and is optimized for insertions at the back end of container. For example:

```
V : Vector_Subtype;
Push_Back (V, New_Item => X);
```

An unbounded vector is implemented as a pointer to an internal array, that automatically grows as necessary to accommodate new items. When more storage is necessary, a new array is allocated that is two times the length of the existing array. (Array growth is always a function of internal size rather than the number of active elements in the vector. If it were otherwise, then insertion complexity would be quadratic rather than linear.)

Insertion at other positions within the vector is permitted:

```
Insert (V, Before => I, New_Item => X);
```

However, the time complexity is linear, because all the elements from the insertion position to the back of the vector have to slide up to make room for the new item. This is why there is no `Push_Front` operation for vectors (although the effect of such an operation can achieved by `Insert`'ing an item before the position `Index_Type'First`).

The key benefit of a vector is that it supports random access of elements in constant time:

```
X := Element (V, Index => I);
Replace_Element (V, Index => J, By => Y);
```

Here we have a selector function that returns the value of the element at the indicated position. However, returning a copy of the element is not a sufficiently general mechanism, as efficiency issues might prohibit copying. And what about modifying the element itself? Consider a hash table implemented as vector of lists – how does a client add an item to the list at a certain index position?

To satisfy this need, all the containers in Charles have an additional generic selector function that accepts an access type as a generic formal type, and which returns an access value designating the element. This gives us a variable-view of the actual element object, which we can then manipulate directly:

```
function To_Access is new Generic_Element (List_Access);
   List : List_Subtype renames To_Access (V, Index => I).all;
begin
   Push_Back (List, New_Item => X);
```

Although a vector automatically resizes as necessary during insertion, if you know the ultimate size of the vector *a priori* then it's always more efficient to perform the allocation in advance. The vector has factory functions for constructing the container at the time of its declaration:

```
procedure Op (N : Natural) is
   V : Vector_Subtype := To_Container (Length => N);
begin
```

There is also a `Resize` operation to explicitly specify the size of the internal array.

## 2.2   Deques

A "deque" (short for **d**ouble-**e**nded **que**ue) also supports random access of elements, but unlike a vector insertion of an element at the front of the container has only *constant* time complexity. It is implemented as an unbounded array of pointers to fixed-size blocks, with each page comprising a fixed number of elements. `Push_Front` works by adding new elements to the first page (working from back to front on the page), and decrementing an offset that keeps track of which element on the page is the first element of the container. When there is no more room on the front page, then a new page is allocated and the offset is reset. The internal array expands as necessary to address all the internal pages. `Pop_Front` works similarly, except that it increments the offset.

As with a vector, the deque allows elements to be inserted (deleted) at any position, but the time complexity is still linear. One difference from a vector is that to make room for the new element, the existing elements are moved toward the end of the deque that is closest to the insertion position.

## 2.3   Lists

Lists are optimized for insertions at any position. All insertions and deletions, in the middle or at either end, have constant time complexity. However, random access is not supported, and advancing from one position to another has a time complexity proportional to the distance between positions.

Rather than referring to elements using a discrete index (as for vectors and deques), for lists we use a separate iterator type:

```
I : constant Iterator_Type :=  Last (List);
E : Element_Type renames To_Access (I).all;
```

There are dedicated operations for pushing an item onto the front or back of the list. To insert an item at a certain position, we specify the position using an iterator value:

```
Insert (List, Before => Iterator, New_Item => Item);
```

The find operation returns an iterator value, which we can test against the distinguished Back iterator value to determine whether the search was successful:

```
I : constant Iterator_Type := Find (List, Item);
begin
   if I /= Back (List) then ... --search succeeded
```

If you need to move elements from one list onto another, a list is optimal because it doesn't require any copying. Instead, only pointers need to be manipulated (the internal node of storage is moved from one list to another, and the respective element counts are adjusted accordingly). For example, calling Splice like this:

```
Target, Source : List_Subtype;
I, J : Iterator_Type;
...
Splice (Target, I, Source, J);
```

moves (*not* copies) the item designated by iterator J in list container Source, onto the list Target just prior to the element designated by iterator I. Additional overloadings of Splice generalize this to move an entire range of elements.

The list container also provides operations for reversing the list, sorting the elements, removing duplicates, filtering, and for merging two lists. The sort operation is stable (the relative order among equivalent items is preserved across a sort), and works by exchanging pointers to nodes, not by copying elements. For example:

```
   function "<" (L, R : ET) return Boolean is ...;
   function Sort is new Generic_Quicksort; --"<" is default
   function Is_Same (E1, E2, : ET) return Boolean is ...;
   function Purge_Duplicates is new Generic_Unique (Is_Same);
begin
   Sort (List);
   Purge_Duplicates (List);
```

The library provides both singly-linked and doubly-linked list containers. Both forward and reverse iteration over a double list is provided, but only forward iteration is possible for single lists. However, if all you need is forward iteration, then the singly-linked list is more space efficient because the overhead per node is smaller. The single list also caches a pointer to the last node, so that pushing an item onto the back of the list can be performed in constant time. (It would be linear otherwise, because it would necessary to traverse the list from the front, just to locate the last item. [4])

Note that the list containers are "monolithic" data structures, meaning that there is no structure sharing (except what is implied by the use of iterators). List assignment is by value, not by reference. The intent of the name "list" is to emphasize that this container is optimized for constant-time insertion and deletion of elements at arbitrary positions. The semantics of this abstraction are *not* the same as for the similarly-named "polylithic" [6] data structure in functional languages as LISP.

# 3    Associative Containers

In addition to sequence containers, the Charles library has "associative" containers that associate a "key" with each element, and which order the keys for efficient retrieval. For a set container an element is its own key, but a map container allows you to index an item by some other arbitrary type (in this sense a map is a generalization of an array).

Both hashed and sorted associative containers are provided. A hashed container has possibly linear time complexity in the worst case, but the average time complexity is *constant*. A sorted container guarantees *logarithmic* time complexity even in the worst case.

In Charles the sorted containers are implemented as red-black trees [8], but this does not preclude other choices — an AVL tree could have been used just as easily. Indeed, multiple implementations would be possible. However, all the components have an identical[2] container interface (because of static polymorphism), so once a component (whether sorted or hashed) has been instantiated, the implementation more or less disappears.

A hashed associative container orders keys by comparing hash values for equality. Keys that hash to the same value are stored in the same bucket, which is implemented as a singly-linked list. (This implies that reverse iteration is not available for hashed containers.) Keys for a sorted associated container are compared using an *equivalence* relation, not by equality. This is called "strict weak ordering." You can think of equivalence as being implemented this way:

```
function Is_Equivalent (L, R : Key_Type) return Boolean is
begin
   return not (L < R) and then not (R < L);
end;
```

To store an item in a sorted associative container the client must provide an order relation for the keys. The canonical example would be a set of employee records, ordered by social security number:

```
type Employee_Type is
   record
      SSN : SSN_Type;
      ...
   end record;

function "<" (L, R : Employee_Type) return Boolean is
begin
   return L.SSN < R.SSN;
end;
```

---

[2] Strictly speaking they aren't really identical, because there are operations that make sense for some implementations, but not others. For example, a hashed container has a **Resize** operation to specify the size of the hash table, but this obviously wouldn't be applicable for a tree-based implementation.

```
package Employee_Sets is
   new Charles.Sets.Sorted.Unbounded (Employee_Type, "<");
```

Now you can add a new employee to the database like this:

```
E : Employee_Type := ...;
begin
   Insert (Set, New_Item => E);
```

and it will be inserted according to the value of E.SSN. If we want to change some information for an employee (let's say his address changed), then we can use Find (really, an instantiation of Generic_Find), which returns an iterator designating the employee record object:

```
I : constant Iterator_Type := Find (S, Key => SSN);
E : Employee_Type renames To_Access (I).all;
begin
   E.Address := New_Address;
```

There is only a subtle difference in how sets and maps are implemented. For a set, the element is its own key, and the set doesn't allocate space for anything other than the element. For a map, the key is separate piece of data, and the map is implemented as key/element pairs. In either case storage is ordered according to the key (per the relational operator for sorted containers, or the hash value for hashed containers).

Note that equality for maps is implemented in terms of the equality operator for elements; keys do not participate in determining whether two maps are equal. Set equality is also defined in terms of the element equality rather than the relational operator used to compute equivalence.

In order to store a key in a map, it must have a definite subtype. One issue is that some keys are more naturally represented using indefinite subtypes, e.g. a name having type String. To use the map the client would first have to convert the key to the definite subtype used to instantiate the package, and then call the map operation. However, this is neither syntactically graceful, nor very efficient (temporaries are created only to be immediately destroyed). For these reasons, and because maps with string keys are nearly ubiquitous, the library provides dedicated map abstractions having type String as the key. The string-key maps also accept a key comparison operation as a generic formal subprogram; for example this would allow key comparisons to be case-insensitive.

Except for Storage_Error, the associative containers do not raise exceptions. The search operation, for example, does *not* raise an exception if the key was not found. Rather, the value of the Back iterator is returned as a nonce, to indicate failure. For example:

```
I : constant Iterator_Type := Find (Map, Key);
begin
   if I /= Back (Map) then ...
```

A similar approach is used for insertions into unique-key maps, and in particular insertion of a key into a map that already contains the key *does not raise an exception*. Rather, a Boolean value is returned indicating whether the insertion was successful:

```
declare
   Iterator : Iterator_Type;
   Success  : Boolean;
begin
   Insert (Map, Key, Item, Iterator, Success);
   if Success then ...
```

If Key is already in the Map, then Success returns False, and the iterator designates the key/element pair whose key matches Key. Otherwise Success returns True and the iterator designates the newly-inserted pair.

Conditional insertion is a necessary feature of any efficient[3] map abstraction. Consider an application that counts the frequency of words in a text file, implemented as a map (the histogram) with key type String and element type Natural. The algorithm can be implemented by simply always trying to (conditionally) insert a count of 0 for a word:

```
procedure Insert (Word : String) is
   Iterator : Iterator_Type;
   Success  : Boolean;
begin
   Insert (Histogram, Word, 0, Iterator, Success);

   declare
      Count : Natural renames To_Access (Iterator).all;
   begin
      Count := Count + 1;
   end;
end Insert;
```

Here we don't care about the whether the insertion was successful; our interest is only that the insertion be *conditional*. If the word is already in the map, the insertion fails, and we simply increment its current count value. If the word was not already in the map, the insertion succeeds, and the associated count is initialized to the value 0 (and then immediately incremented).

For many applications an unconditional replacement of the existing value is appropriate, analogous to replacing the value of a component of an array. The associative containers overload the `Insert` operation with versions that either create a new element if the key does not exist or simply replace the current element if it does.

---

[3] It's efficient because the test to determine whether the key is already in the map, and the insertion if it isn't, is an atomic operation, and therefore the tree traversal is only done once.

Multiset and multimap containers allow multiple keys to be equivalent. Therefore there is no conditional version of Insert because all insertions are *unconditional*. Unless memory is exhausted, insertions always succeed, by allocating a new key/element pair in the map. Elements with equivalent keys are stored in adjacent positions, as a contiguous range. To interrogate elements whose keys are the "same," operations return a pair of iterators designating the ends of the range. For example, suppose we want to look up all the people whose last name is "Heaney":

```
Iter, Back : Iterator_Type;
begin
    Equal_Range (People_Map, "Heaney", Iter, Back);

    while Iter /= Back loop ...
```

Here, Equal_Range returns an iterator pair specifying a half-open range of elements; this permits iteration in the normal way. The Lower_Bound and Upper_Bound operations can be used to return just the respective endpoint.

It is a necessary feature of any general-purpose container library that it provide a way to modify container elements in-place. However, for an associative container, we must take special care to ensure that a key is not modified accidently. This would break the abstraction, which depends on key values being in equivalence order. For a set, the Generic_Element operation imports the access type as access constant rather than access all, so it returns a constant view of the object instead of a variable view, and thus prevents the set element from being modified. The Generic_Key operation for the map similarly returns only a constant view. (Note that map correctness only depends on keys, not elements, so its Generic_Element operation doesn't need any special treatment.)

However, we accept that some applications have a legitimate need to modify set elements. In our employee set example, we modified the element by changing the address component of the employee record. That was a perfectly safe (and reasonable) change, because we didn't modify the key. To accommodate this need, the set container provides a special operation, Generic_Modify_Element, that imports the generic formal access type as access all in the normal way, and which therefore returns a variable view. The spelling is deliberately different, so that if the container instantiation is changed to a set from some other component, then any elements that had been modified via Generic_Element (as for other containers) would be immediately flagged at compile time. The function To_Access that we showed in the example was an instantiation of the special operation. There is a corresponding operation for the map container, Generic_Modify_Key, to admit benign changes to key values.

If you do need to change a key, you *must* delete the element from the container first, make the change and then re-insert the element. For example:

```
I : Iterator_Type := Find (Set, Key => SSN);
E : Employee_Type := Element (I); --make a copy
begin
```

```
Delete (Set, Iterator => I);
E.SSN := New_SSN;
Insert (Set, New_Item => E);
```

The hashed associative containers import a hash function as a generic formal subprogram. Charles provides hash functions for common types as String and Integer. The containers are implemented as an unbounded array of buckets (the hash table), with each array component a linked list of elements. The array itself always has a length that is a prime number, which produces a better scatter when mod'ing the hash value.

The hash table automatically resizes itself as items are inserted, in order to maintain a load factor of 1; this is how it can guarantee that average time complexity is constant. When the load factor is exceeded, a bigger array is allocated and the existing elements are rehashed onto the new hash table. This means that occasionally (whenever the array needs to grow) there will be spikes in the execution time of insert.[4] However, as with a vector, there is a resize operation that allows you to preallocate the buckets array to a specified size.

## 4   Iterators

Containers are important to the extent that we care how quickly we can find an element in the container, or how efficient it is to insert an element, but ultimately it is the *elements* in the container in which we are interested. An iterator is a structured way of gaining access to the elements, without exposing the representation of the container.

A reusable abstraction should be as flexible, efficient, and safe as possible, but these goals are often in conflict. This tension is particularly acute in the design of iterators, because they essentially violate the encapsulation provided by the container in order to allow access to its elements. How type safe should an iterator be? Should a container keep track of all the iterators currently designating items in the container? What should happen if you try to remove an item from the container while it's being designated by an iterator? Charles doesn't try to solve these problems, as that would have grossly compromised the design. Where there has been tension among these goals, I have in general traded type safety for flexibility and efficiency. A client can always build a safe layer on top of a flexible and efficient abstraction, but the opposite is not true.

I prefer to program close to the machine, and therefore iterators have been closely modelled on raw access types. In particular, to be sufficiently general and flexible the iterator type *must* be non-limited and definite. To discover what iterator operations are needed, we can examine how access types are used to traverse a simple linked list:

---

[4] It's not clear whether this would be acceptable in real-time programs, which demand predictable behavior and must be designed around worse case execution times. There is another technique called "incremental hashing" that can be used to smooth out the spikes, and I'm currently debating whether to implement hash tables this way instead.

```
declare
   Node : Node_Access := List.Head;
begin
   while Node /= null loop
      Process (Node.Element);
      Node := Node.Next;
   end loop;
end;
```

From this example we can extrapolate what operations need to be provided by a high-level container:

- An operation to return an iterator designating the "first" element.
- An operation to return a sentinel, to which the iterator is compared to determine whether it has "fallen off the end" of the sequence.
- An operation to inspect the "current" element designated by the iterator.
- An operation to return an iterator designating the "next" element.

If we implement an iterator as a thin wrapper around an access value, this allows us to hide the representation of the container, but preserves the flexibility and efficiency of raw access types. We can then rewrite the example like this:

```
declare
   I : Iterator_Type := First (List);
   B : constant Iterator_Type := Back (List);
begin
   while I /= B loop
      Process (Element (I)); --or Process (To_Access (I).all)
      I := Succ (I);
   end loop;
end;
```

This schema can be generalized to work for any container besides a list, and this is in fact how iterators are implemented in Charles. The fact that the iterator is non-limited and definite means you can implement any other type in terms of the iterator. This is important because real systems are built from the bottom up, by assembling complex abstractions from simpler primitives.

The library has been designed to be flexible, so it is easy to compose abstractions (one container can have another container as its element type), and this is no less true for iterators (so you to can have a container of iterators, too). Consider a set, which orders its elements according to the relation specified when the set package was instantiated. Suppose we want to sort the elements according to some other criterion? One way would be to copy the elements into a list, and then sort the list. However, if the elements are large, or there are a large number of elements, efficiency issues might prohibit this approach. A more efficient technique would be to create a list of set iterators, that designate the elements in the original set, rather than creating a list of set elements. (This assumes, of course, that set iterators are small compared to set elements.) You could then sort the list of iterators according to whatever criterion you desire, irrespective of how the original elements are ordered in the set.

## 5   Conclusion

Charles satisfies the need for a general-purpose container library. It tries to find the right balance between flexibility and ease of use, and resolve the tension among the orthogonal goals of efficiency, safety, and flexibility. Writing complex abstractions that have even modest container needs is a lot simpler with Charles than without.

## References

1. Plauger, P.J., Stepanov, A.A., et al.: The C++ Standard Template Library. Prentice Hall PTR (2001)
2. Musser, D.R., Derge, G.J., Saini, A.: STL Tutorial and Reference Guide. 2nd edn. Addison-Wesley Publishing Company (2001)
3. Stroustrup, B.: The C++ Programming Language. 3rd edn. Addison Wesley Longman, Inc. (2000)
4. Austern, M.H.: Generic Programming and the STL. Addison Wesley Longman, Inc. (1999)
5. Gamma, E., et al.: Design Patterns. Addison-Wesley Publishing Company (1995)
6. Booch, G.: Software Components With Ada. The Benjamin/Cummings Publishing Company, Inc. (1987)
7. Weller, D., Wright, S.: Ada95 Booch Components. World Wide Web, http://www.pushface.org/ (1995)
8. Cormen, T.H., Leiserson, C.E., Rivest, R.L.: Introduction To Algorithms. The MIT Press (1990)

# A Quality Model for the Ada Standard Container Library*

Xavier Franch and Jordi Marco

Dept. Llenguatges i Sistemes Informàtics, Universitat Politècnica de Catalunya
c/ Jordi Girona 1-3 (Campus Nord, C6) E-08034 Barcelona (Catalunya, Spain)
{franch,jmarco}@lsi.upc.es

**Abstract.** The existence of a standard container library has been largely recognized as a key feature for improving the quality and effectiveness of Ada programming. In this paper, we aim at providing a quality model for making explicit the quality features (those concerning functionality, suitability, etc.) that determine the form that such a library might take. Quality features are arranged hierarchically according to the ISO/IEC quality standard. We tailor this standard to the specific context of container libraries, by identifying their observable attributes and establishing some tradeoffs among them. Afterwards, we apply the resulting model to a pair of existing container libraries. As main contribution of our proposal, we may say that the resulting quality model provides a structured framework for (1) discussing and evaluating the capabilities that the prospective Ada Standard Container Library might offer, and (2) analyzing the consequences of the decisions taken during its design.

## 1 Introduction

Most important object-oriented programming languages include some standard libraries of reusable components as part of their definition. Among them, we focus on container libraries. A *container* (also known as *collection*) may be defined as an object that contains (i.e., stores) other objects. Some examples of containers are sets, maps and sequences. Object-oriented programming languages that offer container libraries are: Java, with the Java Collections Framework (JCF) [1]; C++, with the Standard Template Library (STL) [14]; and Eiffel, with the Eiffel Base Library [10].

Unfortunately, this is not the case for the Ada language, in spite of various attempts and claims in this direction, among which we mention:

- Some existing widespread container libraries, such as the Booch Components [3] and the Charles Container Library [4].
- Some events, such as the *Standard Container Library for Ada* workshop held during the Ada Europe 2002 conference.

---

* Work partially supported by the Spanish research project CICYT TIC2001-2165.

J.-P. Rosen and A. Strohmeier (Eds.): Ada-Europe 2003, LNCS 2655, pp. 283–296, 2003.

- Wide initiatives, such as the Application Standard Components Library [2] or the prospective Working Group mentioned during the workshop above.
- Some opinions and claims, such as those in the `comp.lang.ada` discussion list or the *ACM SIGAda* Chair Message for March 2002 *Ada Letters*: "Such a [Container] Library could be an excellent addition to the Ada International Standard".

Needless to say, the existence of a standard container library for Ada would clearly contribute to the quality of the final Ada artifacts and the effectiveness of the software development process itself. For this reason, and also because once deployed, its modification would be certainly difficult and should be avoided whenever possible, having a comprehensive, structured and precise framework for assessing the design and implementation of this library becomes utterly important.

With this objective in mind, we present in this paper a quality model aimed at making explicit all the quality criteria (e.g., reliability, functionality, efficiency etc.) that should be considered when building the Ada Standard Container Library. We use the ISO/IEC 9126-1 quality standard [9] as initial point. It is a very general standard, and so the main goal of the paper is tailoring the quality model proposed therein to the specific context of container libraries and also showing its use by means of the evaluation of two existing container libraries.

The rest of the paper is structured as follows. Section 2 presents the ISO/IEC 9126-1 quality standard, which introduces six groups of software characteristics that will drive the discussion. Sections 3 and 4 analyze some of the form that these characteristics take in the domain of container libraries, and Sect. 5 discuss their relationships. Section 6 applies the resulting quality model to two existing container libraries, an Ada one (Booch Components) and a standard library from other programming language (Java Collection Framework). Finally, Sect. 7 provides the main conclusions of our work.

## 2    The ISO/IEC 9126-1 Quality Standard

The ISO/IEC Quality Standard 9126-1 [9] provides a good framework for determining a quality model for a given domain of software components. An ISO/IEC-9126-1-based *quality model* is defined by means of general *characteristics* of software, which are further refined into *subcharacteristics*, which in turn are decomposed into *attributes*. Attributes collect the properties that software components exhibit. Intermediate hierarchies of subcharacteristics and attributes may appear making thus the model highly structured.

The ISO/IEC 9126-1 standard fixes six top level characteristics: functionality, reliability, usability, efficiency, maintainability and portability (see Table 1). It also fixes their further refinement into subcharacteristics but does not elaborate the quality model below this level, making thus the model flexible. The model is to be completed based on the exploration of the particular software domain and its application context; because of this, we may say that the standard is very versatile and may be tailored to domains of different nature, such as the one of container libraries.

**Table 1.** ISO/IEC 9126-1 characteristics and subcharacteristics

| Functionality | |
|---|---|
| suitability | presence and appropriateness of a set of functions for specified tasks |
| accuracy | provision of right or agreed results or effects |
| interoperability | capability of the software product to interact with specified systems |
| security | prevention to (accidental or deliberate) unauthorized access to data |
| compliance | adherence to functionality-related standards or conventions |
| **Reliability** | |
| maturity | capacity to avoid failure as a result of faults in the software |
| fault tolerance | ability to maintain a specified level of performance in case of faults |
| recoverability | capability of reestablish level of performance after faults |
| compliance | adherence to reliability related standards or conventions |
| **Usability** | |
| understandability | effort for recognizing the logical concept and its applicability |
| learnability | effort for learning software application |
| operability | effort for operation and operation control |
| attractiveness | capability of the product to be attractive to the user |
| compliance | adherence to usability related standards or conventions |
| **Efficiency** | |
| time behavior | response and processing times; throughput rates |
| resource utilization | amount of resources used and the duration of such use |
| compliance | adherence to efficiency related standards or conventions |
| **Maintainability** | |
| analysability | identification of deficiencies, failure causes, parts to be modified, etc. |
| changeability | capability to enable a specified modification to be implemented |
| stability | capability to avoid unexpected effects from modifications |
| testability | capability to enable for validating the modified software |
| compliance | adherence to maintainability related standards or conventions |
| **Portability** | |
| adaptability | opportunity for adaptation to different environments |
| installability | effort needed to install the software in a specified environment |
| co-existence | capability to co-exist with other independent software in a common environment sharing common resources |
| replaceability | opportunity and effort of using software in the place of other software |
| compliance | adherence to portability related standards or conventions |

In our paper, we are interested in considering the model as a means for obtaining quantitative measures of the attributes of the final product. Qualitative metrics would be defined later, when considering how the different attributes should be weighted for taking into account the requirements over the Ada Standard Container Library.

In the next sections, we adapt this quality model to the domain of container libraries. To do so, we apply a methodology presented elsewhere [6] that extends the hierarchy and makes explicit the relationships between the different quality features.

# 3    Quality Attributes for Functionality

Functionality is probably the most relevant quality characteristic in the domain of container libraries. Success of the Ada Standard Container Library requires exhibiting the appropriate functionality once considered its design requirements. It should be noted that "appropriate" does not necessarily mean "exhaustive", because an excess of functionality would impact negatively in other criteria such as usability or operability. Tradeoffs among quality factors, which are part of the quality model, would help to make this clear, see Sect. 5.

This section is structured into five subsections, one for each functionality subcharacteristic. For each subcharacteristic, we present a table with the attributes that play a part on them, together with some explanation.

## 3.1    Suitability

Not only in the domain of container libraries but also in others we have analyzed before [6,5], *Suitability* is perhaps the largest and more complex subcharacteristic. For this reason, it is worth to decompose it into groups of attributes, i.e., new subcharacteristics. In our case, we identify two of them:

- *Core Suitability*. Addresses the types of containers offered and their implementations. These types are mainly characterised by the operations for adding, removing, modifying and searching elements.
- *General Suitability*. Keeps track of additional functionalities offered by (most of) the containers of the library, such as support for concurrent access or iterators.

**Core Suitability.** Table 2 introduces the attributes for *Core Suitability*. Some comments follow:

- The concept of *category* stands for huge groups of behaviour-related containers. Some categories such as sequences or maps will surely be present somehow in the Ada Standard Container Library, while others such as graphs may be a matter of discussion (in fact, most of the standard container libraries present in other languages do not offer this container category).
- Containers and their implementations are kept separated from the very beginning. A container represents an abstract data type, which may have (and probably will have) some different implementations in the library.
- The concept of *operation* should be viewed from the abstract-data-type point of view. In addition to operations, there could be generic algorithms that use those operations (container traversal, merging, etc.). These algorithms are represented by an attribute in the *General Suitability* subcharacteristic.
- It is important to include in the model itself the different types of container elements that are present in the library, and also their relationships, which are not shown in the table for lack of space.
- The second, the third and the fourth attributes are in fact families of attributes, one for each type of category or container.

**Table 2.** Attributes for *Core Suitability*

| Attribute | Definition | Examples |
|---|---|---|
| Category variety | Range of different categories of containers offered by the library | Sequences, maps, sets, trees, graphs |
| Container variety | Range of different containers provided by every category | For sequences: stacks, queues, lists |
| Implementation variety | Range of different implementations provided by every category | For maps: closed hashing, red-black trees |
| Operation variety | Range of different operations provided by every container | For stacks: empty, push, pop, top, isEmpty |
| Element variety | Types of container elements | Universal, comparable |

**General Suitability.** Table 3 lists the attributes for *General Suitability*. It should be mentioned that other attributes could also be incorporated, but we focused on the most usual ones:

- *Position* is a direct access path for elements in the container. *Iterators* provide a means to obtain and possibly manipulate the elements in the container. These two features are present in most widespread container libraries, with these names or others (e.g., iterators in STL, items in LEDA [13]).
- As some libraries do (e.g., STL and LEDA), positions and iterators may be implemented using the same mechanism. But it is important to keep both attributes separate, since their semantics are different.
- It is expected that these attributes will behave uniformly in the whole library, e.g., the same error management mechanisms used throughout the library (see 4.1).

**Table 3.** Attributes for *General Suitability*

| Attribute | Definition | Examples |
|---|---|---|
| Direct access by position | Types and operations for supporting direct access to elements in containers | Type *position*; operation for deletion by position |
| Iterators | Types and operations for supporting traversal of containers | Bidirectional, unidirectional; read, read/write |
| Concurrent access | Mechanisms for managing concurrent access to containers | Semaphores, synchronization |
| Persistency | Mechanisms for storing container elements in a persistent manner | Serialization; operations for writing to disk; file types |
| Algorithmic variety | Range of generic algorithms present in the library or in particular containers | Sorting, merging For arrays: binary search |
| Error management | Mechanisms available for error management | Use of contracts, exceptions, messages |
| Sizeability | Strategies supported for managing the size of the container | Bounded, unbounded and resizeable containers |

- An exception to the last rule is the *Sizeability* attribute, because some implementation strategies may prevent the use of some kind of sizeability strategies (e.g., heap implementations for resizeable containers).

## 3.2  Accuracy

In Table 4 we present the attributes for *Accuracy*. We highlight the following:

- It is necessary to distinguish among the *Error Management* mechanisms available (part of *General Suitability*) and how they are used in container operations. The first attribute in Table 4 addresses the last topic.
- Absence of ambiguity is analyzed at the specification level, not at the implementation one. For instance, it is important to know the policy that the container follows when elements with the same value are found in an ordered traversal. But on the other hand, it is not relevant to know the detailed behaviour of a concrete implementation (e.g., how a hashing strategy handle collisions), provided that it keeps the intended specification.
- Access by position and iterators must be well defined. A survey of these mechanisms in some widespread libraries shows that there are certain conditions that may affect the accuracy of the results and may compromise the integrity of the container. This is the reason why these new attributes have been introduced in this subcharacteristic.

**Table 4.** Attributes for *Accuracy*

| Attribute | Definition | Examples |
|---|---|---|
| Trusted operations | Policies that ensure right results when executing operations | Deleting non-existing elements will not harm the state of the container |
| Absence of ambiguity | Certainty about the behaviour of a container | Priority queue: FIFO ordering of elements with the same priority |
| Accurate access by position | Policies and artifacts that ensure right results when accessing by position | Operation for knowing if a position is bound to the right element |
| Accurate access by iterator | Policies and artifacts that ensure right results when accessing by iterator | Operation for knowing if the current element during traversal has changed |

## 3.3  Interoperability

Table 5 introduces the attributes for *Interoperability*. The second attribute refers to the possibility of using a container in a distributed system with some kind of middleware. To do so, some actions must be taken; e.g., the interface of the container should be defined in some specific Interface Description Language (IDL). Although a great deal of applications would avoid the remote use of containers for efficiency reasons, we think that the attribute must be included in the model for its analysis when designing the Ada Container Standard Library.

**Table 5.** Attributes for *Interoperability*

| Attribute | Definition | Examples |
|---|---|---|
| Language interoperability | Ability to be invoked by programs written in a language different from Ada | Invocation from C++ and Eiffel |
| Component interoperability | Ability to be integrated in heterogeneous systems | CORBA, DCOM, RMI middleware |

## 3.4 Security

Table 6 presents the attributes for *Security*. We remark that:

- *Concurrent Access Security* is closely related to the *Concurrent Access* attribute which belongs to the *General Suitability* characteristic. In other words, as it happened with error management in Sect. 3.2, it is important to distinguish among the available policies and how they are implemented from the security point of view.
- Since positions and iterators provide additional access mechanisms to containers (additional with respect to the access schemes bound to the type of container), it is important to establish their conditions of correct behaviour. Obviously, the two resulting attributes are closely related to the ones appearing in *Accuracy* but it is necessary to distinguish among accuracy of results and data security.

**Table 6.** Attributes for *Security*

| Attribute | Definition | Examples |
|---|---|---|
| Concurrent access security | Policies and mechanisms that ensure safe concurrent access to elements in the container | Facilities for duplicating parts of the container; blocking during an iterator traversal |
| Direct access by position security | Policies and artifacts that ensure safe use of positions when accessing the container | A position bound to an element is the same while it is in the container |
| Iterator security | Policies and artifacts that ensure safe use of iterators when accessing the container | Read-only iterators may not be used in an odd manner |

## 3.5 Functionality Compliance

Table 7 introduces the attributes for *Functional Compliance*. In general, the concept of compliance comes from two different sources. The first one are regulations clearly stated in a document, such as the reference manual of the Ada programming language. The second one stems from the current practices of the community, which have lead to a common foundation widely accepted. Some

**Table 7.** Attributes for *Functionality Compliance*

| Attribute | Definition | Examples |
|---|---|---|
| Domain compliance | Adherence to conventions of names of containers, operations and other artifacts | Function-like containers should be named something close to *map* or *table* |
| Language compliance | Adherence to conventions in the language community | In Ada: *To_List* for transforming a set into a list |

different communities must be taken into account, mainly the Ada community (e.g., use the term *package* instead of *module*), the software libraries community (e.g., use the expression *browsing* when performing a tool-supported search in the library) and the data structure community (e.g., use names such as *stack* and *hashing*).

## 4    Other Quality Attributes in Container Libraries

In the previous section we have analysed all the subcharacteristics belonging to the *Functionality* characteristic. The same should be done with the rest of characteristics of the ISO/IEC 9126-1 quality standard, but this is not possible in this paper for lack of space. So, we have focused on two other subcharacteristics that would play also an important part in the success of the Ada Standard Container Library.

### 4.1    Understandability

Table 8 introduces the attributes for *Understandability*. Some comments follow:

- A clear distinction among types of containers and their implementations is crucial for supporting understandability. Information hiding is the principle that should rule the design of the library. Therefore, no assumptions about implementation policies should appear when defining the type of container.
- Uniformity could be defined as a family of attributes, one for each type of concept in the library. Thus, we may talk about uniformity of error management, uniformity on the way of using generic algorithms, and so on.
- The third attribute is different from the compliance ones (see Sect. 3.5), although some relationships exist.
- Documentation has to be considered here from the user's point of view. This means that we are not addressing project documentation, such as code comments, but documentation for understanding the product.
- The quality of the design affects the understandability of a library. Bad designs may hide concepts and may place features in the wrong place.
- Complexity is a concept that has to be mainly with two factors: size (number of packages, number of methods, etc.) and conceptual difficulty of the implementations, algorithms, strategies, etc.

**Table 8.** Attributes for *Understandability*

| Attribute | Definition | Examples |
|---|---|---|
| Separation between type of container - implementation | Degree of distinction among the semantics of a container type and its available implementations | No assumptions on the available implementations |
| Uniformity | Same strategies and level of detail when dealing with the same concept in different parts of the library | Access by position available to all types of containers |
| Name appropriateness | Behaviour of library features accordingly to their name | The *getCurrent* operation of an iterator does not change the current element |
| Quality of documentation | Appropriateness and comprehension of the documentation to make easy the use of the library | UML diagrams for describing the packages; browsing capabilities |
| Quality of design | Quality of the design of the library | Use of design patterns |
| Complexity | Size of the library and conceptual difficulty of the offered features | Use of advanced implementation techniques |

### 4.2  Changeability

Table 9 presents the attributes for *Changeability*. Issues worth to remark:

- Changeability may be seen from different points of view, namely: extension of the library with new types of containers, implementations, generic algorithms and so on; specialization of existing types of containers with new features, new specific algorithms, etc.; modification of existing features. We could then split the subcharacteristic into three. However, for the sake of brevity, we do not proceed this way.
- Modularity and internal reusability are perhaps the key two factors in this subcharacteristic. Of course, whatever final form the library takes, some degree of modularity and reusability will exist.
- The last two attributes were previously introduced in the *Understandability* subcharacteristic. This situation illustrates the fact that the hierarchy has a graph-like form. However, it should be remarked that the focus of the attribute vary depending on the subcharacteristic. For instance, complexity in *Understandability* has been considered from the user point of view (basically, number of concepts to be understood) while complexity in *Changeability* is considered from the developer point of view (basically, how easy is the design and the code of the library to be modified). In other words, metrics for this two attributes would be definitively different.

## 5  Stating Relationships among Quality Attributes

We have already mentioned that a fundamental point when building a quality model is making explicit the tradeoffs among the different quality factors that

**Table 9.** Attributes for *Changeability*

| Attribute | Definition | Examples |
|---|---|---|
| Modularity | Extent of the decomposition of the library into modules | One package for container type |
| Internal reusability | Degree of reusability of the code inside the library | Use of abstract classes |
| Programming practices | Adoption of best programming practices in-the-small | Avoid global variables; adopt name conventions |
| Decoupling | Independence of the different packages that are in the library | Use the *Template Method* pattern |
| Quality of design | Quality of the design of the library | Use of design patterns |
| Complexity | Difficulty of analysing the internal structure of the library | Intensive use of object-oriented features |

have been identified. The purpose of this section is to identify and characterize the types of relationships that appear in the model.

Relationships may be classified according to two different concepts:

- Which are the types of quality factors related. The most usual case is relating a couple of attributes, but also subcharacteristics may be related. Also, a subcharacteristic may affect an attribute, or the other way round.
- Which is the kind of relationship. We distinguish among three types:
  - Dependency. If $A$ depends on $B$, then $A$ is an attribute that makes sense only if $B$ satisfies some condition on its value.
  - Collaboration. If $A$ collaborates with $B$, growing of $A$ (from the metrics point of view) implies growing of $B$. Often, collaboration is symmetric.
  - Damage. If $A$ damages $B$, growing of $A$ (from the metrics point of view) implies shrinking of $B$.

As an example Table 10 enumerates some representative relationships among the quality factors presented in Sects. 3 and 4. Some explanation is given below:

- The first relationship is a dependency among two attributes. It states that talking about accuracy of iterators is useless when there are no iterators.
- The second one is a damage from one subcharacteristic to another. It reflects that the more suitable the library is, the more difficult to understand.
- The third relationship is a symmetric collaboration among two attributes. Modularity of the library clearly supports separation of concerns, while having a conceptual distinction among type of container and implementation will favour the ability to structure the library in a modular manner.
- Finally, the fourth relationship is a collaboration from one attribute to another. Internal reusability clearly has a positive influence on uniformity of the library, since inherited features will appear in the heirs without any difference.

It should be remarked that the analysis of the relationships among quality attributes requires adopting a qualitative point of view of the quality model. As

**Table 10.** Relationships among quality factors

| Dependee | Depender | Type of relationship |
|---|---|---|
| Iterators | Accurate access by iterator | Dependency |
| Suitability | Understandability | Damage |
| Modularity | Separation type of container – implementation | Symmetric collaboration |
| Internal reusability | Uniformity | Collaboration |

stated in the introduction, qualitative measures are supposed to be introduced when considering a specific context for the library, after analysis of the requirements for its design and implementation. The relationships stated here could be then a great help when considering the implications of some requirements over the quality of the library.

# 6    Applying the Quality Model

In this section we apply the resulting quality model to two existing container libraries: an Ada one, Booch Components (BC) and the standard library from Java, the Java Collections Framework (JCF). Due to lack of space, we do not analyze other interesting Ada libraries (such as Charles) or standard libraries of other languages (e.g., STL).

Table 11 summarizes the evaluation of the attributes in both libraries. The first column identifies the subcharacteristic, the second one the attribute, and the third and fourth the values in the libraries. The table is commented in the rest of the section.

**Core Suitability.** It can be observed that BC provides a richer variety of categories and containers, but there is not a significant difference in implementation strategies. Operations on containers are usual ones: insertions, deletions, etc. Some containers and implementation strategies require comparisons of elements and so both libraries offer comparable elements.

**General Suitability.** Both libraries are not well-suited for access by position. BC does not provide this kind of access, while JCF just in lists. Also, iterators are not as powerful as in other widespread libraries. In fact, we have a proposal [11] enriching BC with access by position and more powerful iterators.

It must be also remarked that the direct access provided by JCF by means of list iterators is in fact the position on the list, so in case of *LinkedList* implementation it is highly inefficient. This property would have appeared if efficiency were included in the part of the quality model developed in this paper. On the other hand, the list iterator are bidirectional and read/write.

Other general suitability attributes are more or less covered by both libraries.

**Accuracy.** Exception handling is widely used for assuring correct results. Concerning iterators and direct access, JCF has a safe although somehow restrictive policy of invalidating iterators when other operations interfere.

**Table 11.** Applying the quality model to the BC and JCF libraries

| | Attribute | BC | JCF |
|---|---|---|---|
| Understandability | Separation type of container – implem. | Yes | Yes |
| | Uniformity | Yes | Not entirely |
| | Name appropriateness | Yes | Not entirely |
| | Quality of documentation | Good | Good |
| | Quality of design | Yes | Not entirely |
| | Complexity | Low | High |
| Changeability | Modularity | Yes | Yes |
| | Internal reusability | Yes, but causes coupling | Yes, but causes inefficiency |
| | Programming practices | Good | Good |
| | Decoupling | No | Yes |
| | Quality of design | Coupling causes lost of quality | Good |
| | Complexity | High | High |
| Core Suitability | Category variety | Sequences, maps, trees, sets, graphs | Sequences, maps, sets |
| | Container variety | Sequences: collections, lists, stacks, rings, queues; Maps: maps; etc. up to 12 varieties | Sequences: lists; Maps: sorted maps; Sets: sets, sorted sets |
| | Implementation variety | One strategy for each container | Lists and maps, two strategies; others, one |
| | Operation variety | Most operations for each container | Most operations for each container |
| | Element variety | Universal, comparable | Universal, comparable |
| General Suitability | Access by position | Not supported | Only in lists, using iterators |
| | Iterators | Unidirectional, read-only | Unidirectional (except for lists), read-and-remove |
| | Concurrent access | Guarded and synchronized | Not specific mechanisms |
| | Persistency | Not supported | Serialization |
| | Algorithmic variety | No generic algorithms | No generic algorithms |
| | Error management | Exception mechanism | Exception mechanism |
| | Sizeability | Bounded, unbounded and resizeable | Unbounded, resizeable |
| Accuracy | Trusted operations | Yes, using exceptions | Yes, using exceptions |
| | Absence of ambiguity | Yes | Yes |
| | Accurate direct access by position | Direct access not supported | Lists invalidate direct access as soon as the container is modified |
| | Accurate access by iterator | Yes, using concurrent access mechanisms | An iterator is invalidated when the container is modified by other mean than the iterator itself |
| Interopera. | Language interoperability | No language interoperability | No language interoperability |
| | Component interoperability | No component interoperability | No component interoperability |
| Security | Concurrent access security | Yes, using concurrent access mechanisms | Controlled by exception mechanisms |
| | Direct access by position security | Direct access not supported | Positions provided by lists not persistent |
| | Iterator security | Safe read-only iterators | Control of modifications |
| F.C.[1] | Domain compliance | Good enough | Good enough |
| | Language compliance | Good enough | Good enough |

[1] Functionality Compliance

**Interoperability.** None of the libraries offer interoperability capabilities. It could be thought of building wrappers, using either the *Export* pragma and *Remote Call Interface* in the case of Ada, or the CORBA and RMI standards in the case of Java.

**Security.** In both cases, the security aspects are well covered.

**Functionality Compliance.** In both cases, the libraries are compliant enough with respect to functionality, although from time to time some odd terms appear.

**Understandability.** It is worth to remark some lack of uniformity in the JCF library. For instance, JCF bidirectional iterators and access by position are only available for lists. Also, some of the JCF method names are not according to their functionality (e.g. the *next* iterator method return the current element and then advances to the next).

The design of the JCF library is not as clear as it should be. For instance, there are two separate hierarchies; this makes more difficult to get the general picture.

With respect to complexity there are different tradeoffs between the two libraries. Basically, BC offers few methods that can be combined to obtain more functionalities while JCF has a lot of different methods some of them with similar functionalities and so the library becomes more difficult to understand.

**Changeability.** Both libraries present internal reusability up to some extent, but their solutions are not optimal. In the case of BC, some dependencies among packages arise and makes changes more difficult to implement (see [11] for details). This coupling also impacts on other attributes. In the case of JCF, some methods of its abstract classes are implemented using only the interface of the class (avoiding coupling) but as a consequence they are not as efficient as possible, and the size of each category abstract class grows.

## 7   Conclusions

In this paper, we have proposed a framework based on the concept of quality model for assessing the quality of the prospective Ada Standard Container Library. The quality model is based on the ISO/IEC 9126-1 quality standard which fits well with the fact that Ada is an ISO standard too. The use of such a quality standard to design an Ada Standard Container Library is an important aspect to make easier the standardization of the new Ada language extensions and evolutions. In the paper, we have tailored the highly abstract quality model proposed in this standard to our specific needs, by adding some quality factors and exploring their relationships. Also, we have checked the applicability of the model to a pair of existing libraries. For the sake of brevity, we have not presented the whole quality model, but a (representative enough) part.

Our paper tries to be a contribution in the design and implementation of the Ada Standard Container Library. It is our believe that the success of this potential library requires as a first step a deep knowledge of the features that must be considered for their inclusion, the different ways on offering these features, and also the consequences of including them. Mechanisms such as shortcuts [7], specific design patterns [8,12] and others may be then more thoroughly evaluated. The quality model may act as a cornerstone in these activities.

But adopting a quality-model-based approach in not only useful for assessment. It provides a structured and precise way of describing the features of the library, making its use more appealing. It also can be used as a first step for getting a certification of the resulting product, issued by a third-party organization; there are currently some organizations that use the ISO/IEC quality standards for issuing such certifications. Last, a quality model supports traceability of decisions taken during the construction of the library, which is specially interesting when considering maintenance of the library.

# References

1. K. Arnold, J. Gosling and D. Holmes. *The Java Programming Language*. Addison-Wesley, 3$^{rd}$ edition, 2000.
2. Application Standard Components Library (ASCL).
   `http://ascl.sourceforge.net/`.
3. G. Booch, D.G. Weller and S. Wright. The Booch Library for Ada 95 (version 1999). Available at
   `http://www.pogner.demon.co.uk/components/bc/`.
4. M. Heaney. Charles Container Library. At home.earthlink.net/~matthewjheaney/.
5. J.P. Carvallo, X. Franch and C. Quer. Defining a Quality Model for Mail Servers. In *Proceedings of the 2$^{nd}$ International Conference on COTS-Based Software Systems (ICCBSS)*, volume 2580 of *LNCS*. Springer-Verlag, 2003.
6. X. Franch and J.P. Carvallo. Using quality models in software package selection. *IEEE Software*, **20(1)**, 2003.
7. X. Franch and J. Marco. Adding Alternative Access Paths to Abstract Data Types. In *Challenges of Information Technology Management in the 21$^{st}$ Century (IRMA'2000)*, pages 283–287. Idea Group Publishing, 2000.
8. N. Gelfand, M. T. Goodrich and R. Tamassia. Teaching Data Structure Design Patterns. In *Proceedings of ACM SIGCSE '98*, 1998.
9. ISO/IEC Standards 9126-1 Software Engineering – Product Quality – Part 1: Quality Model, June 2001.
10. B. Meyer. *Reusable Software: the base object-oriented component libraries*. Prentice Hall, 1994.
11. J. Marco and X. Franch. Reengineering the Booch Component Library. In *Reliable Software Technologies Ada-Europe 2000*, volume 1845 of *LNCS*, pages 96–111. Springer-Verlag, 2000.
12. J. Marco and X. Franch. Shortcuts for the Ada Standard Container Library. Presented at the *Workshop for Standard Container Library for Ada*. Held during the Ada-Europe 2002 Conference, Wien (Österreich). Available at
    `http://www.auto.tuwien.ac.at/AE2002/resources.html`.
13. K. Mehlhorn and S. Näher. *The LEDA Platform of Combinatorial and Geometric Computing*. Cambridge University Press, 1999.
14. D.R. Musser and A. Saini. *STL Tutorial and Reference Guide*. Addison-Wesley, 1996.

# Experiences on Developing and Using a Tool Support for Formal Specification

Tommi Mikkonen

Tampere University of Technology
P.O. Box 553, FIN-33101 Tampere, Finland
tjm@cs.tut.fi

**Abstract.** Tool support is a major issue for any novel software engineering approach. This is particularly important for introducing new methodological issues, as only with an adequate embedding of new ideas in a practical tool set it is possible to include new facilities in the software development process. In this paper, we provide an insight to the feedback obtained when developing and using a tool set for easing the use of a formal method. The paper begins by introducing the underlying formal method, its language representation, and the tool set that is currently available. Based on them, the core contribution of the paper is constituted by the observations made when the tool set has been used in practice.

**Keywords**: formal methods, software engineering, development tools

## 1 Introduction

Tools are essential for software engineers in helping to get the practical matters right. Still, tools should (and can) not replace sound semantics of the underlying method. Embedding the sound principles in a practical tool then becomes the advocated approach.

Our experience on embedding sound principles to a practical tool arises from the development and using of the formal method *DisCo* [25], where we have defined a methodological baseline, created a specification language resembling programming languages, and implemented a supporting tool set. This programming-friendly environment has then been used to study the use of the underlying formal specification method in a pragmatic fashion. With this approach, we have gained the best of both worlds – an easy-to-use tool that lets the developers express formal specifications in a language that closely resembles programming languages, especially Ada.

The rationale of this paper is to gather all the experiences we have on putting the approach and associated tools in practice. Currently, we have more than ten years of experience on working with the underlying methodology. During the years, there have been two generations of the specification language ([9,10]) and associated tool environment systems ([23,1]), as well as numerous reported

J.-P. Rosen and A. Strohmeier (Eds.): Ada-Europe 2003, LNCS 2655, pp. 297–308, 2003.

academic (e.g. [22,2]) and industrial (e.g. [5,20]) case studies. The feedback reported in this paper has been gathered from experiments carried out by tool and methodology developers, students using the tool set in their exercises and project assignments, and case studies implemented in connection with industry, with formal specification work carried out by both industry experts and DisCo specialists. While the main conribution of this paper is to discuss the observations made when applying the methodology with the available tools, we will also provide an introduction to the design methodology, associated language support, and the tool set, because the results would be difficult to analyze without a proper background.

The rest of this paper is structured as follows. Section 2 introduces the underlying formalism, related methodological issues, and the language representation used for composing specifications in practice. Section 3 provides an insight to the current DisCo tool set. Section 4 forms the core of this paper by listing the different observations on using and developing the methodology and the tool set in practice. Section 5 finally concludes the paper.

## 2     The DisCo Method

In this section, we will give a brief (and partly simplified) overview of the DisCo method [25]. The discussion starts from semantic principles and related methodological issues. Based on them, the focus is then shifted to language representation used for composing specifications in practice.

### 2.1     Formal Foundation

The DisCo method has based its formal foundation on the Temporal Logic of Actions, TLA [18]. The practical consequence of such a background is that DisCo specifications consist of state variables and relations between them, similar to any TLA specifications. Some relations, referred to as *state predicates*, only discuss a snapshot of the values of variables in a given global state, constituted by the combined values of all the variables. Other relations, referred to *actions*, model the transition from one state to another in terms of state variables. An operational interpretation for a DisCo specification can be given as follows. Actions are execution "steps" that take place in an atomic, interleaved fashion. Then, by giving an initial state and a set of actions one generates a natural representation for a computing system. In addition to the initial state and the actions, one can also give requirements on scheduling of actions.

In contrast to conventional approaches, where the focus of early development phases has been placed on interfaces, we have reconsidered the concept of modularity. Instead of traditional components, we have decided to develop specifications in *layers*, which resemble e.g. program slices [24] or projections used in verification [17], in the sense that they constitute modular structures based on individual (yet potentially connected) variables and operations on them rather than depending on the implementation architecture. We, however, compose new systems with such layers, not decompose those that already exist.

The purpose of layer-wise specification is to enable modeling in an incremental fashion. This has lead to the adoption of *superposition*, a well-known technique first used in [3], as the main modularity mechanism of the method. The traditional purpose of superposition has been the separation of e.g. correctness and termination concerns in the design of algorithms. Therefore, we believe that it can also be used as a mechanism for separating unrelated concerns at specification level. Similar goals can be seen in e.g. *aspect-oriented* development [4]. The close connection between the two has been suggested by e.g. Katz and Gil [13].

With the above flavor of modularity, a system being modeled can be considered to carry out its activities independent of outside activity at any level of details. We will refer to such systems as *closed systems*. They can be observed but not affected from outside. Any real-life system is obviously an *open* one, i.e., it has an interface that separates the system from its execution environment. However, from the theoretical viewpoint, it is often simpler to model things as closed systems, as both environment assumptions and the system behavior will be described using the same formalism. The definition of the implementation interface between the system and its environment then becomes a crucial design decision to be based on the closed system specification. However, this decision can be postponed until the closed-system specification is detailed enough for high-level validation and verification. Moreover, interface refinements that make the interface more precise are possible. Related approaches include e.g. *problem frames*, where the approach relies on similar ideas but at an informal level [6,7].

## 2.2 Language Representation

State variables are included in the DisCo specification language as *classes* that are patterns for objects. A designer might, for example, give a simple class as follows:

```
class A is      -- Class name.
  i : integer; -- Instance variables.
end A;
```

The definition of the initial values of variables is an issue that has been left outside the specification language used for composing designs. Early versions of the DisCo tool set allowed a special configuration part that could be used to assign initial values to variables, but current tools enable a tool-assisted generation of the initial state. On the formal side, the designer is assumed to give initial conditions as explicit expressions in connection with class definitions. For instance, the following initial condition defines that all the values in instances of class *A* are non-negative.

```
initially and/ object : A::object.i => 0;
```

*Actions* are patterns of interaction that can take place between different objects. They can be taken as multi-object methods that determine how the contents of participant objects can change. An action consists of a signature (name of the action and roles for participant objects), an enabling guard (boolean expression), and the effect of the action to the participant objects (collection of assignments executed in parallel). For example, the following action defines that two instances of class $A$ can swap their values.

```
action swap(x, y : A) is      -- Signature (name and roles).
when true do                  -- Enabling condition.
  x.i := y.i || y.i := x.i;   -- Statements that are executed
end swap;                     -- in parallel.
```

If there are several objects that may take the same role in an action, one of them is nondeterministically selected. Similarly, the action to be executed next is not determined by a flow of control of e.g. a process but is selected in a non-deterministic fashion from those that are enabled, i.e., whose enabling condition evaluates to true for some combination of participant objects. One can associate a scheduling requirement in an action by including an asterisk (*) next to a role name. This requires that if enabled, the action must be executed infinitely often for all the potential combinations of participants of that role.

*Assertions* have been included in the language to allow a methodological way to enforce a contract-based design approach. Similarly to conventional methods, where e.g. class invariants can be used, we advocate the use of assertions that determine the design intention. It is then up to the designer to orchestrate the actions, i.e., operations of the system, such that they cannot violate the assertions. However, we usually use assertions at the level of layers, not classes as advocated in object-oriented setting.

To achieve superposition based modularity, the designer is assumed to give a set of related classes and actions in the same syntactic layers in a fashion that obeys superposition. From the practical perspective, this simply means that once a designer gives a state variable, he is also obliged to give the actions that can alter the value of the variable in the same layer. Then, a situation emerges where the first version of a specification produces an abstract behavior, which can be extended with more variables and details in later increments. For the purposes of DisCo, we have selected the variant of superposition so that safety properties (properties of the form "something bad will never happen") are satisfied by construction. Liveness properties (properties of the form "something good will eventually happen"), implemented with scheduling requirements, require additional considerations.

The language representation of a refinement is straightforward. Whenever referring to something that has already been discussed in existing layers, we use ellipses (...) to refer to the existing parts of the system. Moreover, keywords "extend" and "refined" replace "class" and "action" when defining a refinement, respectively. With these conventions, a layer that contains all the above language items would be introduced as listed in Fig. 1. Another layer, adding another dimension of behavior on top of the previously given layer is introduced in Fig. 2. The resulting composition is illustrated in Fig. 3.

```
layer SimpleSwaps is
  class A is i : integer; end A;
  initially and/ object : A::object.i => 0;

  action swap(x, y : A) is
  when true do
    x.i := y.i || x.i := y.i;
  end swap;
end SimpleSwaps;
```

**Fig. 1.** Simple DisCo layer

```
layer PoweredSwaps is
  import SimpleSwaps;

  extend A by pwr : boolean; end A;
  initially and/ object : A::object.pwr = false;

  refined swap(x, y : A) is
  when ... x.pwr and y.pwr do
    ...
  end swap;

  action Switch(me : A) is
  when true do
    me.pwr := not me.pwr;
  end Switch;
end PoweredSwaps;
```

**Fig. 2.** Simple DisCo refinement

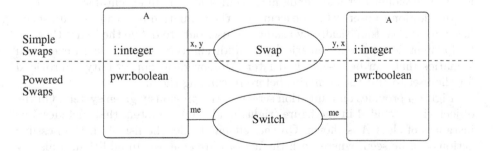

**Fig. 3.** Illustration of a DisCo refinement

# 3    The DisCo Tool Set

The tool set we are currently using has been introduced in [1]. However, some of the ideas can be traced back to a prototype version [23], which was first used to test the soundness of underlying methodological issues as well as usefulness of its language representation.

## 3.1    Compiler

As the starting point of any computer-aided activity in the DisCo tool environment, a compiler is used for reading in completed DisCo specifications, and for verifying syntax and semantics. In particular, this obviously includes checking that the developer does not violate superposition-based specification steps[1].

Layers are used as units of compilation. Internally, the compiler transforms specifications into an intermediate format, where the different base units (e.g. classes and actions) of the specification can be handled separately. Therefore, all the layering included in the specification is "opened" in this phase. However, all the parts of the specification still internally carry an identity tag that allows the tracing of different parts of a completed specification back to their layered origins for debugging purposes.

The architecture described above has resulted in a design where the compiler responsible for reading in the specification and for creating the structure acts as a front-end. Based on this front-end, it is then possible to introduce different back-ends that generate code for different purposes.

## 3.2    Animation Tool

Currently, the most well-established back-end for developing specifications is animation. Any DisCo specification, composed with layers, can be animated so that the designer is responsible for selecting the different actions and the necessary participants for an execution. Then, state changes of different objects can be seen on computer screen. This straightforward animation approach effectively lets the developer test and debug his specifications in an executable fashion.

In addition to user-defined executions, the animation tool can execute specifications semi-automatically by asking the user only to define the issues that the tool cannot determine automatically. In addition, the user can record execution scenarios that can be re-executed later. Moreover, an opportunity is provided for the user to edit the scenarios before re-running them.

Figure 4 provides an animation screen of the previously given system. On the object view included in the figure (right-hand side), a system that includes four instances of class $A$ is shown. On the left-hand side, the list of all the possible actions can be seen, where both the actions are enabled. In addition, a history

---

[1] Early versions of the tool set had an annoying bug that allowed the developer to violate superposition. With developers accustomed to the DisCo principles, it took surprisingly long to identify this very obvious bug.

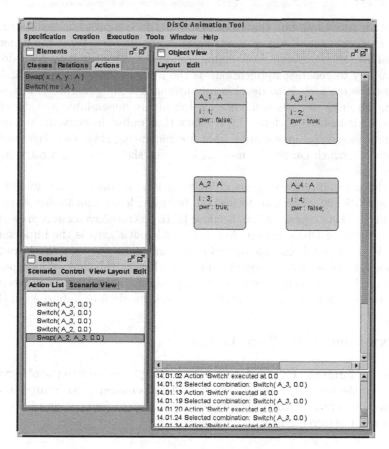

**Fig. 4.** The DisCo animator

window is visible, indicating that several *Switch* actions and one *Swap* action have been already executed. By clicking the button labeled "scenario view", this can be viewed in the traditional graphical format as well. In addition to actions and classes, the action list also includes timing information. This, however, is not utilized by the above specification.

## 3.3   Other Experiments

With formal semantics, one is tempted to offer computer-assisted theorem proving support for formal specification. The experiences with the DisCo method have proven that this is possible but relatively laborious [14]. Therefore, our research activities have aimed at reusable design steps whose proofs could be reused [16], as well as at specification architectures that are verification-friendly [15], rather than at an all-purpose assertion verification support.

Another tool candidate that is easy to identify is a program generator. This has been experimented, but in a setting that can be considered too generic for practical use [19]. For more encouraging results, we currently consider a domain-sensitive way to generate applications as the most prominent approach. A part of the future research is to develop an automatic code generator for a certain domain, the most obvious candidates being highly dependable and distributed systems. This task is hardened by the fact that unlike in conventional setting, where interface definitions can be used for generating skeletons, layers address collaboration which cannot be used as basis for skeletons in a straightforward fashion.

In addition to direct code generation, we have experimented a transition from DisCo to UML [26]. The transformation, however, loses information already included in the DisCo specification, because UML has no convenient representation for e.g. an atomic DisCo action. Another problematic area is the importance of layout. We have not been able to find a convenient way to map the layout used in the DisCo tool set to a conventional case tool. Still, the topic of incorporating some of the DisCo principles in a design process relying on UML is an interesting field of future study, with some preliminary results already published in [12].

## 4     Experiences on Tool Usage

This section forms the core of this paper by giving an insight to the observations we have on the use of the DisCo tool set. The discussion is structured in accordance to the depth of embedding the methodological foundation of the DisCo approach in the development process.

### 4.1   Informal Considerations

Perhaps the biggest surprise for the DisCo development team has been the uniform respect gained by the possibility to animate even partial specifications during the early phases of the development. This has resulted in a pragmatic way to make domain experts and software professionals share a common terminology. With the animation running on the screen, people are tempted to ask questions like "What will happen if the user pushes the button again before the system has reacted?" Animation then lets the designer run the revised scenario, and the result can be read from the screen. This has typically stimulated lengthy discussions, where the client or customer has started to ask questions on how the system should work in different unanticipated situations, even if the client herself should really possess the information, not the specifier, who typically has only formalized the informal specification. Although we have not been gathering any statistical evidence, we feel that no such discussions would have emerged without the animation capabilities that force specifiers and users, as well as external observers, to participate in generating the executions.

The importance of the animation tool has been very clearly stated by Isojärvi in one of our industrial case studies, "Animation is the best part of DisCo [5]."

This issue is particularly obvious when comparing the attitude of industrial developers to the methodological background: Once the formal methodology introduction was finished, and it was the time to use the tool in practice, the events have repeatedly turned from lectures to lively design sessions.

Technically, the above issue is connected to the fact that the method allows the animation of specifications even if the specifications are far from being completed and lack many details needed for the final version. The parts that have not yet been fixed will simply be treated as non-deterministic. In addition, closed-system modeling has established itself as an important mechanism for studying the systems, as even with partial specifications, it is possible to discuss the collaboration of a system and its environment.

## 4.2 Formal Issues

The well-established rationale of formal specification is to avoid unambiquities, misconceptions, and logical errors. In the case of DisCo, we have usually identified such problems in informal specifications when composing DisCo representations based on them. As for the practical cases, unclear issues, potential problems, or clearly identified bugs were found in the majority (if not all) of systems we have studied in detail[2]. As it is possible to advance in increments that take into account one conceptual thing (or concern) at a time, the developers are forced to consider the potential problems in early phases rather than postponing them to later phases. Moreover, practice has shown that it is usually relatively easy to define so small increments that their correct function can be verified and validated by investigating relatively few test cases. This is a strong argument on the positive effect of the use of property-preserving increments in design, achieved due to the use of superposition. In fact, the methodological side of the approach has repeatedly been considered more important than the actual formalization.

Although the majority of logical errors have been found when composing specifications, the compiler and the animation tool have proven themselves valuable for the developers. In addition to being a visualization tool of (semi-)completed specifications as described above, the animation has been used to identify semantical errors in intermediate phases of the specification work. This is comparable to traditional testing and debugging transferred to specification level. The most common errors not related to the domain sensitive logic of the system being modeled have been such that formalized specifications were too relaxed in the sense that some of the necessary restrictions were not imposed.

The downside of animation is that we have witnessed the tendency to think in terms of animations rather than the actual DisCo formalism. While this is relatively natural for people not directly involved in mcthodology development, the same can also be observed within the DisCo team, although to a lesser extent. In particular, during the early phases of the development of the methodology,

---

[2] Categorization of these issues is difficult, as the systems have been at different phases of their life cycle, ranging from a concept study to long-term maintenance

it was not uncommon that a certain instance of the system that was animated on the screen was actually considered as a completed specification, although the formal semantics of the method assume explicitly given assertions and initial conditions of different configurations. Then, when the configuration shown on the screen was altered (e.g. new objects, allowed by the initial condition and assertions, were introduced), some of the informal expectations were invalidated although all the formal aspects included in the rigorous specification were still satisfied. In other words, with the support from an executable specification, the developers overlooked formal descriptions, and took the executable as the semantics of the system. This issue becomes especially obvious when using a theorem prover, as automated reasoning prevents all informal shortcuts.

Similar problems have been encountered with liveness. As validation of liveness properties is impossible with animation only, specifications largely omit scheduling requirements that should be given if the underlying methodology were obeyed in full.

### 4.3   Improved Understanding of Software

As a side-effect of the introduction of concern-separating design style used in the formalism, it is interesting to notice that the designers have transferred some of the main ideas of DisCo into pragmatic use in a more conventional design environment, even if DisCo was not used at all in the project. For instance, [8] introduces a design approach and a software architecture, where the incrementality of DisCo has been ported into a more conventional software engineering setting, together with some design rules to enable preservation of properties when giving new increments. Moreover, in the examples of [21] one can easily identify reflections of the DisCo method. However, applying the approach at the level of UML may make the use of property-preserving increments impractical [11].

In addition, a clear tendency to rely on incremental and iterative design approaches in general can be identified, where designers advance in small steps and aim at early verification and validation. This applies to both the fashion we have used the DisCo method as well as to other software systems designed with more conventional methods.

## 5   Conclusions

In this paper, we have introduced observations concerning the use of a tool-assisted formal method in practice. Based on the experience, it is clear that the facilities offered for studying the specification incrementally yet rigorously at the early phases of the development are helpful. Moreover, facilities offered for animation of specifications have been respected. At the same time, however, we have observed a tendency to think in terms of associated tools rather than the actual method, which implies that the tool should embed the most important properties of the underlying methodology in a practical fashion, to the extent that it is acceptable to use the tool level for reasoning. Incorporating such facilities in a design system developed for e.g. UML-based software processes remains a future challenge. However, also in that environment, we believe that the gains of a state-of-the-art tool incorporating a formal methodology as a baseline for the development will be remarkable.

**Acknowledgements.** The DisCo research team has been partly funded by the Academy of Finland (project 5100005).

# References

1. T. Aaltonen, M. Katara, and R. Pitkänen. DisCo toolset: The new generation. *Journal of Universal Computer Science*, 7(1):3–18, 2001.
2. T. Aaltonen, P. Kellomäki and R. Pitkänen. Specifying cash-point with DisCo. *Formal Aspects of Computing*, 231–232, Vol. 12(4), 2000.
3. E. Dijkstra and C. Scholten. Termination detection for diffusing computations. *Information Processing Letters*, 11(1), 1–4, 1980.
4. T. Elrad, R. E. Filman, and A. Bader. Aspect-oriented programming. *Communications of the ACM*, 44(10):29–32, October 2001.
5. S. Isojärvi. newblock DisCo and Nokia: Experiences of DisCo with modeling real-time system in multiprocessor environment. *Formal Methods Europe Industrial Seminar '97*, Otaniemi, Finland, February 20, 1997.
6. M. Jackson. *Software requirements & specifications: A lexicon of practice, principles and prejudices.* Addison-Wesley, 1995.
7. M. Jackson. *Problem frames: Analyzing and structuring software development problems.* Addison-Wesley, 2000.
8. J. Jokinen, H.-M. Järvinen, and T. Mikkonen. Incremental introduction of behaviors with static software architecture. *Proceedings of the 2002 International Conference on Software Engineering Research and Practice (Eds. H.R. Arabnia and Y. Mun)*, 10–16, CSREA Press, 2002.
9. H.-M. Järvinen. *The Design of a Specification Language for Reactive Systems.* Doctoral dissertation, Tampere University of Technology, 1992.
10. H.-M. Järvinen. *The DisCo2000 Specification Language.* Technical report, Tampere University of Technology, 2000.
11. M. Katara and S. Katz. Architectural views of aspects. *Accepted to AOSD'03, to appear.*
12. M. Katara and T. Mikkonen. Refinements and aspects in UML. *International Workshop on Aspect-Oriented Modeling with UML*, Dresden, September 30., 2002.
13. S. Katz and J. Gil. Aspects and superimpositions. Position paper in Aspect Oriented Programming workshop in ECOOP'99, Lisbon, Portugal, June 1999.
14. P. Kellomäki. Verification of reactive systems using DisCo and PVS. *FME'97: Industrial Applications and Strengthened Foundations of Formal Methods*, (Eds. J. Fitzgerald, C. B. Jones, P. Lucas), 589–604, Springer-Verlag LNCS 1313, 1997.
15. P. Kellomäki. Verification-Friendly Specification of Distributed Systems. *16th World Computer Congress 2000, Proceedings of Conference on Software: Theory and Practice (cds. Y. Feng, D. Notkin and M.-C. Gaudel)*, 480–483, Publishing House of Electronics Industry and International Federation for Information Processing. 2000.
16. P. Kellomäki and T. Mikkonen. Design templates for collective behavior. *Proceedings of ECOOP 2000, 14th European Conference on Object-Oriented Programming (ed. Elisa Bertino)*, 277–295, Springer-Verlag, LNCS 1850, 2000.

308    T. Mikkonen

17. S. Lam and A. Shankar. Protocol verification via projections. *IEEE Transactions on Software Engineering*, SE-10(4), 325–342, July 1984.
18. L. Lamport. The temporal logic of actions. *ACM Transactions on Programming Languages and Systems 16*, 3, 872–923, May 1994.
19. T. Mikkonen. An experimental code generator for implementing formal specifications given as closed systems. *Proceedings of the Fourth Symposium on Programming Languages and Software Tools*, Visegrad, Hungary, 1995.
20. T. Mikkonen. A layer-based formalization of an on-board instrument. Technical Report 18, Software Systems Institute, Tampere University of Techology, 1998.
21. T. Mikkonen and P. Pruuden. Flexibility as a design driver. 52–56, *Computer*, November 2001.
22. M. Setälä. *Formal Modeling of an Avionics System*. MSc. thesis (in Finnish), Tampere University of Technology, 1992.
23. K. Systä. A graphical tool for specification of reactive systems. In *Proceedings of Euromicro'91 Workshop on Real-time Systems*, 12–19, Paris, France, June 12–14, 1991.
24. M. Weiser. Program slicing. *IEEE Transactions on Software Engineering*, Vol. SE-10, No 4, 352–357, 1984.
25. DisCo WWW site. At `http://disco.cs.tut.fi` on the World Wide Web.
26. UML WWW site. At `http://www.rational.com/uml` on the World Wide Web.

# A Behavioural Notion of Subtyping for Object-Oriented Programming in SPARK95

Tse-Min Lin[1] and John A. McDermid[2]

[1] Real-Time Systems Research Group
[2] High Integrity Systems Engineering Research Group

Department of Computer Science
The University of York
Heslington, York   YO10 5DD, UK
{lin,jam}@cs.york.ac.uk

**Abstract** The dynamic aspects of the object-oriented paradigm have prevented the adoption of the latter for the implementation of high integrity systems using the SPARK approach. This paper presents a proposal that allows object-oriented programming in SPARK95, whereas supporting SPARK's static approach for verification by imposing a notion of behavioural subtyping between a type and all its subtypes. Behavioural subtyping supports modular reasoning through supertype abstraction, hence all proofs can be discharged based only on nominal/declared types. An example of proof is also presented.

**Keywords:** behavioural subtyping, modular reasoning, supertype abstraction, object-oriented programming, SPARK

## 1   Introduction

Nowadays, the object-oriented paradigm [3,19] is the programming paradigm most widely used for the implementation of software. However, the dynamic aspects of the object-oriented paradigm (e.g. polymorphism and dynamic/late binding) prevented its use for the implementation of high integrity applications in the SPARK approach [1,2].

In this paper, we present how behavioural subtyping [7,16] can be introduced in the SPARK notation to support object-oriented programming. The behavioural notion of subtyping ensures that a subtype preserves supertype abstraction, i.e. a subtype behaves like its supertype(s), and it is a very important notion because it supports modular reasoning [13,12,22]. Supertype abstraction allows properties to be proved based only on nominal (or declared) types because a supertype stands for all its subtypes.

The overall motivation for the work presented in this paper is the support of object-oriented programming in the SPARK notation, and this is done by:

1. Development of a notion of behavioural subtyping for the SPARK95 language;

J.-P. Rosen and A. Strohmeier (Eds.): Ada-Europe 2003, LNCS 2655, pp. 309–321, 2003.

**Table 1.** Terminology: Object-oriented paradigm and Ada95

| O-O Paradigm | Ada95 |
|---|---|
| class | tagged type + primitive operations |
| state | tagged record |
| method | primitive operation |
| state element/component | component (of a tagged type) |
| method invocation | (dynamic) dispatching |
| subclass/subtype | descendant (derived from a tagged type) |
| superclass/supertype | ancestor (of a tagged type) |
| hierarchy | derivation class (rooted at a tagged type) |
| inheritance | type extension |
| object/instance | value |

2. Proposal of new annotations for the specification of the static properties (or behaviour) of a class and its methods;
3. Definition of syntactical restrictions to ensure that the substitutability principle [5] of the object-oriented paradigm is supported;
4. Presentation of restrictions to forbid the definition of recursive classes, aliasing/sharing of access types, and introduction of subprograms that do not dispatch.

The work presented in this paper is focused on the first objective. A full description of our proposal can be found in [15]. Note that the meaning of the terms used in this paper is drawn from the object-oriented paradigm. Table 1 presents some terms that are commonly used in the object-oriented paradigm and their equivalence in Ada95 [21].

This paper assumes that the reader has knowledge of the object-oriented paradigm and the SPARK approach, and it is organized as follows. The next section presents the definition of a notion of behavioural subtyping for SPARK95. Section 3 illustrates an example where it is proved that the behavioural subtype relation holds for the classes of a points hierarchy. Section 4 presents the related works, conclusions, and suggestions for further work.

## 2    A Behavioural Notion of Subtyping for SPARK95

The inheritance mechanism of the object-oriented paradigm can be used for both specialization and generalization of a type. Inheritance allows the definition of new types (also known as subtypes) from previously defined types (also known as supertypes). The approach adopted by the vast majority of programming languages is to introduce syntactic restrictions to ensure type safe use of polymorphism and late binding. However, very few programming languages provide a mechanism to relate the behaviour of a subtype to the behaviour of its supertype(s). The main consequence of the absence of a behavioral subtype relation [7,

16,22] is that the introduction of a new subtype can result in previously correct program modules to have unexpected (or surprising) behaviour.

If aliasing is constrained (or even forbidden) and each subtype preserves the behaviour of its supertype(s), then modular verification of object-oriented programs is possible [11,12,13,22]. This section presents how the behavioural subtyping approach proposed in [7] can be supported in the SPARK context. In the following we present definitions that are usually found in a denotational semantic model of an object-oriented programming/specification language in a semi-formal way:[1]

1. The notation $\{state\_component \rightsquigarrow value\}$ will be used to represent the "value" of an object. For example, a **Point_2D** object at coordinate (10, 20) is represented as $\{x \rightsquigarrow 10, y \rightsquigarrow 20\}$ whereas a **Point_3D** object at location (101, 2, 54) is represented as $\{x \rightsquigarrow 101, y \rightsquigarrow 2, z \rightsquigarrow 54\}$;[2]

2. The existence of an environment used to record all definitions introduced in a program (identifiers, static and dynamic types, values, syntactical level of declaration, arguments, etc.) is assumed. We are not interested in how it could be represented/defined, but only in that this environment can be used to provide all information of interest.

   This environment is represented by the Greek letter $\varepsilon$ (epsilon). We also use a decorated variation $\varepsilon'$ (and sometimes $\varepsilon''$ or another Greek letter) to represent an environment that is obtained from $\varepsilon$ after the "evaluation" (or execution) of a statement. If the statement does not change anything (such as the statement **null**) then $\varepsilon'$ is exactly the same as $\varepsilon$;

3. $value$ ($[\![id]\!], \varepsilon$) — returns the object that is stored in the identifier **id**. The brackets $[\![\ ]\!]$ are used to denote syntactical elements. If **id** is the name of a class, then $value([\![id]\!], \varepsilon)$ returns the set of all objects/instances **id**. For example, in a pure object-oriented language any type must be introduced by a class, therefore the (built-in) type **Integer** is assumed to be introduced by a class and $value([\![Integer]\!], \varepsilon) = \{\ldots, -1, 0, 1, \ldots\}$;

4. $obj \in value([\![Class\_Name]\!], \varepsilon)$ — means that $obj$ is an element/instance of the set of objects of **Class_Name**. For example, $10 \in value([\![Integer]\!], \varepsilon)$ and $\{x \rightsquigarrow 10, y \rightsquigarrow 20\} \in value([\![Point\_2D]\!], \varepsilon)$;

5. $eval$ ($[\![statement]\!], \varepsilon$) — $eval$ is the evaluation function and it takes two arguments: the statement to be evaluated and the environment to be used in the evaluation. Unlike the $value$ function presented above, the evaluation function can modify the environment and the resultant environment is called $\varepsilon'$, that is, $\varepsilon' = eval([\![statement]\!], \varepsilon)$. For example, if $value([\![x]\!], \varepsilon) = 1$, then $eval([\![x := 100]\!], \varepsilon)$ results in a new environment $\varepsilon'$ such that $value([\![x]\!], \varepsilon') = 100$. SPARK annotations can refer to the previous value of any variable used, therefore, it will be assumed that the environment also records this information. Using the example before, we have that $value([\![x^\sim]\!], \varepsilon') = 1$;

6. $pre$ ($[\![method (obj, arg_2, \ldots, arg_n)]\!], \varepsilon$) — returns true if in the environment $\varepsilon$ the precondition of **method** when invoked with object **obj** and arguments $arg_2$, $\ldots$, $arg_n$ is satisfied, otherwise it returns false;

---

[1] For example: environment, evaluation function, semantic function, value, etc. A reader interested in knowing how such definitions are introduced in the formal semantics of an object-oriented language is referred to [14,23].

[2] **Point_2D** and **Point_3D** are presented in Section 3.

7. *Convert_Sub_to_Super* (*obj*, $\varepsilon$) — the existence of a family of type conversion functions is assumed. In this definition we will assume the existence of a function that converts subtype objects to supertype objects;

8. *static_type* ($[\![\texttt{id}]\!]$, $\varepsilon$) — if **id** is an identifier of a variable, then the function *static_type* returns its static (or declared) type in $\varepsilon$. Once a variable is declared, its static type is fixed — it cannot change in its scope of definition, therefore, for any environment $\varepsilon'$ obtained from $\varepsilon$ the property *static_type* ($[\![\texttt{id}]\!]$, $\varepsilon$) = *static_type* ($[\![\texttt{id}]\!]$, $\varepsilon'$) is true. The introduction of a local variable potentially hides an identifier of the enclosing scope. However, the SPARK approach forbids the redefinition/redeclaration of an identifier that is already in scope. This rule reduces ambiguity;

9. *dynamic_type* ($[\![\texttt{id}]\!]$, $\varepsilon$) — if **id** is an identifier of a variable, then the function *dynamic_type* returns its dynamic type in $\varepsilon$. Due to subsumption, the dynamic type of any variable can be a subtype of its declared/static type. As an example of use of this function, the predicate *dynamic_type* ($[\![\texttt{x}]\!]$, $\varepsilon$) = $\texttt{Point\_3D}$ is true if the dynamic type of **x** in $\varepsilon$ is $\texttt{Point\_3D}$, which means that the predicate *value*($[\![\texttt{x}]\!]$, $\varepsilon$) $\in$ *value*($[\![\texttt{Point\_3D}]\!]$, $\varepsilon$) is also true.

   In the scope of declaration of **id**, the dynamic type of **id** can change during execution to any of the subtypes of its static type. Therefore, the following properties are always true in the scope of declaration of **id**:

$$static\_type\ ([\![\texttt{id}]\!], \varepsilon) = static\_type\ ([\![\texttt{id}]\!], \varepsilon')$$
$$static\_type\ ([\![\texttt{id}]\!], \varepsilon) \sqsubset dynamic\_type\ ([\![\texttt{id}]\!], \varepsilon')$$

   where $\sqsubset$ is the behavioural subtype relation presented below;

10. *id* : *X* is used to indicate that *id* is an identifier of dynamic type *X* in $\varepsilon$.

To guarantee behavioural subtyping the following rules must be enforced in an object-oriented extension of SPARK95:[3]

**Definition 1** *Behavioural Subtype Relation* ($\sqsubset$)

   Given two classes *Super* and *Sub*, *Sub* is a behavioural subtype of *Super*, *Super* $\sqsubset$ *Sub*, if and only if the following properties are satisfied:

1. **Syntactic Rules** — Signature Compatibility

   For all primitive operations **method** in *Super*, that have been redefined in *Sub*:

   a) The number of formal arguments of both methods are the same;

   b) The type of all controlling operands in each method must be the same;

   c) All formal arguments of type *Super* in the supertype method must have type *Sub* in the subtype method;

   d) All formal arguments that do not have type *Super* in the supertype method must have the same type in the subtype method;

   e) The modes of all formal arguments in the supertype method are preserved in the subtype method.

---

[3] These rules describe the verification conditions that must be discharged in order to demonstrate behavioural subtyping. Note that the formulae $(\forall \alpha \mid P \bullet Q)$ and $(\exists \alpha \mid P \bullet Q)$ are equivalent to $(\forall \alpha \bullet P \Rightarrow Q)$ and $(\exists \alpha \bullet P \Rightarrow Q)$, respectively.

The syntactical rules are just what Ada95 requires to ensure that all methods are primitive operations, i.e. they can dispatch. However, they do not tie up the behaviour of a class and its subclasses. In this respect, the most important rules are the semantic rules presented below. They guarantee that a subclass behaves like all its superclasses, or in other words, a class describes the behaviour of all its "unknown" subclasses.[4]

## 2. **Semantic Rules** — Behavioural Compatibility

### a) CORRESPONDENCE OF STATES RULE

$$\forall \, obj_{sub} \in value \, ([\![Sub]\!], \varepsilon) \bullet \exists \, obj_{sup} \in value \, ([\![Super]\!], \varepsilon) \bullet$$
$$Convert\_Sub\_to\_Super \, (obj_{sub}, \varepsilon) = obj_{sup}$$

This rule states that every object of the subtype has a corresponding object in the supertype. If there were only one object $x$ that did not satisfy this rule, then modular reasoning would not be possible because no superclass object could be used as an approximation to the behaviour of $x$.

### b) INITIALIZATION RULE

For every initial state method **Create** in the supertype and subtype:

$$\forall \, id : Super \bullet \exists \, obj \in value \, ([\![Super]\!], \varepsilon); \; \varepsilon' : environment \; |$$
$$eval([\![\texttt{Create} \, (id)]\!], \varepsilon) = \varepsilon' \bullet value([\![id]\!], \varepsilon') = obj$$
$$\forall \, id : Sub \bullet \exists \, obj \in value \, ([\![Sub]\!], \varepsilon); \; \varepsilon' : environment \; |$$
$$eval([\![\texttt{Create} \, (id)]\!], \varepsilon) = \varepsilon' \bullet value([\![id]\!], \varepsilon') = obj$$

This rule ensures that the initial state methods are consistent. That is, there is at least one object that satisfies the invariant of each class. If this rule is not satisfied then the class has no object at all due to a contradictory or false class invariant.

### c) CORRECTNESS OF INITIALSTATE METHODS RULE

For every initial state method **Create** in the supertype that is redefined in the subtype:

$$\forall \, id_1 : Super; \; id_2 : Sub \bullet \exists \, \varepsilon', \varepsilon'' : environment \; |$$
$$eval([\![\texttt{Create} \, (id_1)]\!], \varepsilon) = \varepsilon' \wedge eval([\![\texttt{Create} \, (id_2)]\!], \varepsilon) = \varepsilon'' \bullet$$
$$Convert\_Sub\_to\_Super(value([\![id_2]\!], \varepsilon'')) = value([\![id_1]\!], \varepsilon')$$

This rule states that any object obtained by application of the initial state method in the subtype is related to the object obtained by application of the corresponding initial state method in the supertype. In another words, the initial state of a subclass object is also a (related) initial state of a superclass object. This is important, otherwise a subtype object could follow a history that is not related to a supertype object as both objects would begin their existence in non-related (initial) states.

---

[4] Unknown in the sense that the behaviour of the superclass is an approximation to the behaviour of any of its subclasses, even those that are not defined yet.

d) APPLICABILITY OF SUBTYPE METHODS RULE

For every method **method** in the supertype that is redefined in the subtype:

$$\forall id_1 : Super; \ id_2 : Sub \mid$$
$$Convert\_Sub\_to\_Super \ (value \ (\llbracket id_2 \rrbracket, \varepsilon)) = value \ (\llbracket id_1 \rrbracket, \varepsilon) \bullet$$
$$pre \ (\llbracket \textbf{method} \ (id_1, p_2, \ldots, p_n) \rrbracket, \varepsilon) \Rightarrow pre \ (\llbracket \textbf{method} \ (id_2, p_2, \ldots, p_n) \rrbracket, \varepsilon)$$

where $p_2, \ldots, p_n$ correspond to actual arguments.[5] This rule says that whenever a supertype method is applicable, so is the corresponding subtype method when invoked with the same arguments.

e) CORRECTNESS OF SUBTYPE METHODS RULE

For every method **method** in the supertype that is redefined in the subtype:

i. If **method** is declared as a procedure:

$$\forall id_1 : Super; \ id_2 : Sub \mid$$
$$Convert\_Sub\_to\_Super \ (value \ (\llbracket id_2 \rrbracket, \varepsilon)) = value \ (\llbracket id_1 \rrbracket, \varepsilon) \bullet$$
$$pre \ (\llbracket \textbf{method} \ (id_1, p_2, \ldots, p_n) \rrbracket, \varepsilon) \Rightarrow$$
$$\exists \varepsilon', \varepsilon'' : environment \mid$$
$$eval \ (\llbracket \textbf{method} \ (id_1, p_2, \ldots, p_n) \rrbracket, \varepsilon) = \varepsilon'$$
$$eval \ (\llbracket \textbf{method} \ (id_2, p_2, \ldots, p_n) \rrbracket, \varepsilon) = \varepsilon'' \bullet$$
$$Convert\_dt_{\{p_2, \varepsilon''\}}\_to\_dt_{\{p_2, \varepsilon'\}} \ (value(\llbracket p_2 \rrbracket, \varepsilon'')) = value(\llbracket p_2 \rrbracket, \varepsilon')$$
$$\ldots$$
$$Convert\_dt_{\{p_n, \varepsilon''\}}\_to\_dt_{\{p_n, \varepsilon'\}} \ (value(\llbracket p_n \rrbracket, \varepsilon'')) = value(\llbracket p_n \rrbracket, \varepsilon')$$
$$Convert\_Sub\_to\_Super \ (value \ (\llbracket id_2 \rrbracket, \varepsilon'')) = value \ (\llbracket id_1 \rrbracket, \varepsilon')$$

where each $dt_{\{p_i, \varepsilon'\}}$, $2 \leq i \leq n$, is used to represent the dynamic type of the argument $p_i$ in $\varepsilon'$, i.e. $dt_{\{p_i, \varepsilon'\}} = dynamic\_type \ (\llbracket p_i \rrbracket, \varepsilon')$;[6]

ii. If **method** is declared as a function:

$$\forall id_1 : Super; \ id_2 : Sub \mid$$
$$Convert\_Sub\_to\_Super \ (value \ (\llbracket id_2 \rrbracket, \varepsilon)) = value \ (\llbracket id_1 \rrbracket, \varepsilon) \bullet$$
$$pre \ (\llbracket \textbf{method} \ (id_1, p_2, \ldots, p_n) \rrbracket, \varepsilon) \Rightarrow$$
$$Convert\_dt_{\{res_{sub}, \varepsilon\}}\_to\_dt_{\{res_{sup}, \varepsilon\}} \ ($$
$$value(\llbracket \textbf{method} \ (id_2, p_2, \ldots, p_n) \rrbracket, \varepsilon)) =$$
$$value(\llbracket \textbf{method} \ (id_1, p_2, \ldots, p_n) \rrbracket, \varepsilon)$$

$dt_{\{res_{sub}, \varepsilon\}}$ and $dt_{\{res_{sup}, \varepsilon\}}$ are used to represent the type of the result returned by the methods defined in $Sub$ and $Super$, respectively, using the environment $\varepsilon$.[7]

---

[5] Although $p_2, \ldots, p_n$ occur free in this rule, in practice static type checks ensure that they conform to the signature of **method** as defined in $Super$.

[6] Note that if the dynamic type of $p_i$ is the same in both $\varepsilon'$ and $\varepsilon''$ (i.e. $dt_{\{p_i, \varepsilon'\}} = dt_{\{p_i, \varepsilon''\}}$), then $Convert\_dt_{\{p_i, \varepsilon''\}}\_to\_dt_{\{p_i, \varepsilon'\}}$ is the identity function (that is, no conversion is done). For an argument $p_i$ of mode **in**, SPARK ensures that it cannot be modified, therefore $Convert\_dt_{\{p_i, \varepsilon''\}}\_to\_dt_{\{p_i, \varepsilon'\}} \ (value(\llbracket p_i \rrbracket, \varepsilon'')) = value(\llbracket p_i \rrbracket, \varepsilon')$ holds.

[7] The property to be proved for functions is simpler than the one for procedures because functions in SPARK can neither have side-effects nor change the object and arguments used in the instantiation/invocation.

This rule tells us that:

- If the supertype method is applicable, then rule (2d) ensures that the subtype method is also applicable. However, the modifications done in the arguments of mode **out** and in the objects that received the messages ($id_1$ and $id_2$) must be related. That is, the behaviour of the subtype method is constrained by the behaviour of the supertype method;
- If the supertype method is not applicable, then the subtype method is free to make any modification to the object that received the message and its arguments of mode **out**. That is, the behaviour of the subtype method is not constrained by the behaviour of the supertype method.

□

Although the semantic rules might look intimidating due to the presence of universal and existential quantifiers, we argue that the task of proving that all rules hold is not difficult to be accomplished in practice. This is demonstrated through an example of proof in the next section.

## 3  An Example of Proof

In this section we present a rigorous (or mathematical) proof that Point_3D is a behavioural subtype of Point_2D. Both classes are presented below:

```
1   package Points is
2     type Point_2D is tagged private;
3     --   own x, y;
4     type Point_2D_object is access all
5                       Point_2D'Class;
6     procedure Create (p: access Point_2D);
7     --     access_mode (p: out);
8     --     initialstate;
9     --     derives p from ;
10    --     post p.x = 0 and p.y = 0;
11    procedure Add (p1: access Point_2D;
12                   p2: access Point_2D'Class);
13    --   access_mode (p1: in out; p2: in);
14    --   derives p1 from p1, p2;
15    --   post p1.x=p1~.x+p2.x and
16    --        p1.y=p1~.y+p2.y;
17    function Coord_X (p: access Point_2D)
18                 return Integer;
19    --   access_mode (p: in);
20    --   return p.x;
21    function Coord_Y (p: access Point_2D)
22                 return Integer;
23    --   access_mode (p: in);
24    --   return p.y;
25    ...
26    type Point_3D is new Point_2D with
27                              private;
28    --   own z;
29    type Point_3d_object is access all
30                       Point_3D'Class;
31    function Convert_to_parent (
32                   p: access Point_3D)
33                 return Point_2D_object;
34    --   access_mode (p: in);
```

```
35    --     return r => r.x=p.x and r.y=p.y;
36    procedure Create (p: access Point_3D);
37    --     access_mode (p: out);
38    --     initialstate;
39    --     derives p from ;
40    --     post p.x=0 and p.y=0 and p.z=0;
41    procedure Add (p1: access Point_3D;
42                   p2: access Point_2D'Class);
43    --   access_mode (p1: in out; p2: in);
44    --   derives p1 from p1, p2;
45    --   post p1.x=p1~.x+p2.x and
46    --        p1.y=p1~.y+p2.y and
47    --        (pfIs_Point_3D (p2) =>
48    --         p1.z=p1~.z+
49    --         pfView_in_Point_3D(p2)).z
50    --   function pfIs_Point_3D (
51    --           p: access Point_2D
52    --           return Boolean;
53    --   function pfView_in_Point_3D (
54    --           p: access Point_2D
55    --           return Point_3D_object;
56    function Coord_Z (p: access Point_3D)
57                 return Integer;
58    --   access_mode (p: in);
59    --   return p.z;
60    ...
61  private
62    type Point_2D is tagged
63      record
64        x, y: Integer;
65      end record;
66    type Point_3D is new Point_2D with
67      record
68        z: Integer;
```

```
69      end record;                          95          return Point_2D_object is
70   end Points;                             96      res : Point_2D_object := new Point_2D;
                                             97      begin
                                             98        res.x := p.x; res.y := p.y;
70   package body Points is                  99        return res;
72     procedure Create (p: access Point_2D) is  100    end Convert_to_parent;
73       begin                               101    procedure Create (p: access Point_3D) is
74         p.x := 0; p.y := 0;               102      begin
75       end Create;                         103        p.x := 0; p.y := 0; p.z := 0;
76     procedure Add (p1: access Point_2D;   104      end Create;
77                    p2: access Point_2D'Class) is  105    procedure Add (p1: access Point_3D;
78       begin                               106                   p2: access Point_2D'Class) is
79         p1.x := p1.x + Coord_X (p2);       107      begin
80         p1.y := p1.y + Coord_Y (p2);       108        p1.x := p1.x + Coord_X (p2);
81       end Add;                             109        p1.y := p1.y + Coord_Y (p2);
82     function Coord_X (p: access Point_2D)  110        if p2.all in Point_3D'Class then
83                      return Integer is     111          p1.z := p1.z+
84       begin                               112            Coord_Z (Point_3D_object(p2));
85         return p.x;                        113        end if;
86       end Coord_X;                         114      end Add;
87     function Coord_Y (p: access Point_2D)  115    function Coord_Z (p: access Point_3D)
88                      return Integer is     116                     return Integer is
89       begin                               117      begin
90         return p.y;                        118        return p.z;
91       end Coord_Y;                         119      end Coord_Z;
92       ...                                 120      ...
93     function Convert_to_parent (          121   end Points;
94                        p: access Point_3D)
```

The package Points is used to implement a points hierarchy. The class
Point_2D represents two dimensional points, whereas the class Point_3D represents three dimensional points.

The use of access to class wide types is necessary to support the substitutability principle of the object-oriented paradigm. The annotation **access_mode** is used to make explicit the mode under which the arguments of a subprogram call will be manipulated. The annotation **initialstate** indicates that the subprogram is an initial state operation, and as such it must be the first subprogram to be executed for any object.

The first argument of all primitive operations has a special status in our approach — it corresponds to the object that receives the message. This is necessary because Ada95 does not make explicit the role of each argument of a primitive operation. The proof function pfIs_Point_3D (lines 50–52) is used to verify if the dynamic type of p2 is rooted at Point_3D, whereas the proof function pfView_in_Point_3D (lines 53–55) is used to return a copy of p2 as an object rooted at Point_3D.

Note that we extended the use of the annotation **own** of SPARK. In lines 3 and 28, **own** annotations are used to make visible the state components of Point_2D and Point_3D, respectively. The standard technique in SPARK to make the components of a private record visible is to use proof functions.

Due to space constraints, in this section we only develop the proof that the method Add of Point_3D preserves semantic rule (2e).[8]

**Proof 1** The function Convert_to_parent (lines 31–35) is used to convert 3D points into 2D points, and it corresponds to the semantic function *Convert_Sub_to_Super* used in the definition of behavioural subtyping presented in Section 2. Therefore,

---

[8] The proof of compliance to the other rules can be found in [15, §4.3.4].

$Convert\_Point\_3D\_to\_Point\_2D$ $(value([\![obj_{sub}]\!], \varepsilon))$ in Definition 1 is represented as:

$$value\,([\![\texttt{Convert\_to\_parent}\ (obj_{sub})]\!], \varepsilon) \tag{1}$$

in the points hierarchy.

No method in the points hierarchy changes any of its arguments. That is, for all arguments, if any, they are of mode **in** only.[9] This means that in the formula presented in rule (2e) all predicates of the form:

$$Convert\_dt_{\{p_i, \varepsilon''\}}\_to\_dt_{\{p_i, \varepsilon'\}}\,(value([\![p_i]\!], \varepsilon'')) = value([\![p_i]\!], \varepsilon')$$

for $2 \le i \le n$, are true and do not need to be considered. Furthermore, due to the fact that the preconditions of all methods are true,[10] the property presented in rule (2e) for procedures can be simplified to:

$$\forall\, id_1 : Point\_2D;\ id_2 : Point\_3D\ | \tag{2}$$
$$Convert\_Point\_3D\_to\_Point\_2D\,(value\,([\![id_2]\!], \varepsilon)) = value\,([\![id_1]\!], \varepsilon)\ \bullet$$
$$\exists\, \varepsilon', \varepsilon'' : environment\ |$$
$$eval\,([\![\textbf{method}\ (id_1, p_2, \ldots, p_n)]\!], \varepsilon) = \varepsilon'$$
$$eval\,([\![\textbf{method}\ (id_2, p_2, \ldots, p_n)]\!], \varepsilon) = \varepsilon''\ \bullet$$
$$Convert\_Point\_3D\_to\_Point\_2D\,(value\,([\![id_2]\!], \varepsilon'')) = value\,([\![id_1]\!], \varepsilon')$$

The universal quantifier in (2) requires $id_1$ and $id_2$ to be related. Therefore, we can assume that **p2d** and **p3d** are variables of dynamic types **Point_2D** and **Point_3D**, respectively, with the following property:

$$Convert\_Point\_3D\_to\_Point\_2D\,(value\,([\![\texttt{p3d}]\!], \varepsilon)) = value\,([\![\texttt{p2d}]\!], \varepsilon) \tag{3}$$

which is equivalent, by (1), to:

$$value\,([\![\texttt{Convert\_to\_parent}\ (\texttt{p3d})]\!], \varepsilon) = value\,([\![\texttt{p2d}]\!], \varepsilon) \tag{4}$$

Given that (4) holds, the following properties about **p2d** and **p3d**, in $\varepsilon$, are true:

$$value([\![\texttt{p2d.x}]\!], \varepsilon) = value([\![\texttt{p3d.x}]\!], \varepsilon) \tag{5}$$
$$value([\![\texttt{p2d.y}]\!], \varepsilon) = value([\![\texttt{p3d.y}]\!], \varepsilon) \tag{6}$$

Properties (5) and (6) are true by the return annotation of **Convert_to_parent**. Because **p2d** and **p3d** where selected arbitrarily with the only condition that (3) holds, it is possible to simplify (2) further by using **p2d** and **p3d**, and applying $\forall$-elimination twice:

$$\exists\, \varepsilon', \varepsilon'' : environment\ | \tag{7}$$
$$eval\,([\![\textbf{method}\ (\texttt{p2d}, p_2, \ldots, p_n)]\!], \varepsilon) = \varepsilon'$$
$$eval\,([\![\textbf{method}\ (\texttt{p3d}, p_2, \ldots, p_n)]\!], \varepsilon) = \varepsilon''\ \bullet$$
$$Convert\_Point\_3D\_to\_Point\_2D\,(value\,([\![\texttt{p3d}]\!], \varepsilon'')) = value\,([\![\texttt{p2d}]\!], \varepsilon')$$

---

[9] Recall that when the word "argument" is used, we are not considering the first one because it corresponds to the object that receives the message.

[10] In our proposal, initial state methods (e.g. the **Create** methods of **Point_2D** and **Point_3D**) cannot have preconditions because they must be always applicable. An initial state method must be the first method to be executed for any object.

In (7), the environments $\varepsilon'$ and $\varepsilon''$ are obtained from $\varepsilon$, therefore the following properties are true about **p2d** and **p3d** in $\varepsilon'$ and $\varepsilon''$:

$$value([\![\texttt{p2d}]\!], \varepsilon) = value([\![\texttt{p2d}^\sim]\!], \varepsilon') \tag{8}$$

$$value([\![\texttt{p2d}]\!], \varepsilon) = value([\![\texttt{p2d}^\sim]\!], \varepsilon'') \tag{9}$$

$$value([\![\texttt{p3d}]\!], \varepsilon) = value([\![\texttt{p3d}^\sim]\!], \varepsilon') \tag{10}$$

$$value([\![\texttt{p3d}]\!], \varepsilon) = value([\![\texttt{p3d}^\sim]\!], \varepsilon'') \tag{11}$$

$$value([\![\texttt{p2d}^\sim.\texttt{x}]\!], \varepsilon') = value([\![\texttt{p3d}^\sim.\texttt{x}]\!], \varepsilon'') \tag{12}$$

$$value([\![\texttt{p2d}^\sim.\texttt{y}]\!], \varepsilon') = value([\![\texttt{p3d}^\sim.\texttt{y}]\!], \varepsilon'') \tag{13}$$

Properties (8), (9), (10), and (11) are true by definition of *value*. Property (12) is true by (8), (11), and (5). Property (13) is true by (8), (11), and (6).

All arguments of **Add** are of mode **in**. Therefore, the following properties are true about any actual argument *arg* used to instantiate **Add**:

$\forall \phi, \phi' : environement;$

    $obj : Point\_3D; \; arg : Point\_2D \mid \phi' = eval \, ([\![\texttt{Add} \, (obj, arg)]\!], \phi) \; \bullet$

$$value([\![arg]\!], \phi) = value([\![arg^\sim]\!], \phi') \tag{14}$$

$$value([\![arg]\!], \phi) = value([\![arg]\!], \phi') \tag{15}$$

(14) is true by the definition of *value*, and (15) is true by the definition of **in** mode arguments (they cannot be modified). Now we have all properties necessary to show that **Add** of **Point_3D** complies to the semantic rule (2e). The formula that needs to be proved to hold is (7):

$\exists \varepsilon', \varepsilon'' : environment \mid$

        $eval \, ([\![\texttt{Add} \, (\texttt{p2d}, \texttt{p})]\!], \varepsilon) = \varepsilon'$

        $eval \, ([\![\texttt{Add} \, (\texttt{p3d}, \texttt{p})]\!], \varepsilon) = \varepsilon'' \; \bullet$

$Convert\_Point\_3D\_to\_Point\_2D \, (value \, ([\![\texttt{p3d}]\!], \varepsilon'')) = value \, ([\![\texttt{p2d}]\!], \varepsilon')$

$\equiv$ (one-point rule applied twice)

$Convert\_Point\_3D\_to\_Point\_2D \, (value \, ([\![\texttt{p3d}]\!], \gamma)) = value \, ([\![\texttt{p2d}]\!], \delta)$

$\equiv$ (postcondition of **Add** in **Point_2D**)

$Convert\_Point\_3D\_to\_Point\_2D \, (value \, ([\![\texttt{p3d}]\!], \gamma)) =$

$value \, ([\![\texttt{p2d} \Rightarrow (\texttt{p2d.x=p2d}^\sim\texttt{.x+p.x} \text{ and } \texttt{p2d.y=p2d}^\sim\texttt{.y+p.y})]\!], \delta)$

$\equiv$ (p2d is an instance of **Point_2D**)

$Convert\_Point\_3D\_to\_Point\_2D \, (value \, ([\![\texttt{p3d}]\!], \gamma)) =$

$\{x \rightsquigarrow value([\![\texttt{p2d}^\sim\texttt{.x+p.x}]\!], \delta), y \rightsquigarrow value([\![\texttt{p2d}^\sim\texttt{.y+p.y}]\!], \delta)\}$

$\equiv$ (by (8) and (15))

$Convert\_Point\_3D\_to\_Point\_2D \, (value \, ([\![\texttt{p3d}]\!], \gamma)) =$

$\{x \rightsquigarrow value([\![\texttt{p2d.x+p.x}]\!], \varepsilon), y \rightsquigarrow value([\![\texttt{p2d.y+p.y}]\!], \varepsilon)\}$

$\equiv$ (by (1))

$value \, ([\![\texttt{Convert\_to\_parent} \, (\texttt{p3d})]\!], \gamma) =$

$\{x \rightsquigarrow value([\![\texttt{p2d.x+p.x}]\!], \varepsilon), y \rightsquigarrow value([\![\texttt{p2d.y+p.y}]\!], \varepsilon)\}$

$\equiv$ (postcondition of Add in Point_3D)

$value\ (\llbracket$Convert_to_parent (p3d $\Rightarrow$

(p3d.x = p3d$^\sim$.x+p.x and p3d.y = p3d$^\sim$.y+p.y and

(pfIs_Point_3D (p) =>

p3d.z=p3d$^\sim$.z+(pfView_in_Point_3D(p)).z)))$\rrbracket, \gamma) =$

$\{x \rightsquigarrow value(\llbracket$p2d.x+p.x$\rrbracket, \varepsilon), y \rightsquigarrow value(\llbracket$p2d.y+p.y$\rrbracket, \varepsilon)\}$

$\equiv$ (return annotation of Convert_to_parent)

$value\ (\llbracket$r $\Rightarrow$ (r.x = p3d$^\sim$.x+p.x and r.y = p3d$^\sim$.y+p.y)$\rrbracket, \gamma) =$

$\{x \rightsquigarrow value(\llbracket$p2d.x+p.x$\rrbracket, \varepsilon), y \rightsquigarrow value(\llbracket$p2d.y+p.y$\rrbracket, \varepsilon)\}$

$\equiv$ (r is an instance of Point_2D)

$\{x \rightsquigarrow value(\llbracket$p3d$^\sim$.x+p.x$\rrbracket, \gamma), y \rightsquigarrow value(\llbracket$p3d$^\sim$.y+p.y$\rrbracket, \gamma)\} =$

$\{x \rightsquigarrow value(\llbracket$p2d.x+p.x$\rrbracket, \varepsilon), y \rightsquigarrow value(\llbracket$p2d.y+p.y$\rrbracket, \varepsilon)\}$

$\equiv$ (by (11) and (15))

$\{x \rightsquigarrow value(\llbracket$p3d.x+p.x$\rrbracket, \varepsilon), y \rightsquigarrow value(\llbracket$p3d.y+p.y$\rrbracket, \varepsilon)\} =$

$\{x \rightsquigarrow value(\llbracket$p2d.x+p.x$\rrbracket, \varepsilon), y \rightsquigarrow value(\llbracket$p2d.y+p.y$\rrbracket, \varepsilon)\}$

$\equiv$ (by (5) and (6))

$\{x \rightsquigarrow value(\llbracket$p2d.x+p.x$\rrbracket, \varepsilon), y \rightsquigarrow value(\llbracket$p2d.y+p.y$\rrbracket, \varepsilon)\} =$

$\{x \rightsquigarrow value(\llbracket$p2d.x+p.x$\rrbracket, \varepsilon), y \rightsquigarrow value(\llbracket$p2d.y+p.y$\rrbracket, \varepsilon)\}$

$\equiv$ (tautology)

true

$\square$

# 4 Conclusion

In this paper we presented how the static approach for verification/analysis advocated by SPARK can be extended to support object-oriented programming through the notion of behavioural subtyping. The latter supports supertype abstraction because a type stands for all its subtypes.

Supertype abstraction supports modular reasoning which is, in our opinion, one of the most important benefits of the object-oriented paradigm. We argue that the only feasible approach to support object-oriented programming, whilst preserving the SPARK approach of relying on static analysis to discharge verification conditions, is through supertype abstraction. The advantage of supertype abstraction is twofold: proofs can be discharged based solely on declared types and the introduction of new subtype(s) in a type hierarchy does not require retesting/reverification of previous correct modules [11,22].

Since release 6.1, on July 2002, limited support for tagged types and type extensions have been added to the SPARK language [6]. However, class-wide operations including dynamic dispatching are not allowed, hence object-oriented programming is still not supported in the SPARK language.

It is interesting to note that Eiffel [19] advocates an approach, known as *Design by Contract* (DbC) [9,10,18], that is similar, but not equivalent, to the notion of behavioural subtyping presented in this paper. DbC is based on the use of assertions which are part of the implementation and are checked at run-time.[11] In the SPARK approach, tools are provided for static program analysis and proof, and they are used to guarantee that a program will always comply to its annotations — no run-time checks are introduced in the executable code.

The rules of the behavioural subtype relation presented in Section 2 were defined semi-formally. Therefore, as a future development, it would be interesting to update and extend the formal semantics of SPARK83, presented in [17,20] using the Z notation [8], to support SPARK95 and the object-oriented approach presented in this paper. The proof presented in the previous section is long, but it is not complex and tool support can be implemented to help discharge the verification conditions generated by the semantic rules of the behavioural subtype relation.

Note that in the definition of the class `Point_3D` presented in Section 3, it was necessary to introduce two proof functions to verify the tag and return a copy of the argument `p2` as an object in the hierarchy rooted at `Point_3D`. The proof functions are necessary because the FDL language [2, §11.5] does not support the concepts of "tag" and "view" of Ada95. Therefore, another interesting development would be to define an extension of the FDL language to provide built-in support for tags and views, thus eliminating the need for proof functions to access/verify tags and provide support for type conversion between subtype and supertype objects.

The work presented in this paper has dealt with object-oriented program-ming in the SPARK approach. Support for other features of Ada95, namely concurrency and real-time programming, are considered in another paper [4].

**Acknowledgements.** We would like to thank the anonymous referees for their helpful comments, careful reading and valuable suggestions. Lin would also like to thank Alan Burns and CNPq/Brazil.

# References

1. P. Amey and G. Finnie. SPARK95 – differences from SPARK83. Technical Report S.P0468.73.46, Praxis Critical Systems Limited, September 1998.
2. J.G.P. Barnes. *High Integrity Ada: The* SPARK *Approach*. Addison-Wesley, 1997.
3. G. Booch. *Object-Oriented Analysis and Design with Applications*. Benjam-in/Cummings, 2nd edition, 1994.
4. A. Burns and T.-M. Lin. Adding temporal annotations and associated verification to the Ravenscar profile. In *Ada-Europe 2003*, LNCS (to appear). Springer-Verlag, 2003.

---

[11] In Eiffel, assertions can be written using the language itself (e.g. function call) and, as far as we know, no tool exists to verify (or guarantee) conformance of an Eiffel program to its assertions prior to execution. Hence, the fact that a program in Eiffel satisfies its assertions at run-time does not imply that the program is guaranteed to always comply to its assertions.

5. L. Cardelli. A semantic of multiple inheritance. *Information and Computation*, 76(2/3):138–164, 1988.
6. R. Chapman and P. Amey. SPARK toolset release note - release 6.1. Technical Report EXM/RN, Praxis Critical Systems Limited, 2002.
7. K.K. Dhara and G.T. Leavens. Forcing behavioral subtyping through specification inheritance. Technical Report 95-20c, Department of Computer Science, Iowa State University, 1997.
8. A.R. Diller. *Z: An Introduction to Formal Methods*. John Wiley, 2nd edition, 1994.
9. R.B. Findler, M. Latendresse, and M. Felleisen. Behavioral contracts and behavioral subtyping. *ACM SIGSOFT Software Engineering Notes*, 26(5):229–236, 2001.
10. E. Lamm. Adding Design by Contract to the Ada language. In *Ada-Europe 2002*, volume 2361 of *LNCS*, pages 205–218. Springer-Verlag, 2002.
11. G.T. Leavens. Modular specification and verification of object-oriented programs. *IEEE Software*, 8(4):72–80, 1991.
12. G.T. Leavens and D. Pigozzi. A complete algebraic characterization of behavioral subtyping. *Acta Informatica*, 36(8):617–663, 2000.
13. G.T. Leavens and W.E. Weihl. Specification and verification of object-oriented programs using supertype abstraction. *Acta Informatica*, 32(8):705–778, 1995.
14. T.-M. Lin. A formal semantics for MooZ (in portuguese). Master's thesis, Department of Informatics, Federal University of Pernambuco (UFPE), 1993. Available from www.cin.ufpe.br/ mooz/.
15. T.-M. Lin. *Behavioural Subtype and Covariance of (Input) Arguments in Object-Oriented Specification Languages* (submitted). PhD thesis, Department of Computer Science, The University of York, 2002.
16. B.H. Liskov and J.M. Wing. A behavioral notion of subtyping. *ACM Transactions on Programming Languages and Systems*, 16(6):1811–1841, 1994.
17. W. Marsh. Formal semantics of SPARK: Static semantics (Version 1.3). Technical report, Program Validation Ltd., 1994.
18. B. Meyer. Design by contract. Technical Report TR-EI-12/CO, ISE Inc., 1987.
19. B. Meyer. *Object-Oriented Software Construction*. Prentice Hall, 2nd edition, 1997.
20. I. O'Neill. Formal semantics of SPARK: Dynamic semantics (Version 1.4). Technical report, Program Validation Ltd., October 1994.
21. S.T. Taft and R.A. Duff, editors. *Ada 95 Reference Manual: Language and Standard Libraries*, volume 1246 of *LNCS*. Springer-Verlag, 1997.
22. M. Utting. *An Object-Oriented Refinement Calculus with Modular Reasoning*. PhD thesis, Department of Computer Science, The University of New South Wales, 1992.
23. M. Wolczko. *Semantics of Object-Oriented Languages*. PhD thesis, Department of Computer Science, The University of Manchester, 1988.

# Running Ada on Real-Time Linux

Miguel Masmano, Jorge Real, Ismael Ripoll, and Alfons Crespo

Department of Computer Engineering (DISCA)
Technical University of Valencia, Spain
jorge@disca.upv.es

**Abstract.** The Real-Time Linux operating system (RTLinux) is an attractive platform for real-time programming, since real-time tasks can be guaranteed bounded response times, whilst a number of applications, IDEs, GUIs, etc. are also available for the same platform. In RTLinux, real-time tasks are implemented (in C) as kernel modules. Special care must be taken when writing kernel modules: an error in a single task can make the whole system hang or crash, since they are executed in the kernel memory space.

This is clearly an area where Ada can be of great help: Ada's strong typing, consistency checking, robust syntax and readability, and the availability of high quality compilers, encourage the writing of correct software and allow to catch coding errors early in the implementation.

In this paper, we show the state of development of a compilation system for Ada programs on RTLinux, based on the GNAT Ada compiler.

## 1 Introduction

Several real-time kernels have been developed in the recent years that provide support for Ada programs. MaRTE OS [2] and ORK (Open Ravenscar Kernel) [6] are two examples. MaRTE provides full Ada support, whilst ORK implements the Ravenscar subset [5]. However, these kernels provide only the limited support needed for small embedded systems.

Real-time applications require predictability. In the general case, an operating system (OS) is intended to provide a number of complex services (file systems, complex memory management, a number of drivers for the different hardware devices, networking, etc.) and is designed to fairly share the computing resources among processes, to have good average response times and to provide a high throughput. This fact makes it virtually impossible to achieve real-time performance in general purpose OSs. However, new trends in embedded systems consider that some additional services, such as networking or high-level resource management, are crucial for applications in the fields of complex process control, multimedia, robotics, etc.

A solution to this conflict consists in inserting a real-time layer beneath an existing OS, which provides a virtual machine for the OS, that runs in the background, and to run the real-time workload in the foreground. This is the approach followed by Real-Time Linux (RTLinux, for short) [4] and other OSs like RTX [12] on Windows NT. The benefits of such an approach are predictability for the real-time applications, where the complex non real-time functions are performed in the background, and the data and

J.-P. Rosen and A. Strohmeier (Eds.): Ada-Europe 2003, LNCS 2655, pp. 322–333, 2003.
© Springer-Verlag Berlin Heidelberg 2003

control connections between real-time and non real-time parts of the application are achieved by means of fast real-time buffers (RT FIFOs in RTLinux). The OS device drivers can also be inherited for use without hard timing constraints.

Until today, RTLinux only allowed C programming and some support for C++. Our work focuses on generating an RTLinux application fully written in Ada. There exists a precedent to this work by Shen, Charlet, and Baker [11] which describes the implementation of an Ada tasking kernel as a layer of the Linux OS, derived from an early version of RTLinux. This work achieved to run Ada programs at the Linux kernel level. In their approach, the Ada run-time system is executed as a Linux kernel module. Two levels can be identified: the Ada application at the lower level and the Linux OS providing support to other non real time applications.

In our approach, we maintain the RTLinux 3.1 layer, which offers a POSIX.13 subset and we build Ada applications to be executed on top of RTLinux only by slightly modifying the GNAT 3.14 run-time system and by adding the functionality needed to allow the execution of the Ada program as a Linux module. The fact that RTLinux version 3.1 is more POSIX-oriented than previous releases obviously eases the porting.

At first glance, this approach could seem not suitable for small embedded systems, because it requires to install the full Linux OS. However, an important effort has been done in parallel in our group [13] to minimise the Linux footprint and to include only a minimum number of selected services by means of a configuration tool to generate the embedded system.

The structure of this paper is as follows. Section 2 gives a brief overview of the RTLinux structure. Section 3 explains the modifications we have done to some parts of the GNAT run-time system and the functionalities implemented by the new Real-Time GNAT Layer (RTGL). Section4 presents several examples of the Ada features that this work provides support for. In Sect. 5, some performance metrics are provided. Section 6 presents our conclusions and indications for further work on this topic.

## 2  Overview of RTLinux

A general purpose operating system (OS), such as Linux [10] deals directly with all hardware resources such as input-output addresses, memory management and interrupts. This kind of OSs try to provide good average response times, not to guarantee maximum bounds of execution time. This makes general purpose OSs inadequate to achieve hard real-time performance, with guaranteed deadlines.

In RTLinux [4], the Linux OS is patched in order to create the RTLinux layer, between the hardware and Linux itself. This layer captures interrupts before passing them to Linux and intercepts Linux's disabling and enabling of interrupts. In this way, RTLinux takes control of the real interrupts, whilst Linux only receives interrupt requests when RTLinux has already acknowledged them.

Interrupts directed to Linux will be passed only when Linux has interrupts enabled and RTLinux *reissues* them to Linux. When Linux executes the macro for disabling interrupts it really does not disable them, it only informs RTLinux not to pass more interrupts until the next enabling of interrupts.

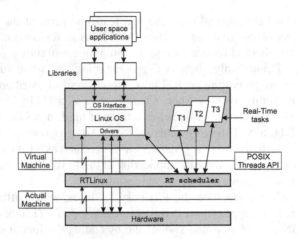

**Fig. 1.** Architecture of RTLinux

Figure 1 shows the architecture of RTLinux. It can be seen as RTLinux having the real view of the hardware and Linux having a virtual view, given by RTLinux.

Real-time applications are implemented as loadable kernel modules. The application's threads may use the RTLinux API, an interface very similar to POSIX for creation and destruction of threads, managing of signals, accessing devices, communication and synchronisation, etc. The different parts of this API are also implemented as modules that can therefore be loaded or not, depending on the real-time application requirements.

Linux modules are dynamically linked with the Linux kernel, forming a single program (the Linux kernel with the added functionalities provided by the modules). Therefore the modules run in the same address space as the kernel. In this situation, an erroneous real-time task can make the whole system crash. According to [4], *the use of the C language aggravates this problem*, since *equivalence of arrays and pointers, type casts make it all too easy to write programs with memory referencing bugs*. This is a good reason to use Ada instead, given its advantages over C to produce programs that are more reliable (e.g., syntax, strong typing, readability...)

It is important to note that the real-time scheduler is also a loadable module, easily replaceable, which allows to test different scheduling policies.

## 3   Creating RTLinux Modules from Ada Sources

RTLinux uses a dynamic module loader from Linux to load and execute real-time modules at the kernel level. These modules are relocatable code in ELF format (Executable and Linking Format) which export several predefined symbols such as the functions `init_module()` and `cleanup_module()`, and other symbols like `author` and `license`. The first two are called when loading and unloading the module, respectively, and implement the desired functionality for such events. Typically, `init_module()` first demands resources and initialises the devices to be used by an associated thread, which is then released. It can be seen as the `main` function in a typical C program. This function is automatically called upon insertion of the module with the Linux command `insmod`. On the other hand, the function `cleanup_module()` is executed at module unload (by means of the Linux `rmmod` command) and it takes care of freeing the previously allocated resources.

RTLinux modules are usually implemented in C, therefore it is straightforward to import these symbols with `include`. But Ada requires a different way of making these symbols visible. We shall distinguish two cases to explain our approach: Ada programs that do not require the run-time system services and programs that do need run-time support (e.g., tasking programs).

### 3.1   Programs with No Run-Time System Support

First, we will describe how it is possible to load an Ada program as a kernel module without considering the run-time system.

There are two ways to load an Ada program as a kernel module; one way is to export the necessary symbols from the Ada program, as suggested by T. Baker in [3]. The other way is to implement a C program, which exports the necessary symbols, and then link it with our Ada program; this is the method of our choice. We have found two reasons for this approach:

- Several RTLinux functions are written as macros, and it is easier to use them in C.
- `init_module()` and `cleanup_module()` call several RTLinux functions (e.g., for the creation of threads). If we adopted the first approach, we should provide an interface to all these functions (Baker suggests to include them in `System.OS_In-terface`). In our approach, we provide an object file that is incrementally linked with the application, therefore hiding RTLinux-specific symbols. This object file defines the Real-Time GNAT Layer, RTGL, that will be explained below.

An Ada program with no run-time can be easily compiled into a loadable kernel module by only providing the appropriate modifiers to the compiler command, such that the program is linked with a simple C object file before the final module is created.

### 3.2   Programs with Run-Time System Support

Programs that make calls to the Ada run-time system require a modified run-time system in order to be executed on RTLinux. In this section, we shall explain the modifications needed to implement our Ada-RTLinux system.

The compiler we are based on is GNAT 3.14 for Linux on an i386 platform, with FSU threads, a POSIX threads library. The applications compiled with GNAT are structured in several layers. The Ada program uses the services of the GNU Ada Runtime Library (GNARL), which interfaces with the underlying OS through a lower level layer called GNULLI (GNARL Lower Level Interface).

We have done a minimum number of changes to GNARL, most of them at the GNULLI level. In particular, the packages `System.OS_Primitives` and `System.OS_-Interface` have been changed to use the interface of RTLinux instead of Linux and to provide adequate support for the ceiling locking operation.

In addition, several packages have been added to provide the following functionalities:

Rtl_IO This package implements a front-end for the rtl_printf() function, which allows to Put objects of several data types on the screen. We have not implemented a Get equivalent yet.

Rtl_Fifo This package provides an interface to use the RTLinux FIFOs. It allows to create and destroy FIFOs, and to send to and receive from them. Rtf_Create and Rtf_Destroy make calls to the Linux Kernel. Since RTLinux modules run in the kernel memory space, special care must be taken to implement these calls. The algorithm we have implemented invokes the FIFO creation primitive by means of an interrupt that is queued to Linux, not a direct call. This avoids random errors in the Linux kernel. This method is not bounded in time, but it is important to note that the FIFOs are created statically at startup, not during the normal program execution.

Rtl_Interrupts This package provides the functionality needed to deal with interrupts. It is basically a binding to the RTLinux functions for interrupt enabling, disabling and handling support.

Rtl_Pt1 This package provides a binding to the RTL Posix Tracing standard (POSIX 1003.1q), version 1.0. This package, in conjunction with the Quivi viewer [1], allows us to graphically monitor the temporal behaviour of the Ada tasks in a program.

### 3.3   The RT-GNAT Layer

The RT-GNAT Layer (RTGL) is a new layer that we have implemented to load an Ada program as a kernel module and which allows us to give support to the GNAT runtime system. It exports kernel symbols (like init_module, cleanup_module, author, license and kernel_version strings), and also the RTLinux interface and library functions.

The services offered by RTGL are:

- Library functions. The GNAT run-time system uses some standard functions (for string management and time conversions) which are not provided by RTLinux or the Linux kernel. RTGL provides them. Some of these functions are taken from the OSKIT project (Utah University, [7]) and from the source code of GNU GCC 2.8.1.
- Exported symbols. Since RTGL is implemented in C, it can import Kernel headers and it can directly export symbols like kernel version, etc. RTGL imports the init_module and cleanup_module functions.
- A dynamic memory manager. RTLinux does not provide dynamic memory management. This feature is provided by the Linux kernel and thus it can not be used from an RT-thread. We need dynamic memory since the Ada runtime uses it for task creation, string creation, etc. When init_module() is executed our memory manager demands a sufficient amount of memory (user-defined) via the bigphysarea module, which allows us to obtain a large amount of contiguous memory and make it available to the GNAT run-time system. This memory is managed with a dynamic memory management algorithm, called DIDMA (Doubly Indexed Dynamic Memory Allocator) that we are currently developing [8]. The main goal of this algorithm is to provide allocation and deallocation of memory in a bounded time.

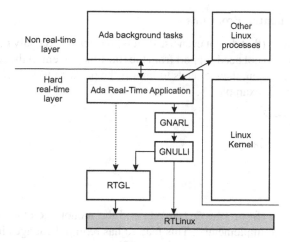

**Fig. 2.** Architecture of the whole system, with an Ada application running on top of RTLinux

The resulting application architecture can be seen in Fig. 2. The Ada application is divided into two parts, one with hard real-time requirements (the *Ada real-time application*) and another part that is scheduled by Linux (*Ada background tasks*), since it has no strict timing requirements. These two parts are two different Ada programs. The link between them is via real-time FIFOs. The real-time FIFO mechanism provides a rapid way of communicating tasks in the real-time and non real-time layers of the system. It also allows the hard real-time tasks to communicate with applications written in a different language. Real-time FIFOs are provided as special devices in the /etc/dev directory.

The Ada real-time application uses the services provided by RTGL only at startup, for the creation of threads. The program then uses the modified run-time system to obtain the tasking services (rendezvous, protected objects, etc). The Linux kernel is scheduled in the background with respect to the Ada real-time application components.

The process of generating the module from the Ada sources consists in calling gnatmake with the appropriate options. A script called rtgnatmake has been written to automate this process.

## 4   Implemented Features

This section explains the Ada features we have already implemented and tested. We have followed an incremental approach, from the simpler to the more complex Ada features. First, we compiled and run a simple sequential *Hello world* program. Then we added delays, exceptions, tasks, protected objects, and asynchronous transfer of control.

To compile an Ada program for RTLinux, the RTGL layer is needed in order to generate the corresponding loadable kernel modules. Besides this layer, some minor modifications have been made to System.OS_Primitives (for non-tasking programs) and System.OS_Interface (for programs that use tasking.)

## 4.1  Simple Sequential Programs

Sequential programs (that do not even use delays) written in Ada only require the RTGL layer, because they do not need support from the run-time system. If the program requires to Put text on the screen, then it needs to import the package RTL_IO. The following is a simple *Hello world* example:

```
with RTL_IO; use RTL_IO;
procedure Hello is
begin
   Put_Line ("Hello World!");
end Hello;
```

## 4.2  Delays

Both the wall clock (from Ada.Calendar) and the monotonic clock (from Ada.Real_Time) have been implemented. This feature has required changes in System.OS_Primitives and in System.OS_Inteface. These changes consist in converting the time types used in RTLinux to the internal representations of time in the GNAT run-time system.

## 4.3  Exceptions

As for today, we have only implemented software generated exceptions, i.e., exceptions generated either by an explicit raise sentence in the program, or generated by the run-time system by means of procedure Raise. Hardware originated exceptions, like FP exceptions, have not been implemented yet. These exceptions generate hardware traps. The default handler in RTLinux just dumps the processor state to the screen and halts the system. We need to investigate further how to treat these exceptions and return to a consistent state. The following sample program can be compiled and executed on RTLinux:

```
with RTL_IO; use RTL_IO;
procedure Exceptions_Example is

   procedure Exception_Error is
      Error : exception;
   begin
      raise Error;
   exception
      when Error =>
         Put_Line ("This is an error");
   end Exception_Error;

   Str: String(1..2);

begin
   Exception_Error;
   delay 1.0;
   Str := "123";  -- This line produces CONSTRAINT_ERROR
exception
   when CONSTRAINT_ERROR =>
      Put_Line ("*** This is a CONSTRAINT_ERROR");
end Exceptions_Example;
```

The screen output of this program is, as expected, the string "This is an error" followed by the message "*** This is a CONSTRAINT_ERROR" one second later.

## 4.4   Tasks

The implementation of tasking has required to modify `System.OS_Primitives` in order to import the RTLinux types and functions used for this purpose. The type `pthread` in RTLinux corresponds to a record, with some components of types defined in the Linux kernel. The required Linux types have been imported to make them visible in the definition of the `pthread` type. Also, the signal masks for threads have been changed to substitute the Linux signals with the RTLinux signals. RTLinux only implements a subset of the Linux signals.

## 4.5   Protected Objects, Ceiling_Locking, and Dynamic Priorities

Protected objects are already implemented in the Linux run-time system of GNAT, but some changes were required to provide an adequate support to the ceiling locking policy. The first requirement has to do with the installation of RTLinux itself. The installation script of RTLinux allows to optionally enable this policy. By default it is disabled, but this policy is necessary to guarantee bounded blocking times. Therefore we need to explicitly enable this option when installing RTLinux.

An important aspect related to the locking policy is the ability to dynamically change a task priority. We have found that the implementation of dynamic priorities in RTLinux does not match the ceiling locking policy correctly. One of the fields of a thread descriptor is the thread's priority. The value of this field represents the priority at which the thread is currently being executed. On the other hand, part of the descriptor of a mutex is a field called *oldprio* that reflects the priority of the thread when it locked the mutex, before the thread inherited the ceiling priority of the mutex. The raising of the thread's priority upon locking the mutex consists in changing the thread's priority in the thread's descriptor. This makes the thread execute at the ceiling priority. When the thread unlocks the mutex, the priority field in the thread's descriptor is overwritten with *oldprio*.

The problem comes when another thread wants to change the base priority of the thread that is locking a mutex. This priority change is also implemented as a change in the priority field of the thread's descriptor. Therefore, this new priority will be lost when the thread unlocks the mutex. We realised this anomaly when we saw some priority changes that did not take effect when the target task was holding a mutex. Actually, we found that no priority change was effective, since setting the priority of a task was always executed whilst holding an internal mutex.

This problem has required to give the adequate support in RTGL for dynamic priorities, that takes into account that `Set_Priority` may be called whilst the target task is running a protected operation. When a thread locks a mutex, we take note that the thread is holding that particular mutex. If the priority of the thread must be changed whilst holding the mutex, then the new priority is written in the *oldprio* field of the mutex. In this way, the priority change will take effect immmediately when the thread unlocks the mutex. For the case of nested calls to protected operations in different protected objects, only the outermost *oldprio* is changed. In fact we have a list of thread descriptors, one of whose fields is the identifier of the mutex held by the thread (or `null`). The access to this list is optimised by means of a simple mapping function that gives access to the mutex descriptor with cost $O(1)$.

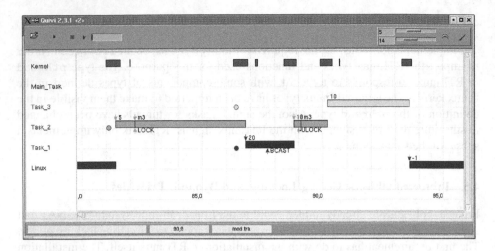

**Fig. 3.** Quivi chronogram of the protected object example with dynamic priorities

In order to check this implementation, we have prepared a test program consisting of:

- A protected object PO with a ceiling priority equal to 18. It has a single protected procedure whose only mission is to busy-wait during 40 $\mu$s.
- Task_1 is a task that alternatively sets the priority of another task (Task_2) to 15, then waits for 500 $\mu$s, then sets Task_2's priority to 5, then waits for another 500 $\mu$s and then repeats the cycle.
- Task_2 is also a cyclic task with a period of 300 $\mu$s. Its priority is initially 15, but it changes due to the execution of Task_1. This task calls the protected procedure of PO once each cycle.
- Task_3 is a cyclic task with a priority equal to 10, i.e., the lowest priority task. It has a period of 150 $\mu$s.

Figure 3 shows a snapshot of the program execution. This chronogram has been obtained with the Quivi tool, mentioned in section 3.2. The left column shows the names of the different entities that can be monitored with Quivi. Kernel represents the kernel activity; Main_Task is the main Ada subprogram; Then we can see the three tasks in this example; and finally the time consumed by the Linux OS can be seen. A dot in a task line represents its activation time, according to its period. When a task starts executing and when there is a context switch, a number indicates the priority of the task gaining access to the processor. Figure 3 shows a time window where Task_2 activates and starts executing with a priority equal to 5. Then it locks the mutex associated with PO and starts executing the protected procedure with the inherited ceiling priority 18. Afterwards, both Task_1 and Task_3 activate. Task_1 gains access to the processor and, as described before, changes Task_2's priority to 15. This change does not take effect immediately, since Task_2 is executing a protected procedure. When Task_1 ends, then Task_2 is chosen to execute. It does it with a priority 18 at the beginning, because it is still running the protected operation, but it still keeps on running after unlocking PO's mutex, despite the fact that Task_3 is also ready. This is only possible because Task_2's priority is effectively 15, which forces Task_3 to wait for Task_2 to complete.

### 4.6 Other Features

We have successfully tested other constructs of Ada such as rendezvous, asynchronous transfer of control, timed entry calls, selective accept and conditional entry calls. These features worked with no more changes required, since the GNAT run-time system already implemented them.

## 5 Performance Metrics

We have done some performance tests for this implementation. Up to now, we know that the maximum frequency of a single real-time task we can achieve is 50 KHz on an Athlon/600 MHz. On the same hardware, MaRTE 1.0 achieves 66 KHz. We did not expect to have better performance, since we have not focused on optimizations yet. For instance, the hardware clocks we are using are the Programmable Interval Timer (PIT) and Time Stamp Counter (TSC) of the ix86 architecture. It takes 3 $\mu$s to program the TSC+PIT timer on a Pentium III/550 MHz. We expect to achieve better results by using the APIC clock, which only requires 324 ns on the same platform. Another important source of performance differences with MaRTE is the fact that MaRTE uses a flat memory approach, with paging disabled, whilst Linux requires paging to be enabled, which also imposes an overhead.

To measure the kernel overhead, we have reproduced the tests presented in [11], based on calculating the maximum achievable utilisation (in practice) of an harmonic task set. The task set is formed by 6 tasks with frequencies 320 Hz, 160 Hz, 80 Hz, 40 Hz, 20 Hz and 10 Hz. We gradually adjust the amount of time consumed by each task and we do it proportionally to the taks' periods, i.e., 1 $\mu$s increments for the most frequent task, 2 $\mu$s for the second, 4 $\mu$s for the third, 8 $\mu$s for the fourth, and so on. The maximum theoretical utilisation in this case is 100%. The difference between the actual utilisation achieved and the theoretical maximum gives a figure of the total overhead imposed by the OS. We have compared an Ada implementation of the test program with a similar program in C, therefore we can measure both the kernel and the run-time system overhead. The measured utilisations are 98.69% for the Ada version and 99.45% for the C version. These numbers indicate a low kernel overhead (1.31% in Ada; 0.55% in C) and a relatively small overhead introduced by the run-time system (99.45% − 98.69% = 0.76%).

These results have been repeatedly obtained, but some times we have had slightly lower utilisations. It is important to note that RTLinux is not in itself a deterministic platform. A Linux process may really disable interrupts (not virtually) by using `ioctl`. For instance, running an X server interferes the results. Our conclusion in this sense is that RTLinux is an excellent platform for soft real-time or teaching purposes, but also that one needs to restrict the running applications to those that respect the RTLinux layer. Obviously, caching is also a source of indeterminism.

# 6   Conclusions and Future Work

In this paper, we have described our approach to run Ada programs on RTLinux, explaining the additions and modifications needed for such a purpose. We have achieved our main goal of running Ada programs on RTLinux. The implementation is still in the development and testing phase, but we have had encouraging results, such as porting a 17-task application that we had previously developed in MaRTE OS for the control of two robotic arms and a chain of conveyor belts with two machining stations. We only had to replace a with clause to make the application run on RTLinux. We also plan to use this approach for teaching real-time systems, since it has interesting advantages: RTLinux is freely available; we can use an instrumented version of RTLinux in combination with the visualization tool Quivi in order to graphically analise timing properties; and we only need one computer for both development and testing.

The porting of GNAT to RTLinux is quite transparent: at the moment, the programmer only needs to know the package Rtl_IO for simple input and output of text. Additionally, Rtl_FIFO provides a binding for using real-time FIFOs and Rtl_Ptl gives an interface to the POSIX tracing facilities.

Another important conclusion from the experiments we have developed is related with the suitability of RTLinux for hard real-time systems. Predictability in RTLinux can only be achieved if we make sure that no running application can gain root privileges to disable interrupts. Otherwise, the timing requirements of the real-time application cannot be guaranteed.

The resulting architecture of the Ada application, with a non real-time layer and a hard real-time one, motivates us to study the applicability of the Distributed Systems Annex of the Ada standard in the future. The whole Ada program can be seen as a multipartition program, one partition running the hard real-time tasks and one or more additional partitions for the non real-time tasks. Ideally, the communication mechanisms between real-time and non real-time tasks (the FIFO buffers) would be hidden under the distributed model of Ada, therefore releasing the programmer to deal with FIFO primitives.

**Acknowledgements.** This work has been partly funded by the European OCERA project [9] and by the Spanish *Comisión Interministerial de Ciencia y Tecnología* project number TIC-99-1043-C03-02. We also want to thank Agustín Espinosa for his work on Quivi and Andrés Terrasa for instrumenting RTLinux.

# References

1. A. Espinosa. Quivi: a Chronogram Viewer in Tcl-Tk (*in Spanish*).
   http://www.dsic.upv.es/~aespinos/quivi.html, 2002.
2. M. Aldea and M. González-Harbour. MaRTE OS: An Ada Kernel for Real-Time Embedded Applications. *Reliable Software Technologies - Ada Europe 2001, Lecture Notes in Computer Science*, 2043:305–316, 2001.
3. T. Baker. Ada and Embedded Real Time Linux. In *SIGAda Conference*, 2000.
4. M. Barabanov. A Linux-based Real-Time Operating System. Master's thesis, New Mexico Institute of Mining and Technology, Socorro, New Mexico, June 1997.

5. A. Burns. The Ravenscar Profile. *Ada Letters*, XIX(4), 1999.
6. J. de la Puente, J. Zamorano, J. Ruiz, R. Fernández, and R. García. The Design and Implementation of the Open Ravenscar Kernel. In *Proceedings of IRTAW10, Ada Letters*, volume XXI, nr 1, pages 85–90, 2001.
7. J. Lepreau, M. Flatt, and et. al. The Oskit Project. University of Utah, http://www.cs.utah.edu/flux/oskit/, 2002.
8. M. Masmano. Doubly Indexed Dynamic Memory Allocator. http://www.ocera.org/download/components/WP4/dynmem.html, 2003.
9. Open Components for Embedded Real-Time Applications. European IST programme 35102. http://www.ocera.org/, 2002.
10. Open Source Development Network. Linux news, information, software, documentation and tutorials. http://linux.com/, 2002.
11. H. Shen, A. Charlet, and T. Baker. A "Bare-Machine" Implementation of Ada Multi-Tasking Beneath the Linux Kernel. *Reliable Software Technologies - Ada Europe 99, Lecture Notes in Computer Science*, 1622:287–297, 1999.
12. VentureCom. RTX. http://www.vci.com/index.asp, 2002.
13. J. Vidal, P. Mendoza, I. Ripoll, and J. Vila. A Tool for Customizing RT-Linux to Embedded Systems. In *Real-Time Linux Workshop, Milano*, 2001.

# A Round Robin Scheduling Policy for Ada

A. Burns[1], M. González Harbour[2], and A.J. Wellings[1]

[1] Department of Computer Science
University of York, UK
{burns,andy}@cs.york.ac.uk
[2] Departamento de Electrónica y Computadores
Universidad de Cantabria, Spain
mgh@unican.es

**Abstract.** Although Ada defines a number of mechanisms for specifying scheduling policies, only one, `Fifo_Within_Priorities` is guaranteed to be supported by all implementations of the Real-Time Systems Annex. Many applications have a mixture of real-time and non real-time activities. The natural way of scheduling non real-time activities is by time sharing the processor using Round Robin Scheduling. Currently, the only way of achieving this is by incorporating yield operations in the code. This is ad hoc and intrusive. The paper proposes a new scheduling policy which allows one or more priority levels to be identified as round robin priorities. A task whose base priority is set to one of these levels is scheduled in a round robin manner with a user-definable quantum.

## 1 Introduction

The Real-Time Systems Annex in Ada 95 defines a number of mechanisms for specifying scheduling policies. It also provides a complete definition of one such policy: `Fifo_Within_Priorities`. This policy, which requires preemptive priority based dispatching, is a natural choice for real-time applications. It can be implemented efficiently and leads to the development of applications that are amenable to effective analysis – especially when combined with the immediate priority ceiling protocol (ceiling locking) on protected objects.

There are, however, application requirements that cannot be fully accomplished with this policy alone. For example, many applications have a mixture of real-time and non real-time activities. The natural way of scheduling non real-time activities is by time sharing the processor, as in most general-purpose operating systems. With the standard policy (`Fifo_Within_Priorities`), some level of rotation can be achieved by giving the set of non real-time tasks the same priority and requiring them to incorporate periodic yield operations (such as delay 0.0). However this is an ad hoc approach, is intrusive, and cannot easily be undertaken with legacy code or when using prewritten or shared libraries. Real-time applications can also benefit from a round robin approach. Although extra task switches increases run-time overheads, round robin execution allows a set of tasks (with the same priority) to make progress at a similar rate.

The aim of this paper is to define a new scheduling policy for the Real-Time Systems Annex, with a view to this definition (or one derived from it) being incorporated into the

J.-P. Rosen and A. Strohmeier (Eds.): Ada-Europe 2003, LNCS 2655, pp. 334–343, 2003.

Annex. This would be one of a set of new policies being proposed for the Annex. For example, a non-preemptive dispatching policy has already been agreed[1].

This paper is organised as follows. First an overview of the POSIX provision is given. Ada is closely associated with POSIX and indeed some Ada run-time systems (e.g. GNAT) are built on top of POSIX compliant kernels. It is important that any proposal for Ada is implementable on POSIX even if the details of the Ada policy are not identical to the Round Robin facility of POSIX. Section 3 then gives the requirements of the Ada policy and Sect. 4 the details of the proposal (including a discussion of some of the priority inheritance problems that ensue). Conclusions are given in Sect. 5.

## 2   The POSIX Policies

The POSIX real-time scheduling model [3,2] is a fixed-priority preemptive model in which there are three compatible scheduling policies defined:

- SCHED_FIFO. It is a priority preemptive policy that uses FIFO ordering for threads of the same priority. It is similar to Ada's FIFO_Within_Priorities policy.
- SCHED_RR. It is also a priority-preemptive policy that uses a round-robin execution quantum to share the processor among threads of the same priority level. In this policy, when the implementation detects that a running thread has been executing for a time period of the round robin quantum or longer, the thread is placed at the tail of the scheduling queue for its priority, and the head of that queue is removed and made the running thread. While a round robin thread is preempted by higher priority threads it does not consume its unused portion of round robin quantum. This policy ensures that if there are multiple threads at the same priority, one of them will not monopolize the processor.
- SCHED_SS. It is the sporadic server scheduling policy, intended for processing aperiodic threads with a guaranteed bandwidth, and with predictable effects over lower priority threads.

These policies can be set on a thread by thread basis, because the effects of mixing them are well defined. For example, when a round robin thread is running, its execution time is limited to its time quantum. After the quantum is elapsed, the thread is sent to the tail of the ready queue for its priority. If a FIFO within priorities thread now comes into execution, it runs until completion (possibly preempted by higher priority threads during its execution). It is the responsibility of the application developer to make sure that no mixture of round robin and FIFO threads is made at the same priority level, if the round robin semantics is to be preserved.

The ranges of valid priorities for the SCHED_FIFO and SCHED_RR may coincide, overlap, or be disjoint. Each of these ranges is required to have at least 32 distinct priority levels. There are functions that allow the application to obtain such priority ranges in a portable way.

In order to portably define a set of priorities that is suitable for mixing real-time and non real-time threads, a new requirement is being proposed in the revision of the POSIX.13 real-time profiles [4], which in essence states that there should be at least one round robin priority level that is below the first 31 priority levels of the SCHED_FIFO policy. In this way, in a portable application, the non real-time threads would use that round robin priority level, while the real-time threads would use the top 31 values of the SCHED_FIFO policy.

Real-time POSIX defines the mutex as the mechanism to achieve mutual exclusive synchronization for shared data and resources. Mutexes have an optional creation attribute, the protocol, that can be used by the application to specify a real-time synchronization protocol, if desired. Two such protocols are defined:

- PTHREAD_PRIO_INHERIT: This is the basic priority inheritance protocol, in which a thread that is blocking one or more threads due to the use of the mutex inherits their priorities.
- PTHREAD_PRIO_PROTECT: This is the immediate priority ceiling protocol, in which each mutex has an attribute called its priority ceiling; while a thread holds the lock on one or more mutexes, it inherits the priority ceilings of those mutexes. This is basically the same as Ada's Ceiling_Locking protocol for protected objects.

One problem that affects real-time behavior is the relationship between the round robin scheduler, and the mutexes. Ideally, to minimize blocking, expiration of the round robin quantum should be delayed if the thread is holding a mutex with one of the PTHREAD_ PRIO_INHERIT of PTHREAD_PRIO_PROTECT protocols. The standard does state that "While a thread is holding a mutex that has been initialized with the PRIO_INHERIT or PRIO_PROTECT protocol attributes, it shall not be subject to being moved to the tail of the scheduling queue at its priority". But the standard does not list the expiration of the round robin quantum as one of the circumstances to which this statement applies, thus leaving the behavior unspecified.

## 2.1 Overheads of Implementation

Depending on the degree of precision that the implementtaion gives to round robin scheduling, the execution time quantum can be measured in a coarse way, for example counting ticks of a particular clock (and accounting for partial ticks as full ones), or in a more precise way by accounting for actual execution time.

As an example, we have implemented a precise round robin scheduler in MaRTE OS [6], and we have measured its overhead relative to the overhead of the SCHED_FIFO policy. We have mixed the implementation of this scheduler with the execution-time budget mechanism also available in MaRTE OS. For example, the expirations of round robin quantum are programmed as execution time events, similar to those of other budgets associated with execution-time timers. The implementation of execution time budgets has taken approximately 66 lines of code, and has incremented the standard context switch from 0.42 microseconds to 0.44, a negligible 20 ns increase (as measured on a 1.1GHz Pentium III). The implementation of the round robin policy itself has taken 30 lines of code, of which 10 lines are shared with the sporadic server scheduling policy, also specified in POSIX and available in MaRTE OS. The context switch time due to the exhaustion of a round robin quantum, 0.75 us, is very similar to the context switch caused by the expiration of a relative sleep algorithm, 0.75 us. Both include the time required to handle the timer hardware interrupt, which was not necessary for the regular context switch. Although these numbers are those inside the OS and do not take into account the time spent by the Ada Runtime System, it is expected that the overheads there will be comparable.

## 3 Requirements for Round Robin Scheduling

Although the basic requirements for Round Robin Scheduling are straightforward, and many examples exist in general purpose operating systems (as well as the POSIX interface definition), there is more than one approach possible to providing this dispatching behaviour; see Aldea Rivas and González Harbour for a discussion [5]. Clearly a non-intrusive method is needed. Hence a policy must be defined that forces Round Robin scheduling on a set of tasks without the need to make any code changes to the tasks themselves. One approach, following POSIX, would be to assign Round Robin status on a per task basis. This is a very general facility that is appropriate for an OS interface definition. However, from the perspective of the Ada Language, introducing a per-task scheduling policy represents a substantially different model than the one currently defined in the Real-Time Annex. In addition, allowing a mixture of tasks with Round_Robin and FIFO_Within_Priorities at the same priority level introduces a potential for mistakes that would degrade real-time performance. The following requirements are therefore necessary and sufficient.

- Round Robin (RR) scheduling should co-exist with standard preemptive priority based scheduling.
- All tasks at any priority level should be dispatched according to a single policy (either Fifo_Within_Priorities or a round robin).
- At a RR priority level, each task is given the same quantum of CPU resource.

This different uses of priority are illustrated in Fig. 1.

When a task exhausts its quantum of CPU resource, it is placed at the back of the dispatching queue of tasks at that priority level.

Hence, we take a priority-centred rather than a task-centred view. A full proposal (see Sect. 4) must, therefore, cater for the use of dynamic priorities that may move a task into or out of a RR priority level. It must also give a meaning to a protected object being assigned a RR priority.

A significant requirement is that the proposal must deal with the situation in which a task's quantum is exhausted while it has an inherited priority. In particular

- No extra lock must be needed to ensure mutual exclusive execution with protected objects.

The most straightforward use of Round Robin scheduling is to require it only at the lowest priority level (as with the new POSIX proposal [4]). All other priorities use the normal preemptive scheme, but the spare capacity that becomes available to the lowest priority level is shared between a set of non real-time tasks. Scheduling analysis could be used to give a minimum value to this spare capacity (over a relatively long time period) and hence, with knowledge of the quantum size and number of tasks at the RR level, a measure of progress could be calculated.

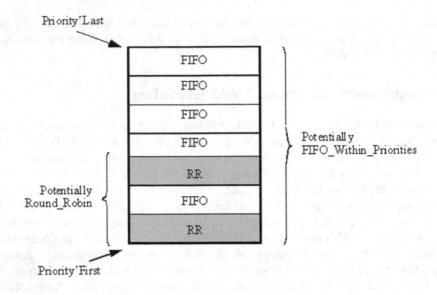

**Fig. 1.** Priority Levels

Round Robin scheduling can also be used at priority levels other than the lowest one. Here, the goal is to reduce the output jitter of a set of periodic tasks which all have the same period and deadline. Without Round Robin scheduling, the variation in the scheduling of these tasks results in a large variation in their response times. If the tasks are outputting to the environment, it may be necessary to bound the jitter on that output. This use of Round Robin scheduling within a real-time program explains our proposal to place the new policy in Annex D (the real-time annex).

A final observation concerns the size of the quantum. Ideally this should be programmable (at least during elaboration) so that each RR priority level can choose the size of its quantum. However, some implementations, for example those built upon POSIX, may need to impose restrictions. A compromise is therefore needed in the proposal.

In a current AI (Ada Issue) a proposal is being considered for adding CPU-Time budgeting capabilities to the Ada standard. This AI defines a new package, `Execution_Time` (as a child of `Ada.Real_Time`), that contains operations to measure the execution time of the different tasks in the system, and to limit the execution time of the tasks to desired execution time budgets, with facilities to detect execution time overruns. As part of this package, a private type called `CPU_Time` is defined to represent periods of execution time. This type seems the most appropriate for specifying the round robin quantum values, and therefore we have used it in this proposal.

## 4  Proposal

An extension is proposed for Annex D of the Ada Reference Manual (ARM). First a new policy-identifier is defined for task dispatching.

```
pragma Task_Dispatching_Policy(Priority_Specific);
```

When Priority_Specific is used, the dispatching policy is defined on a per priority level. This is achieved by the use of a new configuration pragma:

```
pragma Priority_Policy (Policy_Identifier, Priority
                        [,Policy_Argument_Definition]);
```

The Policy_Identifier shall be Fifo_Within_Priorities, Round_Robin or an implementation-defined identifier. For Policy_Identifier Fifo_Within_ Priority there shall be no Policy_Argument_Definition. For Policy_ Identifier Round_Robin, the Policy_Argument_Definition shall be a single parameter, Round_Robin_Quantum, of type CPU_Time (from the Execution Time proposal).

For other policy identifiers, the semantics of the Policy_Argument_Definition are implementation defined. At all priority levels, the default Priority_Policy is Fifo_Within_Priorities. The locking policy associated with the Priority_Specific task dispatching policy is Ceiling_Locking.

An implementation that supports Round Robin Scheduling must provide the following package:

```
with Ada.Real_Time.Execution_Time; use Ada.Real_Time;
package Round_Robin_Dispatching is
   Default_Quantum : constant Execution_Time.CPU_Time;
   Minimum_Quantum : constant Execution_Time.CPU_Time;
   Maximum_Quantum : constant Execution_Time.CPU_Time;
   function Nearest_Supported_Quantum
            (Q : Execution_Time.CPU_Time)
            return Execution_Time.CPU_Time;
   Maximum_Priority_Level : constant System.Priority;
private
   -- definition of constants, implementation defined
end Round_Robin_Dispatching;
```

Here a default quantum is given together with the minimum and maximum values supported. Note although the type is private, a value in seconds can be obtained via the procedure Split defined in Execution_Time. If an implementation only supports one value then the minimum and maximum are set to the default. The function is provided to deal with implementations that only support discrete values for the quantum. If the user calls this function with a desired budget then the function returns the actual value that will be used at run-time (this will be at least the minimum and no more than the maximum).

The final constant in the package gives the maximum priority that can be specified for RR dispatching. Note this is of type Priority and hence cannot be at an interrupt priority level. The lowest value this can be is Priority'First; in which case only the lowest priority is available for RR scheduling.

Use of pragma Priority_Policy will be checked with the following rules applying:

– If the same priority is given in more than one pragma, the partition is rejected.

- If the Policy_Identifier is Round_Robin and the value of Priority is greater than Round_Robin_Dispatching.Maximum_Priority_Level then the partition is rejected.
- If the Policy_Identifier is Round_Robin and no quantum is given, the default in Round_Robin_Dispatching applies.
- If Task_Dispatching_Policy is not Priority_Specific for the partition in which a pragma Priority_Policy appears, then the partition is rejected.

It is now necessary to consider the dynamic semantics of the proposal. If all priority levels have Priority_Policy Fifo_Within_Priorities then this is equivalent to Task_Dispatching_Policy Fifo_Within_Priorities. The dynamic semantics defined in D.2.2 of the ARM for Fifo_Within_Priorities apply to any priority level with Priority_Policy Fifo_Within_Priorities.

For Policy_Identifier Round_Robin, the same rules for Fifo_Within_Priority apply with the additional rules and modifications:

- When a task is added to the tail of the ready queue for a priority level with Priority_Policy Round_Robin, it has an execution time budget set equal to the Round_Robin_Quantum for that priority level. This will occur when a blocked task becomes executable again.
- When a task is preempted (by a higher priority task), it is added to the head of the ready queue for its priority level. If this is a RR priority level then it retains its remaining budget.
- When a task with a base priority at a RR priority level is executing, its budget is decreased by the amount of execution time it uses.
- When the implementation detects that a task with a round robin priority has been executing for a time larger than or equal to its round-robin quantum, the task is said to have exhausted its budget. When a running task exhausts its budget, it is moved to the tail of the ready queue for that priority level. The semantics of this move is equivalent to the task with priority Pri executing Set_Priority(Pri); see D.5(15) of the ARM. Hence, for example, it will continue to execute within a protected operation.

The last rule together gives the important details of the proposal. First it is the base priority of a task that is significant. If a task's base priority is at a RR level then it will consume its budget whenever it is executing even when it has inherited a higher priority (i.e. its active priority is greater than its base priority). The final point deals with the key question of what happens if the budget becomes exhausted while executing in a protected object. To ensure mutual exclusion, without requiring a further lock, it is necessary to allow the task to keep executing within the PO. It will consume more than its quantum but the expected behaviour of system is maintained. The usual programming discipline of keeping the code within protected objects as short as possible will ensure that quantum overrun is minimised. Further support for these semantics comes from observing that execution within a PO is abort-deferred. Quantum exhaustion is a less severe state than being aborted; deferred behavior thus seems appropriate.

To complete the definition, a few details have to be covered. A task that has its priority changed, via the use of Set_Priority, may move to, or from, a round-robin priority

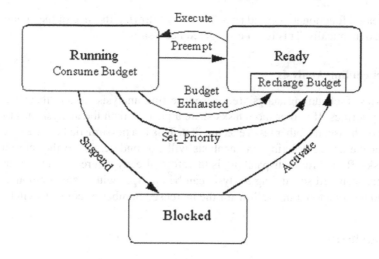

**Fig. 2.** State Transistion Diagram

level. If it is moved to a RR level then it is placed at the tail of the ready queue and given a full quantum. If it moves to a Fifo_Within_Priorities scheme then it is again placed at the tail of the ready queue but no quantum is set. Figure 2 gives an illustration of the rules defined above.

There are no additional rules concerning a protected object with a priority assigned that is a round robin priority. A task does not obtain a budget by executing with an active priority at a round robin level – it is only the base priority of a task that determines its scheduling policy.

An example of usage is as follows. Assume package Ada.Real_Time is visible. To set the default priority level to round-robin with a quantum of 50 milliseconds:

```
pragma Task_Dispatching_Policy (Priority_Specific);

pragma Priority_Policy (Round_Robin, Default_Priority,
                Execution_Time.Time_Of(0,Milliseconds(50)));
```

By setting the quantum to be very large (if allowed by Round_Robin_Dispatching. Maximum_Quantum) a 'run until blocked' at the lowest priority level can be requested:

```
pragma Task_Dispatching_Policy (Priority_Specific);

pragma Priority_Policy (Round_Robin, Priority'First,
                Execution_Time.CPU_Time_Last);
```

This proposal introduces round-robin scheduling by assuming that such a scheme would be used at a particular priority level (or levels) and that a single quantum value is appropriate for each level. Mixing round-robin and non-round-robin at the same priority level, or having quanta defined on a per-task basis is not considered necessary – and could be achieved by using supervisor tasks and execution time budgeting.

Increased functionality could be achieved by allowing the size of the quantum to be changed dynamically. This is not considered necessary.

### 4.1  Scheduling Analysis

Most forms of scheduling analysis (e.g. response-time analysis) assume that all tasks have distinct priorities. If two or more tasks share a priority then the analysis has to assume that, for each task, the other tasks execute first. This is a pessimistic but safe assumption; whatever the actual ordering, the analysis will not underestimate the interference of other tasks. Round robin dispatching is therefore already catered for by this approach, and hence standard scheduling analysis can be applied (with a small amount of extra overhead been factored in to allow for the increased number of context switches).

## 5  Conclusions

This paper has introduced a proposal for adding Round Robin scheduling to the portfolio of policies supported by the Real-Time Systems Annex. Although priority based preemptive scheduling is the policy of choice for most real-time applications, it is increasingly the case that large have both more complete real-time requirements and non real-time components. With just one non real-time task it is easy to construct the code of the task as a non-blocking infinite loop, and give it the lowest priority in the system. But if there are two or more such tasks it is not possible to share this base priority without injecting 'relinquish CPU' statements. Round Robin scheduling is the standard way of time sharing such tasks. Such sharing is also of benefit to real-time code by allowing a set of tasks to make progress at a similar rate.

The motivation for the proposal given here is to allow applications to simultaneously use priority based scheduling and Round Robin. It is intended to forward this proposal (or variant of it) as an AI to the ARG for possible inclusion in the next Ada revision.

**Acknowledgements.** The topic of Round Robin scheduling has been discussed at the Real-Time Workshops, IRTAW – we acknowledge the input from people attending these workshops. The proposal reflects work undertaken as part of the EU funded FIRST project.

## References

1. A.Burns. Non-preemptive dispatching and locking policies. In M. Gonzále Harbour, editor, *Proceedings of the 10th International Real-Time Ada Workshop*, pages 46–47. ACM Ada Letters, 2001.
2. IEEE. Information Technology - Standardized Application Environment Profile - POSIX Realtime and Embedded Application Support (AEP) (POSIX): 1003.13:1998, 1998.
3. IEEE. Information Technology – Portable Operating System Interface (POSIX): 1003.1:2001, 2001.
4. IEEE. Information Technology – Standardized Application Environment Profile - POSIX Realtime Application Support (AEP) Draft 1003.13 (2), 2002.

5. M. Aldea Rivas and M. González Harbour. Extending Ada's real-time systems annex with the POSIX scheduling services. In M. González Harbour, editor, *Proceedings of the 10th International Real-Time Ada Workshop*, pages 20–26. ACM Ada Letters, 2001.
6. M. Aldea Rivas and M. González Harbour. MaRTE OS: An ada kernel for real-time embedded applications. In *Reliable Software Technologies, Proceedings of the Ada Europe Conference, Leuven*. Springer Verlag, LNCS 2043, 2001.

# A Proposal to Integrate the POSIX
# Execution-Time Clocks into Ada 95[*]

Javier Miranda[1] and M. González Harbour[2]

[1] Applied Microelectronics Research Institute, Univ. Las Palmas de Gran Canaria
35017 Las Palmas de Gran Canaria, Spain
jmiranda@iuma.ulpgc.es
[2] Departamento de Electrónica y Computadores, Univ. Cantabria
39005 – Santander, Spain
mgh@unican.es

**Abstract.** In this paper we present a proposal to integrate the POSIX.1 execution-time clocks and execution-time timers into the Ada 95 language. This proposal defines a new package named *Ada.CPU_Time* and describes the modifications done to the GNAT front-end and run-time to support it. Additionally this proposal discusses some usage schemes of this new interface.

**Keywords:** Scheduling, Hard Real-Time, Ada 95, Execution-Time, GNAT

## 1 Introduction

Traditional real-time systems were built (and still are) using executive schedulers [4]. In these systems, if a particular task or routine exceeded its budgeted execution time, the system could detect the situation. Basically, whenever the minor cycle interrupt came in, it could check whether the current action had completed or not. If not, that meant an overrun. Unfortunately, in concurrent real-time systems built with multitasking preemptive schedulers, there is no equivalent method to detect and handle execution-time overruns. This is the case for systems built using the Ada tasking model and the associated Real-Time Annex [9].

In addition to detecting budget overruns, many flexible real-time scheduling algorithms require the capability to measure execution time and be able to perform scheduling actions when a certain amount of execution time has been consumed (for example, sporadic servers in fixed priority systems, or the constant bandwidth server in EDF-scheduled systems). Support for execution-time clocks and timers is essential to be able to implement such flexible scheduling algorithms.

[*] This work has been partially funded by the Spanish Research Council (CICYT), contract numbers TIC2001–1586–C03–03 and TIC99-1043-C03-03, and by the Commission of the European Communities under project IST-2001-34820 (FIRST).

J.-P. Rosen and A. Strohmeier (Eds.): Ada-Europe 2003, LNCS 2655, pp. 344–358, 2003.
© Springer-Verlag Berlin Heidelberg 2003

In recognition of all these application requirements, the Real-Time extensions to POSIX have recently incorporated support for them. Real-Time POSIX supports execution-time clocks and timers that allow an application to monitor the consumption of execution time by its tasks, and to set limits for this consumption. The next revision of the Ada language should support this functionality in order for the language to continue to be the best for programming real-time applications.

There has been a previous proposal to include execution-time clocks and timers into the Ada language [6,3]. That proposal defines a new package that includes all the operations required to access the execution-time monitoring functionality. Some of those operations belong to a protected object type that represents execution-time timers. That approach does not require modification of the compiler, nor of the run-time system (provided that the POSIX execution-time functionality is available), but its execution-time type is not integrated in the Ada language as a Time type.

In this paper we present a proposal to integrate the POSIX.1 execution-time clocks and timers into the Ada 95 language. In this proposal we try to minimize the implementation impact on the compiler and run-time system, and therefore we will not include any new language construct. It is composed of a new package named *Ada.CPU_Time* and the proposed modifications to the GNAT [5] front-end and run-time to efficiently support the execution-time clocks and timers. The proposal is being submitted to the Ada Rapporteur Group (ARG) for possible inclusion in a future revision of the Ada standard.

This paper is organized as follows. In Sect. 2 we present the interface of our proposed *Ada.CPU_Time* Ada package. In Sect. 3 we present five basic usage schemes of the execution-time clocks and timers. In Sect. 4 we briefly present the modifications done to the GNAT compiler to allow the use of execution-time timers in the Ada 95 timed sentences. We close with some conclusions and references.

## 2  Ada.CPU_Time

Our proposed interface to handle CPU_Time is based on the standard Ada.Real-Time interface. The major differences between our Ada.CPU_Time interface and the standard Ada.Real_Time interface are:

- The new data type *Clock_Id* is used to represent execution-time clocks. A value of this type represents the execution-time clock of a given task. According to the POSIX definition of execution time [7], it is implementation defined to whom will be charged the effects of interrupt handlers and run-time services on behalf of the system.
- The *Time* type represents absolute values of execution time as measured by a given execution-time clock. Values of this type have an internal Clock_Id value that ties the value of execution time to its associated clock. If a variable of type Time is not initialized, the value of its internal Clock_Id is undefined.

The type Time is a time type as defined by ARM95, Sect. 9.6, and thus values of this type may be used in a delay_until_statement.

- The *Time_Span* type represents length of execution-time duration, and its values are not dependent upon any particular execution-time clock (or task).
- The new function *CPU_Clock* is used to get the execution-time clock identifier associated with each Ada task.
- The function *Clock* has a new parameter used to specify the execution-time clock to be read. In order to keep compatibility with the Ada *Calendar* and *Real_Time* packages, if no parameter is passed then the execution-time clock of the calling task is returned.
- The new function *Clock_Id_Of* returns the identifier of the execution-time clock associated with the Time parameter T.
- The function *Time_Of* has a new parameter used to associate the time to an execution-time clock.
- The exception *Time_Error* is raised by the function Clock if the Clock_Id parameter is not valid. This exception is also raised by operators "+" and "-", and the function Clock_Id_Of, if an execution-time parameter is not valid (for example, if it is not initialized). In addition, it is also raised by a delay_until_statement if a Time value corresponding to the task executing the statement is used, because otherwise this situation would cause a deadlock.
- The exception *Incompatible_Times* is raised by operator "-" if the execution-time parameters correspond to different execution-time clocks.

```
with Ada.Task_Identification;
package Ada.CPU_Time is
   type Clock_Id is private;

   type Time is private;
   Time_First : constant Time;
   Time_Last  : constant Time;
   Time_Unit  : constant := implementation-defined-real-number;

   type Time_Span is private;
   Time_Span_First :  constant Time_Span;
   Time_Span_Last  :  constant Time_Span;
   Time_Span_Zero  :  constant Time_Span;
   Time_Span_Unit  :  constant Time_Span;

   Tick : constant Time_Span;

   function CPU_Clock
     (Task_Id : Ada.Task_Identification.Task_Id
        := Ada.Task_Identification.Current_Task) return Clock_Id;
   function Clock (C : Clock_Id) return Time;
   function Clock_Id_Of (T : Time) return Clock_Id;

   function "+"  (Left : Time;      Right : Time_Span) return Time;
   function "+"  (Left : Time_Span; Right : Time)      return Time;
   function "-"  (Left : Time;      Right : Time_Span) return Time;
   function "-"  (Left : Time;      Right : Time)      re-
turn Time_Span;
   function "<"  (Left, Right : Time) return Boolean;
   function "<=" (Left, Right : Time) return Boolean;
```

```
   function ">"   (Left, Right : Time) return Boolean;
   function ">="  (Left, Right : Time) return Boolean;
   function "+"   (Left, Right : Time_Span) return Time_Span;
   function "-"   (Left, Right : Time_Span) return Time_Span;
   function "-"   (Right : Time_Span) return Time_Span;
   function "*"   (Left : Time_Span; Right : Integer) re-
turn Time_Span;
   function "*"   (Left : Integer;   Right : Time_Span) re-
turn Time_Span;
   function "/"   (Left, Right : Time_Span) return Integer;
   function "/"   (Left : Time_Span; Right : Integer) return Time_Span;
   function "abs" (Right : Time_Span) return Time_Span;
   function "<"   (Left, Right : Time_Span) return Boolean;
   function "<="  (Left, Right : Time_Span) return Boolean;
   function ">"   (Left, Right : Time_Span) return Boolean;
   function ">="  (Left, Right : Time_Span) return Boolean;

   function To_Duration  (TS : Time_Span) return Duration;
   function To_Time_Span (D : Duration)   return Time_Span;
   function Nanoseconds  (NS : Integer) return Time_Span;
   function Microseconds (US : Integer) return Time_Span;
   function Milliseconds (MS : Integer) return Time_Span;

   type Seconds_Count is new Integer range -Integer'Last ..
           Integer'Last;
   procedure Split (T : Time; SC : out Seconds_Count; TS :
           out Time_Span);
   function Time_Of (SC: Seconds_Count; TS : Time_Span; C : Clock_Id)
       return Time;

   Time_Error          : exception;
   Incompatible_Times  : exception;
private
   . . .
end Ada.CPU_Time;
```

## 3   Usage Schemes for Ada.CPU_Time Timers

Using package *Ada.CPU_Time*, described above, we can design some basic usage schemes that depend on the particular needs of the application task whose execution time is being monitored. We have identified five major schemes:

- **Handled** [6]. This is the case in which an execution-time overrun is detected, but the task is allowed to complete its execution. This is applicable to systems under testing, or for tasks that have a high degree of critically (an thus cannot be stopped) or for which an occasional execution-time overrun can be tolerated, but needs to be reported.

  In this scheme, the application task uses a single variable T to remember the value of the execution-time clock and to evaluate the execution time of the work. If the execution time is higher than MAX_TIME then it handles the execution-time error.

```
task body Periodic_Handled is
   C : Clock_ID := CPU_Clock; -- My execution-time clock identifier.
   T : Time;
begin
   loop
      T := Clock (C);
      do task useful work;
      if Clock (C) - T > MAX_TIME then
         Handle the error;
      end if;
      delay until Next_Start; -- Global clock delay.
   end loop;
end Periodic_Handled;
```

- **Priority Change** [6]. In this scheme when the overrun is detected the priority of the task is lowered or increased (depending on the application requirements). A simple implementation of this scheme uses two nested tasks: the *Worker* task and the *Supervisor* task. The Supervisor task sleeps until the execution time of the Worker task reaches the instant of the priority change. When this happens the supervisor lowers (or increases) the priority of the worker task. If the Worker task completes the work before the instant of the priority change then it aborts the Supervisor task.

```
task body Worker is
   task Supervisor;
   task body Supervisor is
      C : Clock_ID := CPU_Clock (Worker'Identity);
      T : Time := Clock (C) + To_Time_Span (TIME_OF_PRIORITY_CHANGE);
   begin
      delay until T;
      Lower (or increase) the priority of the worker task;
   end Supervisor;
begin
   do useful work;
   abort Supervisor;
end Worker;
```

An alternative implementation of this scheme does not require *abort*. It can be done by means of the Ada *select* statement and one *entry* (i.e, *Work_Done*). When the worker completes its work calls Work_Done. The supervisor can then be implemented by means of a timed selective accept. If the entry call is received before the instant of priority change is reached then the work has been successfully completed in time; otherwise the time-budget has expired and the Supervisor lowers (or increases) the priority of the worker task. In addition the Supervisor task must accept the call to Work_Done that will be issued by the Worker at the end of its work.

```
task Worker;

task Supervisor is
   entry Work_Done;
end Supervisor;

task body Worker is
```

```
begin
   do useful work;
   Supervisor.Work_Done;
end Worker;

task body Supervisor is
   C : Clock_ID := CPU_Clock (Worker'Identity);
   T : Time := Clock (C) + To_Time_Span (TIME_OF_PRIORITY_CHANGE);
begin
   select
      accept Work_Done;
   or
      delay until T;
      Lower (or increase) the priority of the worker task;
      accept Work_Done;
   end select;
end Supervisor;
```

- **Stopped** [6]. This is the case in which if an execution-time overrun is detected, the associated task execution is stopped to allow lower priority tasks to execute within their deadlines. The task itself waits until its next activation and then proceeds normally.

  In the implementation of this scheme an asynchronous select statement is used to abort the task's work if an execution-time overrun is detected.

```
task body Periodic_Stopped is
   C : Clock_ID := CPU_Clock;   -- My execution-time clock identifier
   T : Time;
   Next_Start : Duration;
begin
   loop
      T := Clock (C) + To_Time_Span (WORST_CASE_EXEC_TIME);
      select
         delay until T;         -- Execution-time clock
         Handle the error;
      then abort
         do useful work;
      end select;
      Next_Start := ...;
      delay until Next_Start;   -- Global clock
   end loop;
end Periodic_Stopped;
```

- **Imprecise** [6]. This scheme corresponds to the case in which the task is designed using the imprecise computation model [10], in which the task has a mandatory part (generally short and for which it is easier to estimate a worst-case execution time), and an optional part that refines the calculations made by the task. Since the worst-case execution time of this optional part is usually more difficult to estimate, this part will be aborted if an execution-time overrun is detected. This technique is also valid for cases in which the optional part continuously refines the quality of the results; we can let the optional part run until it exhausts its execution-time budget, and then use the last valid result obtained.

The implementation of this scheme consists of using the "handled" approach for the mandatory part of the task, and the "stopped" approach for the optional part. After the optional part, whether it is aborted or not, another mandatory part may exist to cause outputs of the task to be generated. Therefore this scheme can be implemented as follows:

```
task body Periodic_Imprecise is
   C : Clock_ID := CPU_Clock;   --  My execution-time clock identifier
   T : Time;
   Next_Start : Duration;
begin
   loop
      T := Clock (C) + To_Time_Span (WORST_CASE_EXEC_TIME);
      do mandatory part I;
      select
         delay until T;          --  Execution-time clock
      then abort
         do optional part;
      end select;
      do mandatory part II;
      Next_Start := ...;
      delay until Next_Start;    --  Global clock
   end loop;
end Periodic_Imprecise;
```

# 4  Detailed Description of the Integration of the Execution-Time Timers into the GNAT Compiler

In this section we describe the modifications done to the GNAT compiler to support the POSIX execution-time timers in Ada. For each Ada timing statement (delay until, timed entry call, and timed selective accept) we present the modifications done to the GNAT front-end and run-time.

## 4.1  Delay Until Statement

**Front-End**

- **Semantics.** The subprogram *Analyze_Delay_Until* has been modified to allow the use of the *Ada.CPU_Time.Time* type in the Ada 95 *delay until* statement.
- **Expander.** When an *Ada.CPU_Time.Time* type variable is used to specify the timeout, the expander has been modified to transform the *delay until* statement as follows:

```
Original Ada Code          Expanded Code
---------------            -------------
delay until T;             Ada.CPU_Time.Delays.Delay_Until (T);
```

## Run-Time

- Package **System.Tasking.** A POSIX execution-time timer and two flags have been added to the Ada Task Control Block (ATCB). The flags are used to remember if the timer has been created (and therefore the execution-time timer field is valid), and if the ATCB timer is currently in use.
- Package **Ada.CPU_Time.Delays.** The subprogram *Delay_Until* has been programmed to do the following actions:
  1. Defer the abortion of the calling task.
  2. Lock the ATCB of the calling task.
  3. If the ATCB has not been created then create, program, and arm the ATCB execution-time timer; if the ATCB had been created but it is not in use then program and arm the ATCB execution-time timer; otherwise create, program and arm a new execution-time timer. In all these cases the address of a *Delay Block* register composed of the following fields is associated with the execution-time timer:
     - The ATCB address of the calling task.
     - A boolean field (*Timed_Out*) initialized to false. This field will be used by the timed statements to differentiate the case of the timeout expiration from the case in which the blocked task is awakened by some other task (i.e. the acceptor of a timed entry call, or the caller of a selective wait).
  4. Pass the calling task to the *Delay_Sleep* state.
  5. Stop the calling task until the timeout expires. This is done by blocking the calling task using the caller *mutex* and *condition variable* declared by GNAT in the ATCB for this purpose.
  6. Pass the calling task to the *Runnable* state.
  7. If the ATCB timer was re-used then mark it as "not in use". Otherwise, remove the execution-time timer.
  8. Unlock the ATCB of the calling task.
  9. Verify if a request to abort the calling task has been received during this delay. If true then abort the task; otherwise undefer its abortion.

A task is used to handle the signal associated with all the execution-time timers. This task does following actions:
1. Block all the signals.
2. Activate the signal associated with all the execution-time timers.
3. Await for the execution-time timers signal.
4. Get the address of the *Delay Block* register associated with the execution-time timer.
5. Set to *True* the field *Timed_Out* of this *Delay Block* register.
6. Awaken the task that programmed this execution-time timer.
7. Go to step 3.

**Behavior.** The calling task programs an execution-time timer and becomes blocked until this timer expires.

## 4.2   Timed Entry Call

**Front-End**

- **Semantics.** No modification was required.
- **Expander.** When an *Ada.CPU_Time.Time* variable is used to specify the timeout the expander has been modified to transform the Ada timed entry call statement in the following way:

```
Original Ada Code            Expanded Code
---------------              -------------
select                       declare
  T.E                          P : params := (param, param, param);
  <<S1>>;                      B : Boolean;
or                           begin
  delay until <<CPU_TIMEOUT>>;  CPU_Timed_Entry_Call
  <<S2>>                         (Acceptor => <Acceptor-Task_ID>,
end select;                       Entry_Id => <Entry_Index>,
                                  Uninterpreted_Data => P'Address,
                                  Timeout  => <<CPU_TIMEOUT>>,
                                  Mode        => Absolute_CPU_Mode,
                                  Successful => B);
                               if B then
                                 <<S1>>;
                               else
                                 <<S2>>;
                               end if;
                             end;
```

**Run-Time**

- Package **System.Tasking.Rendezvous** The new *CPU_Timed_Entry_Call* subprogram is based on the GNAT *Timed_Entry_Call subprogram*. The main differences with the original GNAT version are:
  - The data type of the *Timeout* parameter is *Ada.CPU_Time.Time* (instead of *Duration*).
  - Its body calls *CPU_Wait_For_Completion_With_Timeout* instead of the GNAT *Wait_For_Completion_With_Timeout* version.
- Package **System.Tasking.Entry_Calls** The new subprogram *CPU_Wait_For_Completion_With_Timeout* is based on the GNAT subprogram *Wait_For_Completion_With_Timeout*. The only difference is that it calls *CPU_Timed_Sleep* (instead of the GNAT subprogram *Timed_Sleep*).
- Package **System.Task_Primitives.Operations** The new subprogram *Ada.CPU_Timed_Sleep* calls *Ada.CPU_Time.Delays.Delay_Until*.

**Behavior.** If the rendezvous can be immediately accepted the subprogram *CPU_Timed_Entry_Call* completes the rendezvous and returns *True* in the out mode parameter *Successful*. Otherwise it programs an execution-time timer by calling the subprogram *Ada.CPU_Time.Delays.Delay_Until*.

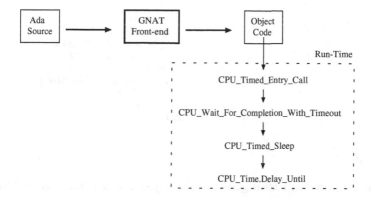

**Fig. 1.** Run-time calls to implement the CPU timed entry call

- If the call is accepted before the timeout expires then the receiver task unblocks the caller by calling *Wakeup*, the same subprogram called by the CPU_Time signal handler. The unblocked calling task detects this state by evaluating the *Signaled* field associated with its execution-time timer (still false because the execution-time timeout has not expired). Therefore the calling task removes its execution-time timer and returns *True* in its out mode parameter *Successful*.
- Otherwise (the timeout expires) the CPU_Time signal handler sets to *True* the Signaled field associated with the execution-time timeout, and the unblocked calling task removes its execution-time timer and returns *False* in its out mode parameter *Successful*.

### 4.3  Timed ATC

**Front-End**

- **Semantics.** No modification required.
- **Expander.** When the Ada.CPU_Time time variable is used to specify the timeout the GNAT expander has been modified to generate the following block of code:

```
Original Ada Code              Expanded Code
---------------                --------------
select                         declare
  delay until <T>;                D : aliased Delay_Block;
  <<S1>>;                      begin
or                               Abort_Defer;
  <<abortable_part>>             CPU_Arm_Timer (<T>, D'access);
end select;                      begin
                                   begin
                                     Abort_Undefer;
                                     <<abortable_part>>
                                   at end
                                     Async_Cancel_Timer (D'access);
```

```
                              end;
                         exception
                            when _abort_signal =>
                               Abort_Undefer;
                         end;
                         if Timed_Out (D) then
                            <<S1>>;
                         end if;
                      end;
```

## Run-Time

- Package **System.Tasking.Async_Delays.** The new CPU_Arm_Timer subprogram does the following actions:
  1. Increments the ATC nesting level of the calling task.
  2. Initializes all the fields of the Delay_Block.
  3. Calls the subprogram *Ada.CPU_Time.Delays.Arm_Timer.*
- Package **Ada.CPU_Time.Delays.** The new CPU_Arm_Timer subprogram
  arms the execution-time timer and returns.

**Behavior.** First of all let's briefly explain the semantics of the GNAT "at
end" statement. It is a handler which provides a common way out of a block of
statements even when an exception is propagated.

In the above code, after the execution-time timer is armed the abortable part
is executed. If the abortable part completes its execution before the execution-
time timer expires then the *Async_Cancel_Timer* is called to disarm the timer.
Otherwise the execution-time timer expires and the signal catcher calls the run-
time subprogram *Locked_Abort_To_Level* which defers the abortion of the blocked
task, and cancels all the nested ATC (if any) done in the abortable part by
raising the internal exception *_abort_signal.* After the abortion is undefered (in
the exception handler) if the timeout had expired then the block of statements
$<< S1 >>$ is executed.

### 4.4   Timed Selective Accept

### Front-End

- **Semantics.** It has been modified to disallow the simultaneous use of
  *Ada.CPU_Time* and *Ada.Real_Time* time variables to specify multiple timeouts in the Ada 95 timed selective accept[1].
- **Expander.** When the Ada.CPU_Time time variable is used to specify the
  timeouts of the following Ada 95 timed-selective statement the GNAT expander has been modified to generate the following block of code:

---

[1] "If a selective_accept contains more than one delay_alternative, then all shall be
delay_relative_statements, or all shall be delay_until_statements for the same time
type." ARM95 [9], Sect. 9.7.1(13).

```
Original Ada Code
---------------
select
    accept E1 ...;
or
    accept EN ...;
or
    delay until <CPU_TIMEOUT_1>;
or
    delay until <CPU_TIMEOUT_N>;
end select;
```

```
Expanded Code
-------------
declare
    S : Entry_Barriers := (others => True);
    P : params := (param, param, param);
    D : time_array (1 .. N) :=  (<CPU_TIMEOUT_1>, <CPU_TIMEOUT_N>);
    E_Index : integer := 0;
    D_Index : integer := 0;
begin
    CPU_Timed_Selective_Wait
         (Open_Accepts        => S'address,
          Select_Mode         => delay_mode,
          Uninterpreted_Data  => P'Address,
          Timeout             => D,
          Mode                => Absolute_CPU_Mode,
          Index               => E_Index,
          CPU_Time_Index      => D_Index);
    if E_Index = 0 then
             Some timeout has expired.
        case D_Index is
            when 1 => <<CPU_S1>>
                ...
            when N => <<CPU_SN>>
        end case;
    else
        --  Some entry call has
        --  been accepted.
        case E_Index is
            when 1 => <<S1>>
                ...
            when N => <<SN>>
        end case;
    end if;
end;
```

## Run-Time

- Package **System.Tasking.Rendezvous.** The new *CPU_Timed_Selective_-
  Wait* subprogram is based on the GNAT *Timed_Selective_Wait* subprogram.
  The main differences with the original GNAT version are:

  • The data type of the *Timeout* parameter is an array where all the time-
    outs specified in the Ada *select* statement are passed by the front-end.

- Its body calls a variant of the *CPU_Timed_Sleep* which receives the time-outs array and returns the index of the expired CPU timeout. This index is returned in the *CPU_Time_Index* parameter. The possible values of the out mode parameters *Index* and *CPU_Time_Index* are:

|                              | Index           | CPU_Time_Index    |
| ---------------------------- | --------------- | ----------------- |
| Some entry call was accepted: | \<entry index\> | 0                 |
| Some deadline expired:       | 0               | \<timeout index\> |

- Package **System.Task_Primitives.Operations.** The new variant of the subprogram *Ada.CPU_Timed_Sleep* calls *Ada.CPU_Time.Delays.Multiple_Delay_Until*.
- Package **Ada.CPU_Time.Delays.Multiple_Delay_Until.** This subprogram does the same actions as *Delay_Until* (described in Sect. 4.1). However, instead of using a single execution-time timer, it programs as many execution-time timers as the number of *delay until* alternatives specified by the programmer in the Ada 95 timed selective accept.

  In order to identify the execution-time timer which expired, an array of *Delay Block* registers containing the address of the caller's ATCB and the *Timed_Out* field is used. When one execution-time timeout expires the unblocked task traverses this array to look for the execution-time timeout which has its *Timed_Out* field set to *True*. If no execution-time timer has its *Timed_Out* field set to true it means that some entry call was accepted, and therefore the task was unblocked by the caller.

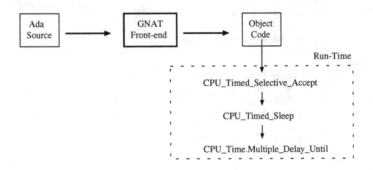

**Fig. 2.** Run-time calls to implement the CPU timed selective accept

# 5    Discussion

A prototype implementation has been developed for the execution-time budgeting proposal presented in this paper, using the MaRTE operating system [1,2] that provides a POSIX.13 [8] minimal real-time system interface and includes execution-time clocks and timers. As we have shown, the implementation requires small modifications to the compiler and to the run-time. The modifications were implemented in a short period of time and should not represent a large effort to current compiler implementors. We have compared this implementation with the one presented in [3] defining support for the execution-time clocks and timers as a library package. This latter implementation is very simple, because it does not require modifications to the compiler nor to the runtime system. The overheads are quite similar in both implementations, and are small.

For both approaches, implementations on bare machines or systems without the POSIX execution-time clocks and timers would be a bit more complex because the underlaying execution-time monitoring functionality would have to be implemented in the scheduler. Reference [6] describes one such implementation and it can be seen that it is relatively simple, and that it does not introduce any significant overhead into the scheduler.

Both approaches were discussed at the International Real-Time Ada Workshop in 2002. The conclusion from the Workshop was to recommend the approach presented in this paper, with execution-time functionality integrated in the Ada language, because it provides a simpler model to programmers. However, the group felt that there is a strong need to have the execution-time functionality in Ada, so if the second proposal, with a library implementation, has more probability of success, the group would also recommend its adoption.

Both proposals have been submitted to the Ada Rapporteur Group (ARG) for the inclusion of the execution-time clocks in the next revision of the Ada language. At the time of writing this paper the ARG has expressed preference for the package solution because it has less implementation impact, but a final decision has not been made, and discussions will continue.

## 6   Conclusions

We have presented a proposal to integrate the POSIX.1 execution-time clocks and timers into the Ada 95 language. This proposal defines a new package named *Ada.CPU_Time* and describes the modifications done to the GNAT front-end and run-time to allow the use of the execution-time timers in the timed Ada statements. We have also discussed some usage schemes of the *Ada.CPU_Time* interface.

As a proof of concepts we have modified the GNAT sources to support the execution-time timers with all the Ada 95 timed statements: delay until, timed entry call (to tasks and to protected objects), timed asynchronous transfer of control (ATC). and timed selective accept. We have a modified version of the GNAT compiler which implements all the proposals presented in this paper, and which can be used on top of the MaRTE Operating System [1,2]. Among the different options for implementing execution-time clocks in the Ada language, this proposal represents an easy to use alternative with limited implementation impact. It is now the task of the ARG and the Ada community to decide which of the alternatives is best for inclusion in the next revision of the Ada language.

# References

1. Aldea Rivas M. and González Harbour M. *MaRTE OS: Minimal Real-Time Operating System for Embedded Applications* Departamento de Electrónica y Computadores. Universidad de Cantabria. *http://marte.unican.es/*
2. Aldea Rivas M. and González Harbour M. *MaRTE OS: An Ada Kernel for Real-Time Embedded Applications.* Proceedings of the International Conference on Reliable Software Technologies, Ada-Europe-2001, Leuven, Belgium, Lecture Notes in Computer Science, LNCS 2043, May, 2001, ISBN:3-540-42123-8, pp. 305,316.
3. Aldea Rivas M. and González Harbour M. *Extending Ada's Real-Time Systems Annex with the POSIX Scheduling Services.* IRTAW-2000, Las Navas, Avila, Spain.
4. Burns A. and Wellings A. *Real-Time Systems and Programming Languages.* 3rd edition. Addison-Wesley, 2001.
5. Comar, C., Gasperoni, F., and Schonberg, E. *The GNAT Project: A GNU-Ada9X Compiler.* Technical report. New York University. 1994.
6. González-Harbour M., Aldea Rivas M., Gutiérrez García J.J., Palencia Gutiérrez J.C. *Implementing and using Execution-Time Clocks in Ada Hard Real-Time Applications.* International Conference on Reliable Software Technologies, Ada-Europa'98, Uppsala, Sweden, in Lecture Notes in Computer Science No. 1411, June, 1998, ISBN:3-540-64563-5, pp. 91,101.
7. IEEE Std. 1003.1:2001, Information Technology – Portable Operating System Interface (POSIX).
8. IEEE Std. P1003.13-1998, Information Technology – Standarized Application Environment Profile – POSIX Realtime Application Support (AEP). The Institute of Electrical and Electronics Engineers.
9. Intermetrics, Inc. *Ada 95 Language Reference Manual.* Intermetrics, Inc., Cambridge, Mass., January, 1995.
10. J. Liu, K.J. Lin, W.K. Shih, A. Chang-Shi Yu, J.Y. Chung, and W. Zhao. *Algorithms for Scheduling Imprecise Computations.* IEEE Computer, pp. 56–68, May 1991.

# A Test Environment for High Integrity Software Development

Alejandro Alonso, Juan Antonio de la Puente, and Juan Zamorano

Departamento de Ingeniería de Sistemas Telemáticos
Universidad Politécnica de Madrid, E-28040 Madrid, Spain
{aalonso,jpuente}@dit.upm.es, jzamora@fi.upm.es

**Abstract.** The paper describes the architecture and implementation of the dynamic analysis tools that are part of the DOBERTSEE environment. DOBERTSEE is a low-cost, flexible software engineering environment which enables different development processes and methods to be supported by integrating different tools. An XML-based language is used as a basis for integration. The current version of the environment supports the HRT-HOOD method and Ravenscar Ada, and includes an extensive set of static and dynamic analysis tools.

## 1 Introduction

Software engineering environments have hardly lived up to their promises in supporting extensive development of high-quality software along the whole life cycle of software products. In spite of the sound technical approaches that can be found in some commercial systems, such factors as the monolithic nature of many of them, the difficulties that are often found when trying to adapt them to the particular methods and idioms of a particular development, and the lack of flexibility in supporting different kinds of development processes have limited their use in many application areas, in particular in the high-integrity systems field.

The DOBERTSEE (Dependable On-Board Embedded Real-Time Software Engineering Environment) project [1] was launched by the European Space Agency (ESA) in 2000 with the aim of developing a new, open software engineering environment for on-board software which supports up-to-date technology with a low cost, in such a way that new methods and tools can easily be integrated into it. The project builds on the results of the previous ECSS-PMod project, in which the requirements of a software engineering environment (SEE) for supporting the ECSS[1] software standards [2] were developed and a prototype implementation was built [3].

It can be expected that such a SEE can be easily integrated in a company's practice, by adapting it to the particular processes, methods, and tools used by the project teams, while keeping the required investment costs low enough to be affordable even for small or medium size projects. The technical approach is based on the extensive use of an XML based language, called CASEML, as the glue between different tools and representations of software. The environment is targeted to on-board spacecraft software, and is intended

---

[1] European Co-operation for Space Standardization.

J.-P. Rosen and A. Strohmeier (Eds.): Ada-Europe 2003, LNCS 2655, pp. 359–367, 2003.

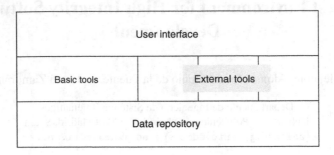

**Fig. 1.** SEE architecture

to be flexible enough to cover different process models and methods in the framework of the ECSS-E40 standard.

The main focus of this paper is on the dynamic analysis tools which have been developed as part of the DOBERTSEE environment. Section 2 describes the main features and the overall architecture of DOBERTSEE. Section 3 discusses the role of testing in the DOBERTSEE software design process and the general approach to testing that has been followed in the project. The tools themselves are described in Sect. 4. Finally, some conclusions are presented in Sect. 5

## 2   The DOBERTSEE Environment

The SEE environment provides a platform in which different tools can be integrated, together with a basic set of services and a common user interface (figure 1). The basic services include support for life cycle definition, enactment of the ECSS process model, configuration management, documentation, traceability, and distributed cooperative work. A common data repository is also part of the core environment.

The SEE can thus be seen as a framework into which a variety of software tools can be integrated. The DOBERTSEE project aims at building a particular instance of SEE for on-board embedded real-time systems by selecting one such set of tools which are appropriate for the application domain. The implementation of SEE is based on open standards such as ASIS [4] and Tcl/Tk [5], so that other tools can be easily integrated into it.

On board spacecraft software systems typically have high integrity and hard real-time requirements, for which the development methods and tools have to be used. Vardanega [6] proposed a design approach for this kind of systems based on extensive use of static and dynamic analysis as early as possible in the development process. The approach is based on the use of HRT-HOOD [7,8] as a design method, and a subset of Ada 95 including the Ravenscar profile [9] for implementation, in such a way that static and dynamic analysis techniques, including response time analysis, [10,11] can be used in the design process. Figure 2 shows a scheme of the overall design process used in DOBERTSEE, which explicitly adds dynamic analysis to the original proposal.

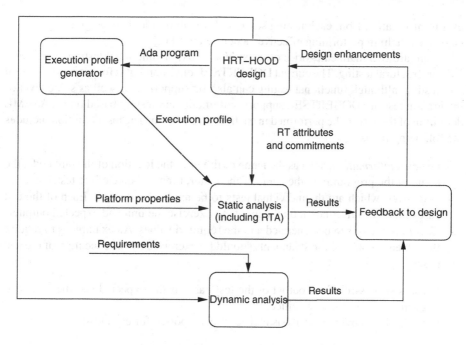

**Fig. 2.** DOBERTSEE design process

The set of tools integrated with the basic SEE in DOBERTSEE is targeted to the design process, and currently includes:

- STOOD, a graphic HRT-HOOD tool [12].
- GNAT/ORK, an Ada 95 compilation system and run-time system supporting the Ravenscar profile for ERC-32 computers [13].
- The TSIM simulator for ERC-32 computers [14]
- The RTA tool for response time analysis [15].
- Static analysis tools developed by OEI-UPM.
- Dynamic analysis tools developed by DIT-UPM.

The static and dynamic analysis tools were developed specially for the DOBERTSEE environment. The next sections describe the dynamic analysis tools and some related issues.

## 3   Testing in DOBERTSEE

In spite of the importance of static analysis techniques for the validation and verification of high integrity systems, dynamic analysis or testing is still of paramount importance in order to get confidence on the correctness of software systems [10]. Of course testing has well known limitations, and a systematic approach has to be used in order to get useful information from each testing session. Tool support and coverage analysis may be of great help in performing effective, accurate tests [16].

There are two main categories of tests: *black-box*, or functional testing, and *white-box*, or structural testing. The current DOBERTSEE environment is focused on structural unit testing, although functional testing can also be supported, as well as some level of integration testing. DOBERTSEE supports automatic unit testing based on a CASEML definition of the tests to be performed in each testing session. The test definition includes the following items:

- *General information*, such as the name of the unit, the location of the unit code, the name of the programmer, the name of the tester, time and date of the test, etc.
- *Test cases*, which are the individual tests to be applied to the unit. Each of the test cases includes information on the inputs to exercise the unit and expected outputs.
- *Coverage metrics* to be generated and the required values. An example is to require that at least a 90% of the statements should be exercised by the execution of the test cases.

A test is successful if the output of the test cases is the expected and the produced coverage metrics met those specified.

The following coverage analysis metrics are supported for each test:

- *Statement coverage*: check that at least a significant number of the statements in the code have been executed and identify those which have not been executed.
- *Module coverage*: check that every procedure, function, task, entry, and package initialization sequence has been executed at least once.
- *Loop coverage*: check if every loop in the code has been executed zero times, once, or more than once.
- *Decision coverage*: check that every decision in the code has been executed with values *true* and *false* at least once.
- *Condition coverage*: check that every boolean element has been evaluated to *true* and *false* at least once.
- *Basis path coverage*: check the number of independent paths which are executed in the test. This metric is closely related to the cyclomatic complexity of a procedure [17].

A unit is considered to have been successfully tested if the results of the execution of all the test cases are the expected ones, and the coverage metrics, that have been specified in the test, are met.

Figure 3 shows the data flow of the operations required to perform automatic unit testing based on the above principles. The main operations are:

- *Code analysis*: in order to compute the coverage metrics some information about the code execution is required (e.g. how many times a loop has been executed, which statements have been exercised, etc.) This information is obtained by *instrumenting* the code, i.e. by inserting additional statements in the original code that store the required information in appropriate data structures. The information can then be recovered and analyzed when the instrumented code is executed.

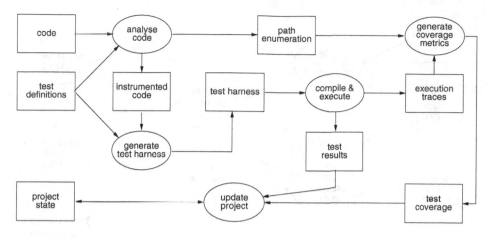

**Fig. 3.** Data flow for automatic unit testing

- *Test harness generation*: a *test harness* is a program that executes the required test cases by invoking the tested unit with the appropriated data, and generates a trace of the execution results for future evaluation. It is important to note that a test harness may have to be executed on an embedded computer, which makes the whole process a bit more difficult.
- *Code compilation and execution*: once a test harness has been produced, it has to be compiled and run on the target platform or on a simulator. The output of this process is the result of the execution of all test cases that were included in a test harness, together with the trace information produced by the instrumented code.
- *Coverage metrics generation*: The coverage metrics requested by the test definition are generated out of the execution traces produced by code instrumentation and the path information provided by the code analysis process.
- *Project update*: The DOBERTSEE environment keeps information about the state of the software project under development. The project state has to be updated with the results of the test execution and the coverage measurements. In particular, the test definition is updated by including information on the test execution. Other project information may also be updated as a consequence of the test results.

This set of operations constitute the core of the dynamic analysis activity of the design process (figure 2).

## 4   Tool Architecture and Design

### 4.1   Overview

The automatic testing process described in Sect. 3 is implemented in DOBERTSEE by means of a driving tool, the *test harness tool*, which is integrated as a vertical tool in the SEE architecture (figure 1). The test harness tool is launched from the common SEE

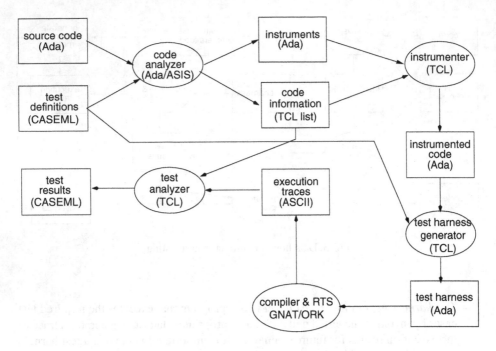

**Fig. 4.** Architecture of dynamic analysis tools

interface and in turn invokes a number of dynamic analysis tools. Figure 4 shows the overall architecture of the dynamic analysis tools, which closely follows the data flow diagram depicted in figure 3.

The tools are implemented in a variety of languages, including Ada, Tcl, and CASE-ML. The GNAT ASIS interface [18] has been used in order to recover semantic information from the Ada compiler, and a new ASIS binding has been developed for the Tcl tools. The implementation languages have been chosen in order to make it easier to interface the tools with the basic SEE infrastructure and the GNAT compiler.

## 4.2   Tcl/ASIS Binding

The Ada Semantic Interface Specification (ASIS) [4] is a standard interface between an Ada environment and any tool that requires information from it. An Ada environment includes semantic and syntactic information on Ada units, whose exact contents and internal representation is compiler-dependent. The purpose of ASIS is to provide an open and published callable common interface.

For all these reasons, ASIS is a fundamental asset for code analysis. However, as the main implementation language in DOBERTSEE is Tcl/Tk, it is necessary to call ASIS functions from this environment. The implementation of a binding for this purpose was not difficult, as it is possible to call C functions from a Tcl/Tk interpreter. Therefore, it is possible to call ASIS from a Tcl/Tk interpreter by using the Ada capabilities for interfacing with C. There are only some restrictions on the data types to be used, which are not related to the use of Ada.

Two kinds of capabilities were developed in order to simplify the use of ASIS from Tcl scripts:

- *Generation of Tcl lists*: As previously described, the output of the code analyzer is used by a Tcl program to instrument the code. Therefore, if the analysis results are output in a format suitable for Tcl, the instrumentation tool code is much simpler. According to this consideration, the results of the code analysis are stored as Tcl lists, which are a basic data structure in this language.
- *A generic iterator*: One of the most powerful components of the ASIS interface is an iterator, which enables traversing a syntactic tree and performing a given operation on all its elements. A more robust and adapted version of this iterator has been implemented for the DOBERTSEE dynamic analysis tools. It includes an initialization operation which is not part of the standard ASIS iterator, and enables Tcl variables to be retrieved at the output of a function call.

In addition, a number of packages and Tcl dynamic libraries written in Ada have been developed in order to support the Tcl/ASIS binding and make its use easier.

### 4.3   Support for Coverage Metrics

The different kinds of coverage metrics described in Sect. 3 are supported by several tools. The code analyzer provides the basic information required to instrument the code according to the required metrics, and the code information for the analysis of the test results. The test analyzer takes this information and the test execution traces, and produces the analysis results in SEE compatible format (CASEML).

The implementation of the coverage metrics has been divided in four steps, in accordance with the tool structure depicted in figure 3:

- Code analysis.
- Code instrumentation.
- Code compilation and execution.
- Results analysis.

Code analysis is based on the generic iterator previously mentioned. There should be an instantiated package for each supported metric. There are two main procedures that should be provided for this instantiation:

- Postoperation: it is the procedure in charge of analysing the Ada code in order to gather the information required for each metric. This information is stored in a *Tcl* list for future processing. This list includes the location of Ada statements that are relevant for the metric under consideration.
- Initialization: it is the procedure that initializes the variables required in the code analysis for a specific metric.

Code instrumentation is performed by a *Tcl* program, which calls in turn to a procedure for each of the supported metrics. The inputs are the lists previously generated and the source code to be analysed. The goal of the instrumentation is to insert code into the unit under test, in order to gather information relevant to each metric. This information is stored in dedicated variables for each metric during the test execution.

The test harness generator takes into account the required metrics, in order to include the associated packages. When the code is executed, the information produced by the instrumentation code is stored in a file for future processing.

Code analysis is performed by a *Tcl* procedure for each metric. These procedures get the appropriate information from the file with the results and produce the associated metric. Finally, the tool checks whether the coverage metrics produced are within the acceptable bounds specified in the test definition.

An important effort was devoted to make the addition of a new metric as simple as possible. For each of them it is necessary to include the procedures described in this section. Then, it is simply required to insert the appropriate calls from the tool procedures that performs the above mentioned steps. The only detail to take into account is to assign to the new metric an unique name.

## 5   Conclusions

The DOBERTSEE project has shown that it is possible to build an open, low cost environment for high integrity real-time systems. The use of common tools and notations makes it easy to integrate new tools into its basic platform, so tailoring the SEE environment to different application fields.

The test harness tool is a fundamental component of DOBERTSEE. Its design and implementation are a practical demonstration of the power of the SEE approach, and provide the basis for the adaptation and extension of the tool to future project needs.

We believe that this approach can extend the use of integrated software engineering environments to application areas where they are currently seldom used, by allowing project managers and developers to integrate their preferred tools in the environment and to adapt its configuration to their particular needs and to the requirements of different kinds of software processes.

**Acknowledgements.** The DOBERTSEE project has been funded by ESA/ESTEC under contract no. 15133/01/ NL/ND. It has been carried out by a consortium including UPM (Technical University of Madrid), TNI (Techniques Nouvelles d'Informatique), and ES-TEC. We would like to thank Juan Garbajosa, the project leader, and Jorge Amador, the project officer, for the fruitful discussions that we have had on the topics described in this paper.

We would also like to acknowledge the work of Ignacio Martín, Miguel Muñoz, and Miguel Ángel Parra, who have implemented and tested most of the tools that are described in this paper.

# References

1. Garbajosa, J., Alonso, A., Amador, J., Bollaín, M., de la Puente, J.: A low cost software engineering environment for on-board real-time software. In: Workshop on Advanced Real-Time Software Technologies, Aranjuez, Spain (2002)
2. ECSS: ECSS-E-40A Space Engineering — Software. (1999) Available from ESA.
3. GMV-ECSS-UM-01: SEE User Manual. 4.1 edn. (2000) Also available from ESA.
4. ISO: Ada Semantic Interface Specification (ASIS). ISO/IEC- 15291:1999. (1999)
5. Ousterhout, J.: Tcl and the Tk Toolkit. Addison-Wesley (1994)
6. Vardanega, T.: Development of on-board embedded real-time systems: An engineering approach. Technical Report ESA STR-260, European Space Agency (1999) ISBN 90-9092-334-2.
7. Burns, A., Wellings, A.: HRT-HOOD: A design method for hard-real-time. Real-Time Systems 6 (1994) 73–114
8. Burns, A., Wellings, A.: HRT-HOOD(TM): A Structured Design Method for Hard Real-Time Ada Systems. Elsevier Science, Amsterdam (1995) ISBN 0-444-82164-3.
9. Burns, A., Dobbing, B., Romanski, G.: The Ravenscar profile for high integrity real-time programs. In Asplund, L., ed.: Reliable Software Technologies — Ada-Europe'98. Number 1411 in LNCS, Springer-Verlag (1998)
10. ISO/IEC: Guide for the use of the Ada Programming Language in High Integrity Systems. (2000) Technical Report TR 15942:2000.
11. Burns, A., Dobbing, B., Vardanega, T.: Guide for the use of the Ada Ravenscar profile in high integrity systems. Technical Report YCS-2003-348, University of York (2003)
12. TNI: STOOD home page. http://www.tni.fr (2003)
13. de la Puente, J.A., Ruiz, J.F., Zamorano, J.: An open Ravenscar real-time kernel for GNAT. In Keller, H.B., Ploedereder, E., eds.: Reliable Software Technologies — Ada-Europe 2000. Number 1845 in LNCS, Springer-Verlag (2000) 5–15
14. Gaisler, J.: TSIM Simulator User's Manual. Gaisler Research. (2001)
15. de la Puente, J.A.: RTA User's Guide. DIT-UPM. 1.3 edn. (2001)
16. Ghezzi, C., Jazayeri, M., Mandrioli, D.: Fundamentals of Software Engineering. Prentice-Hall, Englewood Cliffs, New Jersey (1991)
17. Watson, A.H., McCabe, T.J.: Structured Testing: A Testing Methodology Using the Cyclomatic Complexity Metric. NIST Special Report 500-235 (1996).
18. Rybin, S., Strohmeier, A., Kuchumov, A., Fofanov, V.: ASIS for GNAT: From the prototype to the full implementation. In Strohmeier, A., ed.: Reliable Software Technologies — Ada-Europe'96: Proceedings. Volume 1088 of Lecture Notes on Computer Science., Springer-Verlag (1996)

# Normalized Restricted Random Testing

Kwok Ping Chan[1,*], Tsong Yueh Chen[2], and Dave Towey[1]

[1] Department of Computer Science and Information Systems
The University of Hong Kong, Pokfulam Road, Hong Kong
{kpchan, dptowey}@csis.hku.hk
[2] School of Information Technology
Swinburne University of Technology, Hawthorn 3122, Australia
tychen@it.swin.edu.au

**Abstract.** Restricted Random Testing (RRT) is a new method of testing software that improves upon traditional random testing (RT) techniques. This paper presents new data in support of the efficiency of RRT, and presents a variation of the algorithm, Normalized Restricted Random Testing (NRRT). NRRT permits the tester to have better information about the target exclusion rate (R) of RRT, the main control parameter of the method. We examine the performance of the NRRT and Original RRT (ORRT) methods using simulations and experiments, and offer some guidance for their use in practice.

## 1 Introduction

It is widely recognised that software testing plays an important role in software development. Exhaustive testing (the checking of all possible input combinations) is seldom feasible, and testers are usually only able to use a small portion of a program's input domain.

A common method of testing software is to draw test cases at random from the input domain. This method, Random Testing (*RT*), has been the subject of much debate in the testing community, and, as Gutjahr [10] points out, few topics in software testing methodology are more contentious than the question of whether or not it is efficient. Starting in the 70's with Myers', "probably the poorest [testing] methodology is random input testing," opinions swung full-circle in the 80's, when Duran and Ntafos [9] concluded that Random Testing might even be more cost effective than Partition Testing, a testing methodology that intuitively, should outperform Random Testing. Surprising as this was, further evidence from Hamlet and Taylor served to confirm the observation [12].

While Random Testing may well be surrounded in some controversy, what is undeniable is that it does offer some distinctively attractive features: In addition to alleviating the overheads associated with partitioning, efficient algorithms exist to generate test cases [11], and reliability estimates and statistical analyses

---

* All correspondence should be addressed to: K.P. Chan, Department of Computer Science and Information Systems, the University of Hong Kong. Email: kpchan@csis.hku.hk

J.-P. Rosen and A. Strohmeier (Eds.): Ada-Europe 2003, LNCS 2655, pp. 368–381, 2003.
© Springer-Verlag Berlin Heidelberg 2003

are also easily calculated [16]. In many situations, it is recommended to apply *RT* in the early stages of software development, since it is then that the program being tested is more error-prone, and Random Testing may be most effective [14].

Although Random Testing consists of no more than repeatedly drawing test cases randomly from a specified input range, it has been found that by requiring the testing candidates to be more evenly distributed, and far separated from each other, the program failure can be more effectively identified under some situations. This forms the basis of the Adaptive Random Testing (*ART*) method proposed by Chen et al. [6] (see also [7]), which has been shown to outperform ordinary *RT* by as much as 50%. Restricted Random Testing (*RRT*) [5] is an alternative methodology which, although motivated by the same goals underlying *ART* of guaranteeing a good spread of the test cases within the input domain while maintaining low overheads, ensures the distribution through the use of exclusion zones around non-failure causing test cases.

In this paper, we present additional experimental data for the *RRT* method, and introduce a variation, Normalized Restricted Random Testing (*NRRT*). Although sharing the same fundamental algorithm, *NRRT* attempts to improve upon the shortcomings of Ordinary *RRT* (hereafter referred to as *ORRT*).

An analysis of the *ORRT* data revealed that the best error detection was obtained when the target exclusion (*R*) was at a maximum. Unfortunately, except in the simplest cases, prior knowledge of the maximum *R* is not available. *NRRT*, on the other hand, permits the tester to know in advance what the maximum *R* should be, and therefore enables a faster and more accurate testing process. It was also found that the shape of the input domain, in addition to determining the maximum possible exclusion, also influenced the effectiveness of the exclusion zones. *NRRT*, by homogenizing the input domain and exclusion zones, improves the effectiveness of the *RRT* algorithm.

We next introduce the background of this study, and describe some of the notation used in this paper. In Sect. 3, we describe a simulation and some experiments that were performed to investigate the effectiveness of the *RRT* methods. The results of these studies are discussed in Sect. 4, and some conclusions and suggestions for when to use the methods are given.

# 2   Preliminaries

## 2.1   F-Measure

Traditionally, the effectiveness of testing strategies have been measured according to the probability of detecting at least one failure (the *P-measure*), or by the expected number of failures successfully detected (the *E-measure*). These measures however, are less than ideal, and Chen et al. [6] propose instead to use the expected number of test cases required to find the first failure (the *F-measure*) as a gauge of how effective the method is. In support of this they point out that, in practice, testing usually stops when a failure is found. Therefore, the *F-measure* is not only intuitively more appealing than either the *E-* or *P- measures*, but also is more realistic from a practical viewpoint. As with [5] the *F-measure* is used to evaluate the testing strategies presented in this paper.

## 2.2  Background

When conducting testing using a random selection algorithm, aside from avoiding repetition of test cases (points located in the input domain and used to test the software for failures), no account is taken of previous, successful test cases. Random test cases are repeatedly generated and tested until one causes a failure. Such a test case is referred to as a failure-causing input [8].

Following the notation of Chen and Yu [8], for an input domain $D$, we let $d$ denote the domain size, and $m$ the size of the failure-causing input regions. The failure rate, $\theta$, is then defined as $m/d$. For Random Testing with replacement of test cases [13], the expected value for $F$ is equal to $1/\theta$, or equivalently, $d/m$. So, when using Random Testing on an input domain with a failure rate ($\theta$) of 0.1%, we expect the *F-measure* to be 1/(0.1%), 1000.

## 2.3  Failure Patterns

Previous research has identified three major categories of failure patterns (input points which cause failures), and reported on how they influence the performance of some partition testing strategies [3]. The categories are: *point*, characterised by individual or small groups of failure-causing input patterns; *strip*, characterised by a narrow strip of failure-causing inputs; and *block*, characterised by the failure-causing inputs being concentrated in either one or a few regions. Figures 1a–1c give examples of each of the categories (the outer boundaries represent the borders of the input domain, and the shaded areas represent the failure-causing regions).

Since detecting failure regions for programs in which the failure-causing proportion of the input space ($\theta$) is large, is a relatively trivial task, and should be fairly easily completed by any reasonable selection strategy, it is those cases where $\theta$ is small that require more attention. Chen et al. [7] suggest that for non-point patterns, by slightly modifying the basic random testing technique,

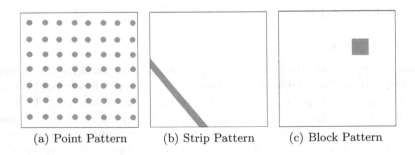

(a) Point Pattern          (b) Strip Pattern          (c) Block Pattern

**Fig. 1.** Types of Failure Patterns

and requiring that the testing candidates be more evenly distributed, and far separated from each other, the failure detection capability can be significantly improved. This suggestion motivates the *RRT* methods.

## 2.4    Method

When testing according to the *RRT* method, given a test case that has not revealed failure, rather than simply generate another test case randomly, we restrict the areas of the input domain from which subsequent test cases may be drawn. To ensure a minimum distance between all test cases, a circular area around non failure-causing inputs is defined, and subsequent test cases are restricted to coming from outside of these regions.

Although the exclusion zone around each test case is the same for all inputs, the area of each zone decreases with successive attempts. The size of the exclusion zones is related to the size of the entire input domain. For example, in two dimensions, with a target exclusion area of $A$, if there are $n$ points around which we wish to generate exclusion zones, then each exclusion zone area will be $A/n$, and each exclusion zone radius will be $\sqrt{A/(2n\pi)}$.

Because of the effects of overlapping zones and of portions of the zones lying outside the input domain, the actual coverage within the input domain may be significantly less than the target coverage. In this paper we refer to the target coverage when discussing our methods and results.

## 3    Empirical Study

To investigate the effectiveness of the *RRT* method, we conducted some empirical studies. In the first case, we simulated a failure region and, by varying the target exclusion area $(R)$, calculated how effective *RRT* was. We also applied the algorithm to several previously studied error-seeded programs [5,6], and found the results to be very favourable. Finally, motivated by some of the shortcomings of the method, we introduce an adapted form of the method and apply this to the programs, again obtaining encouraging results.

## 3.1    Simulation

For certain types of failure patterns, in particular the non-point pattern types, it has been suggested that the failure detection capabilities for *RT* can be improved by ensuring that the test cases are more evenly spread over the input domain [6,7]. To test this, we conducted a simple simulation [5] in which a certain percentage of a square input domain was specified to be a failure zone (a region from which an input drawn would cause the tested program to fail). The failure zone was circular in shape, and its size was set at 0.001 of the entire input area $(\theta = 0.1\%)$. Having advance knowledge as to what portion of the input domain is failure-causing $(\theta)$ enables the expected *F-measure* by Random Testing to be calculated $(1/\theta)$.

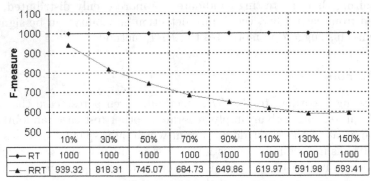

**Fig. 2.** Expected *F-measures* for Random Testing (*RT*) compared with the calculated *F-measure*, averaged over 10,000 trials, for Restricted Random Testing (*RRT*). The failure region ($\theta$) is 0.1% of the entire input domain. The target exclusion (*R*) varies from 10% to 150%

The target exclusion zone (*R*) was varied between 10% and 150%. (even with a target exclusion of 150%, the actual exclusion is often still less than 100%.) The experiments were repeated 10,000 times and the mean *F-measure* for each target exclusion percentage was calculated. The results are reproduced in Fig. 2.

As can be seen from the figure, the experimentally measured *F-measure* for *RRT* shows considerable improvement over the expected values for Random Testing. Additionally, the failure-finding efficiency appears to improve as the target exclusion (*R*) increases. These results support our hypothesis that detection capabilities of Random Testing can be improved by ensuring that the test cases are more evenly spread over the input space.

## 3.2   Adaptive Random Testing

An alternative method of testing, also motivated by the goal of improving the failure detection capabilities of ordinary Random Testing by forcing a more even spread of the input cases, was proposed by Chen et al. [6]. They named this method the Adaptive Random Testing (*ART*) method, and it works as follows: two sets of test cases are maintained: the executed set, a set of those test cases which have been executed but without causing failure; and the candidate set, a set of cases selected randomly from the input domain. Each time a new test case is required, the element in the candidate set with the maximum minimum distances from all the executed set elements is selected. Distance is measured as the Euclidean distance. In the Fixed Size Candidate Set (*FSCS*) version of the method, the candidate set is maintained at a constant size, and is reconstructed each time a test case is selected.

## 3.3   Error-Seeded Programs

In our study, we tested the $RRT$ method against seven error-seeded programs previously studied by Chan et al. [4]. These are published programs [1,15], all involving numerical calculations, which were written in C (converted to C++ for our experiments) and which varied in length from 30 to 200 statements.

Using Mutation Analysis [2], errors in the form of simple mutants were seeded into the different programs. Four types of mutant operators were used to create the faulty programs: arithmetic operator replacement ($AOR$); relational operator replacement ($ROR$); scalar variable replacement ($SVR$) and constant replacement ($CR$). These mutant operators were chosen since they generate the most commonly occurring errors in numerical programs [4]. For each program, after seeding in the errors, the range of each input variable was then set such that the overall failure rate would not be too large [4]. Table 1 summarizes the details of the error-seeded programs.

**Table 1.** Program name, dimension ($D$), input domain, seeded error types, and total number of errors for each of the error-seeded programs. The error types are: arithmetic operator replacement ($AOR$); relational operator replacement ($ROR$); scalar variable replacement ($SVR$) and constant replacement ($CR$)

| Program Name | D | Input Domain From | Input Domain To | AOR | ROR | SVR | CR | Total Errors |
|---|---|---|---|---|---|---|---|---|
| bessj | 2 | $(2.0, -1000.0)$ | $(300.0, 15000.0)$ | 2 | 1 | | 1 | 4 |
| bessj0 | 1 | $(-300000.0)$ | $(300000.0)$ | 2 | 1 | 1 | 1 | 5 |
| cel | 4 | $(0.001, 0.001,$ $0.001, 0.001)$ | $(1.0, 300.0,$ $10000.0, 1000.0)$ | 1 | 1 | | 1 | 3 |
| erfcc | 1 | $(-300000.0)$ | $(300000.0)$ | 1 | 1 | 1 | 1 | 4 |
| gammq | 2 | $(0.0, 0.0)$ | $(1700.0, 40.0)$ | | 3 | | 1 | 4 |
| plgndr | 3 | $(10.0, 0.0, 0.0)$ | $(500.0, 11.0, 1.0)$ | 1 | 2 | | 2 | 5 |
| sncndn | 2 | $(-5000.0, -5000.0)$ | $(5000.0, 5000.0)$ | | | 4 | 1 | 5 |

## 3.4   ART Applied to Seeded Programs

The results of Chen et al.'s experiments [6] using the Fixed Size Candidate Set ($FSCS$) version of Adaptive Random Testing ($ART$) method on the error-seeded programs are given below (Fig. 3). In their version, the candidate set has a fixed size of 10 elements. The results are expressed in terms of their improvement over the $RT$ $F$-measure. For most of the programs, $ART$ offers significant improvement over the calculated $F$-measure for Random Testing ($RT$).

We next present the results for the Restricted Random Testing ($RRT$) methods when applied to the same error-seeded programs. First, we used the original, Ordinary RRT ($ORRT$) method directly on the programs. We then applied an adapted version of the algorithm, the Normalized $RRT$ ($NRRT$) method, which

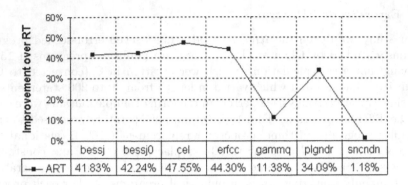

**Fig. 3.** Improvement in F-measures for Adaptive Random Testing ($ART$) compared to Random Testing ($RT$), when applied to the error-seeded programs. (Taken from Chen et al. [6])

attempts to reduce some of the influence that the input domain shape has on the effectiveness of $RRT$.

### 3.5   ORRT Applied to Seeded Programs

In an earlier study [5], we converted seven of the original error-seeded programs to C++ and applied the Ordinary Restricted Random Testing ($ORRT$) method. We varied the target exclusion ratio ($R$) between 10% and 220%, and obtained an *F-measure* for each ratio by calculating the average over 5,000 iterations.

We continued these experiments allowing the target exclusion ($R$) to increase to its maximum. (Maximum target exclusion is defined as the highest target exclusion ratio at which every iteration (5,000) of the method revealed a failure.) The value of the maximum $R$ was found to vary for each program, sometimes, as was the case for the *cel* program, quite dramatically (Table 2). Further investigation revealed that the maximum $R$ was related to the shape of the input domain.

A summary of the results of the comparison between $ORRT$ and Random Testing ($RT$) is given in Figs. 4 and 5. The figures show the percentage improvement of $ORRT$ over $RT$.

For most of the programs, as was the case with the simulation, there appears to be an increase in the improvement over $RT$ corresponding to the increase in target exclusion area. The *gammq* program, although not displaying the relationship between improvement and target exclusion increase as strongly as the other programs, does show overall improvement over $RT$. Only *sncndn*, as was found in the $ART$ results [6], does not show any significant improvement.

When we conducted the experiments using $ORRT$ over the target exclusion ($R$) range 10% to 220%, the *cel* program appeared to perform badly [5]. However, when this range was significantly extended (to 32,000%), it was found that

**Table 2.** Maximum Target Exclusion rates ($R$) for the error-seeded programs, using Ordinary Restricted Random Testing ($ORRT$). The results are averaged over 5,000 iterations

| Program Name | Maximum Target Exclusion (R) |
|:---:|:---:|
| bessj | 220% |
| bessj0 | 100% |
| cel | 32,000% |
| erfcc | 100% |
| gammq | 200% |
| plgndr | 460% |
| sncndn | 150% |

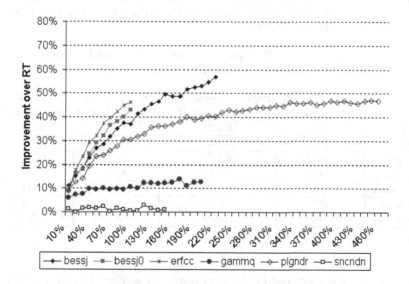

**Fig. 4.** Comparison of the *F-measures* for Ordinary Restricted Random Testing ($ORRT$) when applied to the error-seeded programs. The target exclusion ($R$) varies from 10% to 460%. The figures show the improvement of $ORRT$ over $RT$

the program actually did respond favourably, and did display the characteristic improvement related to the increasing target exclusion ($R$) (Fig. 5).

In almost all cases, the best improvement is obtained when the maximum target exclusion ($R$) is used; this information is summarized in Table 3.

By varying the size and shape of the input domains for the programs, it was discovered that these parameters affect the value of the maximum target exclusion. Since the $ORRT$ results indicate that the best improvements over $RT$ are obtained when the target exclusion ($R$) is at a maximum, prior knowledge of the maximum $R$ would be very useful to testers using the $RRT$ method.

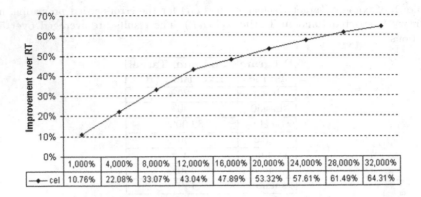

**Fig. 5.** Improvement in the *F-measures* for Ordinary Restricted Random Testing (*ORRT*) compared with *RT*, when applied to the cel program. The target exclusion (*R*) varies from 1,000% to 32,000%

**Table 3.** Program name, Max *R*, Improvement over *RT* at Max *R*, *R* for best improvement, and best improvement over *RT* for the error-seeded programs, using *ORRT*

| Program Name | Max Target Exclusion (R) | Improvement over RT at Max R | R for Best Improvement | Best Improvement |
|---|---|---|---|---|
| bessj | 220% | 56.74% | 220% | 56.74% |
| bessj0 | 100% | 43.03% | 100% | 43.03% |
| cel | 32,000% | 64.31% | 32,000% | 64.31% |
| erfcc | 100% | 46.24% | 100% | 46.24% |
| gammq | 200% | 12.76% | 170% | 13.94% |
| plgndr | 460% | 46.84% | 450% | 46.94% |
| sncndn | 150% | 1.05% | 120% | 3.01% |

Therefore, we adjusted *ORRT* to incorporate a scaling feature enabling the exclusion zones to be normalized to the input domain dimensions. We then applied this Normalized *RRT* (*NRRT*) method to the seeded programs.

## 3.6   NRRT Applied to Seeded Programs

We adapted the *RRT* method so that, instead of a uniform exclusion zone (circle, sphere, etc.) around each non failure-causing test case, we moulded the exclusion zone to the input domain. We defined the exclusion zone initially to be within a unit square/cube/hypercube, and then mapped the points to the actual input domain. We refer to this version of *RRT* as the Normalized Restricted Random Testing (*NRRT*) Method.

The same error-seeded programs as above were tested to calculate the *F-measure*, and, as before, we varied the target exclusion ratio (*R*) and obtained an *F-measure* for each ratio by calculating the average over 5,000 iterations.

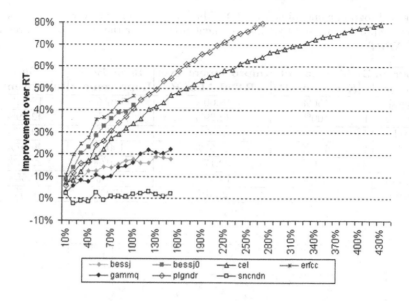

**Fig. 6.** Comparison of the *F-measures* Normalized Restricted Random Testing (*NRRT*) when applied to the seeded programs. The target exclusion (R) varies from 10% to 430%. The figure shows the improvement of *NRRT* over *RT*

Because the initial input domain has been homogenized, programs with the same dimension were now expected to have similar maximum target exclusion rates ($R$), differing now only according to the size of the failure regions.

A summary of the results of the comparison between *NRRT* and Random Testing is shown in Fig. 6. The figure shows the percentage improvement of *NRRT* over *RT*. As before, for most of the programs there appears to be an increase in the improvement over *RT* corresponding to the increase in target exclusion area. In fact, it is only the *sncndn* program that, again, shows no significant improvement over the Random Testing result. As expected, the maximum target exclusion rates for programs with the same dimensionality are similar, and the best results are again obtained when this maximum exclusion ratio is used. These results are shown in Table 4.

A comparison between the best *RRT* and *ART* results is given in Fig. 7, which compares *ORRT*, *NRRT*, and *ART* according to their improvement over the calculated *F-measure* for Random Testing (*RT*).

From the figure, it is can be seen that the *RRT* methods compare favourably with the *ART* method, even for the *sncndn* program. In almost all cases, the *NRRT* results are very similar to, or better than, the *ORRT* results. Only for the *bessj* program does the *NRRT* method perform less well than the other methods.

In the next section, we discuss these results, and some other details regarding the implementation of *RRT*.

**Table 4.** Program name, dimension (D), Max $R$, Improvement over $RT$ at Max $R$, $R$ for best improvement, and best improvement over $RT$ for the error-seeded programs, using $NRRT$

| Program Name | D | Max Target Exclusion (R) | Improvement over RT at Max R | R for Best Improvement | Best Improvement |
|---|---|---|---|---|---|
| bessj | 2 | 150% | 18.00% | 130% | 18.83% |
| bessj0 | 1 | 100% | 42.29% | 100% | 42.29% |
| cel | 4 | 430% | 79.37% | 430% | 79.37% |
| erfcc | 1 | 100% | 46.67% | 100% | 46.67% |
| gammq | 2 | 150% | 22.17% | 150% | 22.17% |
| plgndr | 3 | 270% | 79.55% | 270% | 79.55% |
| sncndn | 2 | 150% | 2.25% | 120% | 3.27% |

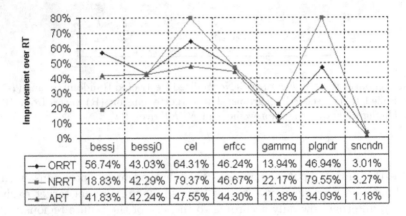

| | bessj | bessj0 | cel | erfcc | gammq | plgndr | sncndn |
|---|---|---|---|---|---|---|---|
| ORRT | 56.74% | 43.03% | 64.31% | 46.24% | 13.94% | 46.94% | 3.01% |
| NRRT | 18.83% | 42.29% | 79.37% | 46.67% | 22.17% | 79.55% | 3.27% |
| ART | 41.83% | 42.24% | 47.55% | 44.30% | 11.38% | 34.09% | 1.18% |

**Fig. 7.** Comparison of best improvement of *F-measures* for Ordinary Restricted Random Testing ($ORRT$), Normalized Restricted Random Testing ($NRRT$), and Adaptive Random Testing ($ART$), when applied to the seeded programs. All figures refer to the improvement over the calculated $RT$

# 4   Discussion

## 4.1   Experiment Results

In the previous section, the results from the various experiments showed that the $RRT$ methods outperform $RT$, and compare favourably to $ART$ when the target exclusion area is at a maximum. With the exception of the *sncndn* program, all the error-seeded programs showed an increase in the improvement over $RT$ as the target exclusion ratio increased for the $RRT$ methods.

The *cel* program, which at first seemed to perform so badly under $ORRT$, and later with a maximum target exclusion ($R$) of 32,000% showed excellent improvement over $RT$ (64.31%), displayed one of the features that motivated the modifications to $ORRT$, which resulted in the $NRRT$ method. The reason for the extreme Max $R$ was traced to the extremely inhomogeneous shape of the input domain: *cel* takes input from a 4D domain whose sides are in proportion 1:300:1,000:10,000. Because *ORRT* makes use of a fixed shape exclusion region, for the *cel* program much of the exclusion zone fell outside the input domain. When we homogenized the input domain by using the *NRRT* method, the improvement over *RT* actually increased to nearly 80%, and the maximum $R$ was reduced to 430%!

As previously explained [5], the poor response of the *sncndn* program to the *RRT* methods is related to the failure pattern, which is characterised as the point pattern type. According to our hypothesis, failure detection capabilities for programs with this failure pattern type are not expected to improve when using *RRT*.

With the exception of the *bessj* program, the results for *NRRT* are similar to, or improve upon, the results for *ORRT*. *NRRT* was motivated by a desire to improve the spread amongst test cases in the input domain, especially for inhomogeneous domains. For those programs with homogeneous input domains (*sncndn*, and the 1-D programs: *bessj0* and *erfcc*), the performance for *ORRT* and *NRRT* is the same.

The *bessj* program, although showing excellent improvement over *RT* with the *ORRT* method, had a poorer performance under *NRRT*. The reason for this appears to be connected to its failure pattern, which is best described as a narrow block type, or an incomplete strip type. This is currently being investigated further.

## 4.2   Complexity and Overheads

The main overhead in the *RRT* method, common to both *ORRT* and *NRRT*, is associated with the generation of a valid test candidate. It is necessary to first generate a random point, and then check if this point lies within any exclusion zone. If it does lie in any of the zones, then it is discarded, and another random point is generated. The process of generating and discarding random points continues until a point lying outside the exclusion zones is produced. The number of attempts to generate a valid test candidate goes up as the target exclusion ratio ($R$) increases.

## 4.3   Summary and Conclusion

Previously, a new method of testing software, Restricted Random Testing (*RRT*), was introduced [5]. In this paper, additional data for this method, and a new adaptation, Normalized Restricted Random Testing (*NRRT*), were presented. The *NRRT* algorithm improves on the Ordinary Restricted Random Testing (*ORRT*) version by adapting the exclusion zones to better suit inhomogeneous domains, and thus more easily enabling a better spread of the test candidates.

The *RRT* methods have been shown to compare favourably with both *RT*, and *ART*, another testing method previously introduced in the literature [6,

7]. Preliminary research has indicated a correlation between the type of failure pattern and which $RRT$ version ($ORRT$ or $NRRT$) performs best. The overheads of the method are related to the generation of the test candidate. When the execution overheads of testing a program exceed those of the test candidate generation, then the $RRT$ methods offer considerable improvement over Basic Random Testing ($RT$).

**Acknowledgement.** We should like to gratefully acknowledge the support given to this project by the Research Grants Council of the Hong Kong Special Administrative Region, China (Project No: HKU 7007/99E).

# References

1. Association for Computer Machinery, *Collected Algorithms from ACM, Vol. I, II, III*, Association for Computer Machinery, 1980.
2. T.A. Budd, "Mutation Analysis: Ideas, Examples, Problems and Prospects", *Computer Program Testing*, B. Chandrasekaran and S. Radicci (eds.), North-Holland, Amsterdam, pp. 129–148, 1981.
3. F.T. Chan, T.Y. Chen, I.K. Mak and Y.T. Yu, "Proportional Sampling Strategy: Guidelines for Software Testing Practitioners", *Information and Software Technology*, Vol. 28, No. 12, pp. 775–782, December 1996.
4. F.T. Chan, I.K. Mak, T.Y. Chen and S.M. Shen, "On the Effectiveness of the Optimally Refined Proportional Sampling Testing Strategy", *Proceedings of the 9th International Workshop on Software Technology and Engineering Practice (STEP '99)*, S. Tilley and J. Verner (eds.), IEEE Computer Society, Los Alamitos, California, pp. 95–104, 1999.
5. K.P. Chan, T.Y. Chen and D. Towey, "Restricted Random Testing", *Software Quality - ECSQ 2002, 7th European Conference*, Helsinki, Finland, June 9–13, 2002, Proceedings. LNCS 2349, Springer-Verlag, Berlin Heidelberg 2002
6. T.Y. Chen, H. Leung and I.K. Mak, "Adaptive Random Testing", submitted for publication.
7. T.Y. Chen, T.H. Tse, and Y.T. Yu, "Proportional Sampling Strategy: A Compendium and Some Insights", *The Journal of Systems and Software*, 58, pp. 65–81, 2001.
8. T.Y. Chen and Y.T. Yu, "On the Relationship Between Partition and Random Testing", *IEEE Transactions on Software Engineering*, Vol. 20, No. 12, pp. 977–980, December 1994.
9. J.W. Duran and S.C. Ntafos, "An Evaluation of Random Testing", *IEEE Transactions on Software Engineering*, Vol. 10, No. 4, pp. 438–444, July 1984.
10. W.J. Gutjahr, "Partition Testing vs. Random Testing: The influence of Uncertainty", *IEEE Transactions on Software Engineering*, Vol. 25, No. 5, pp. 661–674, September/October 1999.
11. R. Hamlet, "Random Testing", *Encyclopedia of Software Engineering*, edited by J. Marciniak, Wiley, pp. 970–978, 1984.
12. R. Hamlet and R. Taylor, "Partition Testing Does Not Inspire Confidence", *IEEE Transactions on Software Engineering*, Vol. 16, No. 12, pp. 1402–1411, December 1990.

13. H. Leung, T.H. Tse, F.T. Chan and T.Y. Chen, "Test Case Selection with and without Replacement", *Information Sciences*, 129 (1–4), pp. 81–103, 2000.
14. P.S. Loo and W.K. Tsai, "Random Testing Revisited", *Information and Software Technology*, Vol. 30, No. 9, pp. 402–417, September 1988.
15. W.H. Press, B.P. Flannery, S.A. Teulolsky and W.T. Vetterling, *Numerical Recipes*, Cambridge University Press, 1986.
16. M.Z. Tsoukalas, J.W. Duran, and S.C. Ntafos, "On some Reliability Estimation Problems in Random and Partition Testing", *IEEE Transactions on Software Engineering*, Vol. 19, No. 7, pp. 687–697, July 1993.

# Testing Safety Critical Ada Code Using Non Real Time Testing

Y.V. Jeppu[1], K. Karunakar[2], and P.S. Subramanyam[3]

[1] Deputy Project Director (IFCS – Flight Control Laws)
Aeronautical Development Agency, Bangalore, India
yvj_2000@yahoo.com
[2] Group Director (Independent Verification and Validation)
Aeronautical Development Agency, Bangalore, India
[3] Project Director (Integrated Flight Control Systems)
Aeronautical Development Agency, PB NO 1718,
Vimanapura PO, Bangalore, India

**Abstract.** Testing the Flight Control Laws and the Airdata Algorithm form a major part of the Flight Control System development cycle. These two safety critical software are validated and verified on very costly rigs in every aircraft development program. The Indian Light Combat Aircraft, with fixed gains and the limited first flight envelope, was successfully test flown without a single software fault starting January 4, 2001. This paper highlights a new methodology followed by the LCA team to test and certify the digital fly-by-wire control law. The philosophy used is to test the quadruplex software in a single strand mode. Only the Control Laws and the Airdata Algorithm are tested against a validated FORTRAN model. The Model code is developed by the Control Law designers and represents the functionality in totality. A fault injection methodology is used on the model, called "Delta Model", to verify the efficacy of the test cases. The paper discusses the methods, typical cases and shares some of the experiences of the Indian LCA Team.

## 1 Introduction

The advantages of having an unstable aircraft and the availability of fast computers have made feasible the highly agile modern day fighter aircraft. The fly-by-wire flight control system on board such aircraft has the ability to tailor the system characteristics at every point in the flight envelope. This is achieved by using the Digital Flight Control Laws (DFCL) scheduled at every flight condition. The scheduling of parameters requires the Air Data System (ADS), which comprises of sensors and a complex algorithm, to give the required parameters. The requirement of a very high mean time between failure (MTBF) necessitates quadruplex redundancy in the on board computer, sensors and actuators that form the flight control system (FCS). The DFCL and

J.-P. Rosen and A. Strohmeier (Eds.): Ada-Europe 2003, LNCS 2655, pp. 382–393, 2003.

the ADS software are designated safety critical and are developed in accordance with strict standards and specifications [1].

The FCS software goes through extensive verification and validation for a fighter aircraft program. A well-documented design process is used to develop the software. An extensive textbook approach to software testing is used to validate the software at every stage of development. The software goes through Unit level testing, Integration testing and Hardware Software Integration [4]. The DFCL and ADS are again tested on ground in a facility called "Ironbird" [1]. The Ironbird is a hardware-in-loop test facility where the real actuators, flight model and the flight computer are tested in a closed loop mode. At the Ironbird rig the DFCL and the ADS are tested against a design model code in an automated fashion. No human intervention is required unless the test fails. As many as 100,000 test cases are executed in this fashion! [1]

## 2    The Indian LCA Software

The FCS for the Indian Light Combat Aircraft (LCA) has gone through a development process as described for a typical fighter aircraft above [2]. It has a quadruplex digital flight control computer (DFCC). The software has been coded in Ada language. The DFCL and ADS have been coded using the BEACON tool, which is a Computer-Aided Control System Design software [5]. The software has gone through a series of testing and finally flown on the initial series of flights without a single error, which could abort the scheduled flights [3].

This paper describes a novel method used by the LCA team to cut short the Ironbird level tests. The testing was carried out in a non real time mode and the method was accepted as a formal test methodology to validate the DFCL and ADS software for certification of onboard Flight Control System software.

## 3    The DFCL and ADS Code

The Control Law Design team designed the DFCL and ADS for the LCA. This design was released for coding as a functionality document. The design is in the form of control block diagrams with specification for each block. The blocks are control elements like filters, nonlinear elements, faders, summations, rate limiters and trigonometric functions. Scheduled gains are specified as lookup tables. The detailed control law and the ADS algorithm are also described in an algorithmic language like FORTRAN. Control law designers and a test pilot, who is a member of the DFCL design team, validate the FORTRAN code in the engineer-in-loop simulator and later at various other real-time simulation platforms. The use of a test pilot in the design phase has provided valuable insight into the DFCL design process to the LCA Team [2].

The software requirement specification (SRS) just specifies that the control law and ADS should behave as specified in the functionality document. It is therefore essential that this specification be executable to validate the DFCC code [6].

## 4    LCA Software Test Process

The DFCC software goes through several tests before its release for the Ironbird level tests. The Computer Software Unit (CSU) and Computer Software Component (CSC) level tests are carried out as per DOD-STD-2167A. These tests are generated using the SRS and the Software Design Document (SDD). The tests cover Equivalence Partitioning, Boundary Value Analysis, Path Coverage, Compound Expression and Loop Tests. Functional tests are carried out using a tool called "Comparator" and white box structural testing with a "Logiscope" tool. The formal code released after these tests undergoes Software Integration Tests (SIT) and Hardware Software Integration tests (HSI). This is carried out on a special test rig called "MiniBird".

The formal configured code is released for Ironbird tests where the DFCC with the simulated avionics, pilot controls and the actuators are tested in open and closed loop mode. This facility is a very costly setup and usage time on this facility is strictly controlled. Moreover, the focus of testing is mainly hardware/software integration issues, redundancy management and fault free flying.

## 5    Requirement for a New Methodology

The time constraints and project pressures lead the LCA team to look for additional schemes to augment the verification and validation of the DFCL and ADS at Ironbird and Minibird levels. Moreover, the error bounds envisaged in the Ironbird, on analysis, were found to be very high to give any meaningful insight into the DFCC code. The question asked in various reviews by the management and the certifying agency was "Has the DFCL and ADS been coded correctly as per the functionality?"

## 6    NRT Philosophy

A new group was formed to look into a formal method to clear the DFCL and ADS very early in the program phase. The group evolved a method to validate and verify the code in a non real time (NRT) mode. The basic concept is to test just the single strand (not the 4 channel operation) Ada code on a target Single Board Computer (SBC). Figure 1 shows the schematic of the NRT philosophy. The Ada code is clipped and stubbed to remove all four channel related code. The DFCL and ADS code are tackled separately. An Ada driver code is added to the clipped code to inject the test inputs and tap out the required outputs. This code with the driver is compiled with the same compiler and settings as the final build of DFCC software. A single board computer using the same processor as the DFCC is used as a target for the Ada executable.

The FORTRAN model code developed by the DFCL and ADS designers is used as a benchmark for the Ada code. This code is a validated code, the validation carried out at various simulation platforms and by another independent agency. The model code is attached to a driver and compiled to run on a PC. The PC software called

PACTS (Platform for Automatic Control law Tests) is used to simulate the test cases. Figure 2 shows the NRT setup.

The test cases are generated in view to test each and every control block element. The inputs are given to the code at the sensor input point like in case of a black box testing. Tap out points are assigned at the input and output of each control element block. This is similar to the white box approach. The test cases are designed to bring out every possible error in each control block. This is verified by introducing faults into the model code called "Delta Model".

## 7  Concept of Delta Model

Fault injection is a common practice in the software testing process. This has been used to design test cases and to estimate the effect of error on the final output [7]. This concept is used to design the test cases for NRT. A test case is designed and tested on a copy of the FORTRAN model code. This code is doctored such that the block under test is perturbed by the addition of error. The error could be errors in the coefficients for the filters, error in the algorithm in case of rate limiters and faders and reversal of sign in case of summation blocks. Figure 3 shows the schematic of the Delta Model.

As an example say the block under test is Filter A and this is called by using a function FILTER() in the code.

```
O = FILTER(COEFFA,COEFFB,DELTAT)
```

Here the functionality document specifies COEFFA to be equal to 10.2345 and COEFFB to 22.4641. The Delta Model has the perturbed value of COEFFA as 10.23461 and the test case has to trap this error.

The test case is executed on the Model and the Delta Model and deemed to be efficient if it can bring out the error in the component under test. The Delta Model is discarded and the executable deleted to avoid confusion once a test case is frozen. Only one error is introduced at a time in a single block for a specific test. Every test case thus identified is brought under configuration management.

## 8  Configuration Control

A high degree of formalism was introduced into the test process in the conception phase. The Ada code under test, the FORTRAN model and the test cases are all Configuration-Controlled Items (CCI). No change is possible on these items without a formal release of change sheet. All test results from Ada code and the Model are CCIs. The test result file has information about the code version, date executed, test case name and issue number. A database and a management system developed inhouse is used to track all the version changes, test executed and pass/fail status.

# 9    NRT Test Process

The NRT test process starts with the Test Plan document with traceablity to the Ground Test Plan Document. This is the base document highlighting the scope of the test, the system under test and broad classification of the items to be tested. This document gives details to prepare the Request for Pre-Tests (RFPT). These define the test cases, which are to be executed on the Delta Model. Each test case has a unique 8-letter identifier, which identifies the test case and all the results, reports etc to follow the test. A RFPT would be converted to Request for Test (RFT) once it is cleared on the Delta Model. The RFT is a CCI and contains a set of test cases, a description of the definition of the system under test, functionality reference, the type of inputs to be injected and the variables to be tapped out for analysis.

Generation of the RFT automatically generates two Input Definition Files (IDF), which are the inputs to the Ada and FORTRAN Model driver. Unlike script-based approach to software testing the IDF specifies the type of input waveform like sine wave or pulse train. This makes the files compact and it is very easy to verify the results based on the test case description. The IDFs are created using software called "PACTS" developed for automated DFCL and ADS tests. This software is a Microsoft Windows based application software running on a PC. The two IDFs are CCI and describe the test case as given in the RFT in a form readable by the respective drives.

The PACTS software has the ability to read the FORTRAN IDF and execute the Model code on the PC. The output is an ASCII file containing columns of data corresponding to variables tapped out. The Ada IDFs are processed in a VAX environment for execution on the target SBC. The output of the SBC is also an ASCII file with suitable headers and data from the Ada code output. These two files are compared in time and frequency domain to generate the pass/fail status automatically based on specific criteria.

Each test case after execution and comparison generates a test report automatically. In case a test has failed user intervention is required for analysis and to continue further tests. Extensive use of Matlab © Mathworks (http://www.mathworks.com) has been used for analysis. A software problem report is raised giving the cause of failure. The testing ends when all the test cases have been executed and all the problems resolved with agreement between the algorithm designers and the software designers. A final report closes the test activity. The schematic of the NRT test process is given in Fig. 4.

# 10    Typical Inputs

NRT emphasizes on stressing the software to its limits. Very large input values over the complete frequency range are injected into the software to stress test the various blocks. Every block is tested with a sine sweep waveform. In case of filters this method gives the frequency response of the block. Rate limiters and non-linearities are excited using sine sweeps to check for frequency dependencies. Large amplitude pulse trains bring out the performance of the rate limiters. Slow ramps are used to characterize the non-linear functions. A combination and ramps and sine waves are used to exercise the gain tables over the entire range. Events are excited with pulse trains with varying widths to check for sub sampling. A single test case will have combination of all these waveforms to check a specific control block.

## 11  Pass / Fail Criteria

The pass/fail criteria for the tests were generated based on the Delta Model tests and preliminary tests on the Ada code. As the error increases with the increase in the input levels percentage errors are computed. In case of very small signal levels this criteria cannot be used, as the percentage errors tend to be very high. The criteria finally used for the error Enrt in NRT is

Enrt = % error, if signal value > 1.0
else
Enrt = the error value itself

A threshold of 0.0002 is used to clear the test cases. This has been found to be optimum for the DFCL and ADS. It has been the team's experience that errors greater than these value normally signify some coding differences between the Ada code and the Model.

## 12  Results

The NRT process has been applied to test various builds of software and this has given the LCA team a lot of experience and insight into what kind of errors pass through various levels of testing. We share a few of them here

- The functionality defines that the scheduled gains should be faded based on certain events. This was not coded in the Ada software. Static tests at Iron bird were clearing the DFCC. In NRT the test case checked for the complete range of input values for the table lookup in combination with the associated events toggling. This rapidly changing test input could trap this error and the software was modified. The functionality was modified to bring clarity to this aspect
- The notch filters, which are very important to prevent structural modes from entering DFCC loop, were coded with lesser precision values. The notch filters are designed after optimization and precise values have to be entered to get the correct frequency characteristics [8]. The errors were in the 4th decimal place and NRT could be used to analyze the effect of this on the system performance. Frequency response analysis from sine sweep inputs was used for accurately determining the performance degradation.
- The integrators were implemented using Tustin transformation in the Ada code and Euler method in the Model code. Errors due to this sort of algorithmic differences could be trapped even though the Ironbird had passed these tests.
- There was an error in the interpolation table indexing. The compiler normally should have trapped this error. The error was trapped in NRT as very high input

values (Mach numbers greater than 3) were injected at the input points and the SBC hung up.

- Table data for interpolation were entered with truncated values instead of rounded off values.
- Limits placed on the inputs were electrical units instead of engineering units. The functionality specifies engineering unit limits on pilot inputs. The software team had placed the scaled limits on inputs.
- Typical errors found in the code are due to the functionality being misunderstood and data values entered wrongly. This is because the DFCL is coded using BEACON tools and the code is generated automatically. The errors in this case will be mostly data entry errors. Certain portions of the code were coded for optimization and here the indexing problems had manifested.

## 13  Advantages of NRT

The team has perceived the following advantages of NRT after extensive rounds of testing.

- NRT tests the software for its functional behavior and performance. It is basically a functional testing and because the tester is tapping out intermediate variables to increase testability it has shades of "grayness" associated with it. The test does not verify the software as a piece of code but treats it as a control system, something similar to a hardware circuit.
- There is a good idea when to stop the tests as the test cases are based on knowledge of the system. Each block has a specific number of tests associated with it and therefore unlike black box testing the number and types of inputs is finite.
- It is easy to analyze the effect of certain errors on the final system performance. As an example, the effect of changed filter coefficients on the system response is very easily decipherable. The question "Okay, we have an error can we live with it?" is answerable avoiding costly changes.
- The test methodology has boosted the confidence of the pilots and the design team. Debugging the code during Ironbird tests has been simplified, as the software design team knows exactly where not to look for bugs.
- The Ironbird and Minibird tests have been reduced by about 70 % as all the DFCL and ADS tests have been covered in NRT. Ironbird level checks are for four-channel operation, path connectivity from sensors to actuators, setting of events physically by failing a hardware item or selecting switches.
- The NRT testing activity can be started very early in the software development cycle and this has saved cost and time to the project.
- Regression testing is easy on every build of software as the testing is completely automatic once the test cases are in place.

# 14  Best Practices

Some of the best practices for embedded DFCL and ADS implementation are as follows

- The DFCL or ADS functionality should be augmented with the Model code. The Model code often is a better representation of the functionality than the text document.
- A functional level stress testing is a must preferably with the participation of the DFCL designer and the software designer. Most of the errors found were due to functionality not being understood by the software designer. If the tester has also understood it in the same way then major design flaw may pass unnoticed.
- Extensive testing is a must at the end of the day in spite of good SQA plans, good compilers and autocode generators being used for software development.
- Static tests do not give out much information about the system. Always test for transient behavior.
- There is a need for synergy between the DFCL designers, Software designers and the Independent Verification and Validation team.
- If the model code were the same as the final embedded code testing would be simpler. There is a requirement to think of a hardware platform where the DFCL designers, software developers and pilots can work together very early in the project phase for a program of this nature.

# 15  Conclusion

Software faults, in safety critical systems, have caused loss of millions of dollars and many lives in the past. The Indian LCA team has test flown the aircraft without a single software fault being encountered. This is a major achievement and speaks volumes of the software development process and the Independent Verification and Validation activity. A method has been described in this paper, which has successfully complemented the various test activities and has been accepted by certification agency as one of the important milestones to certify the software. The NRT test activity with its formalism has paved the way for the flight clearance of the aircraft saving a lot of time, effort and money for the project.

**Acknowledgement.** The authors wish to thank Programme Director, LCA and Director, Aeronautical Development Agency for their encouragement and support for the NRT activity. We wish to thank Mr Sukanta K Giri and Ms Kavitha Rajalakshmi, Scientists, IV&V, ADA for their technical support during the execution of NRT.

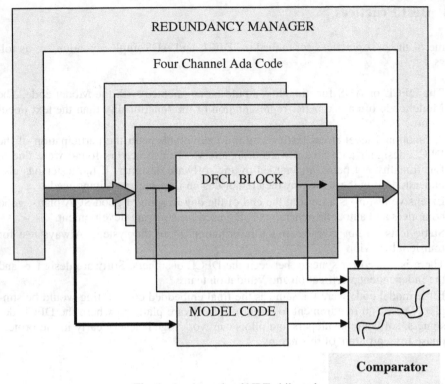

**Fig. 1.** A schematic of NRT philosophy

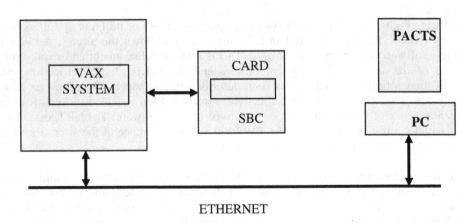

**Fig. 2.** The NRT setup shows the Single Board Computer (SBC) connected to the VAX system. The PACTS software running on the PC collects Test data, simulates the Model code and analyses results

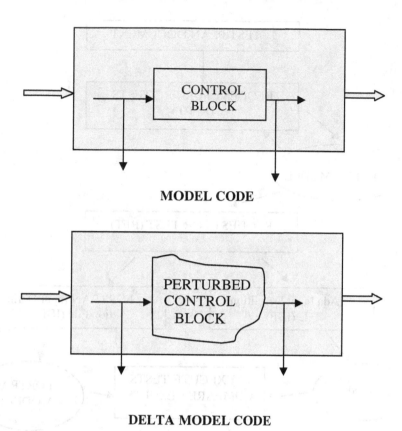

**MODEL CODE**

**DELTA MODEL CODE**

**Fig. 3.** The Model code and the Delta Model are shown as a schematic. A single block under test as defined by the test case is perturbed by changing a constant or logic in the Fortran Model code. This is the Delta model code. The test case is executed on the Model and the Delta Model. A test case is declared adequate if it can find the error in the block under test

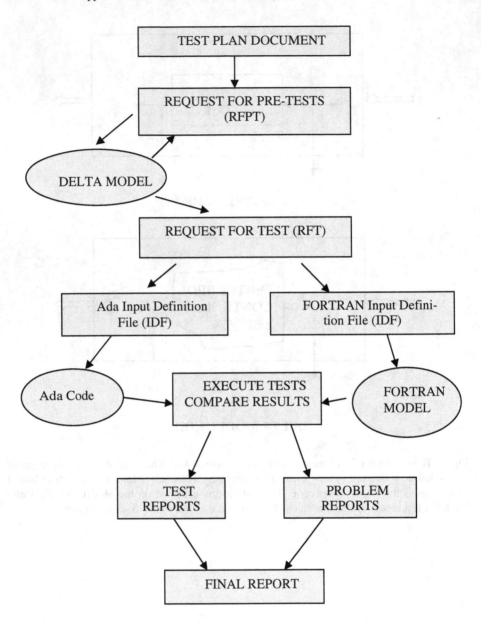

**Fig. 4.** The schematic NRT Test Process shows the various steps involved from Test Plan documentation to the final Test Report.

# References

1. Roger W Pratt: Flight Control Systems, Progress in Astronautics and Aeronautics, Vol. 184, AIAA and IEE (2000).
2. Shyam Chetty, Girish Deodhare, B.B. Misra: Design, Development and Flight Testing of Control Laws for the Indian Light Combat Aircraft, AIAA Guidance, Navigation and Control Conference, Monterey, CA, August 2002
3. Wg. Cdr. R. Kothiyal, Wg. Cdr. T. Banerjee: First Flight of India's Light Combat Aircraft", 45th SFTE Symposium Proceedings, Sept. 2001, 88–95
4. Steven R Rakitin: Software Verification and Validation - A Practitioner's Guide, Artech House (1997)
5. Richard C. Dorf: The Electrical Engineering Handbook, Boca Raton: CRC Press LLC (2000)
6. Alfred R and Theodor T.: Aspects of Flight Control Software – a Software Engineering Point of View, Control Engineering Practice, Vol. 8, 2000, 675–680
7. Voas J: A Tutorial on Software Fault Injection, Spectrum 2000,
   http://www.cigital.com/papers/download/spectrum2000.pdf
8. Animesh C et al.: Design of Notch Filters for Structural Responses with Multi-Axis Coupling, AIAA Journal of Guidance, Control and Dynamics, Vol. 22, No. 2, March 1999, 349–357

# The Standard UML-Ada Profile

Francis Thom

Artisan Software Tools
Stamford House, Regent Street
Cheltenham, UK, GL50 1HN
Francis.Thom@Artisansw.com

**Abstract.** The Unified Modeling Language (UML) has become the *de facto* modelling language for business processes and software intensive systems. However, the UML[1] lacks a rich-set of semantics specific to the Ada Programming language, which would enable a designer to both design an *abstract solution* based on the systems requirements, and an *implementation solution* based on the semantics of the Ada Programming Language. So as to bridge this semantic gap between the modelling language and the programming language the UML provides *Profiles*. A UML profile forms a part of the *extensibility mechanism*, inherent to the UML, enabling the standard UML to be extended to encompass the semantics of an individual programming language. This paper explores the current definition of the ARTiSAN UML-Ada Profile[2], and its usage and flexibility in developing Ada source code from a UML model. The UML-Ada Profile at the design-level not only enables implementation issues to be resolved prior to constructing the code, it also has the advantage of facilitating the automatic generation of Ada Source code.

---

[1]　At the time of writing (09-02-03) the latest publicly available version of the UML specification from the Object Management Group (OMG) is version 1.4. Although Action Semantics have been defined for a UML 1.5, this specification is still in the process of being finalised (a 'proposed' specification is currently available). As for the "UML 2.0", apart from the submissions for the Infrastructure and Superstructure of a "UML 2.0", nothing has yet been agreed and certainly not finalised. Contrary to some arguments about simplifying the UML 1.x series and removing some of its ambiguities, the current submissions are more complex, and more worrying, inconsistently ambiguous, requiring yet more effort (and hence time) from the OMG to resolve these problems. It is possibly to late to inject some 'pragmatic engineering' justification for some of the proposed 'new' features in the "UML 2.0". For those UML end-users (i.e. engineers) who are interested in the "UML 2.0", you should possibly start thinking about "UML 2.1" hopefully, the version of the "UML 2.x" series that removes all the inconsistencies, ambiguities and links the definition of the UML Standard with the evolving MDA initiative. Having said this, ARTiSAN Software Tools have repeatedly stated that we are committed to the 'current issued version' of the UML and we see no reason, barring 'pragmatic engineering' and 'end-user' complaints, for not implementing any finalised "UML 2.0" specification.

[2]　On the 29th of October 2002, Ada-UK convened at their bi-annual conference and discussed the need for a standardised UML-Ada Profile. As a result of this meeting a working group has been set-up to standardise the Ada-Profile. The intention is to make the Ada-Users community the first programming language-specific community to agree on a standard for creating Ada source code from UML.

J.-P. Rosen and A. Strohmeier (Eds.): Ada-Europe 2003, LNCS 2655, pp. 394–404, 2003.
© Springer-Verlag Berlin Heidelberg 2003

# 1    Introduction

## 1.1    The Problem

All programming languages are outside the scope of the UML. The UML, a visual modeling language, is not intended to be a visual programming language, in the sense of having all the necessary visual and semantic support to replace programming languages. The UML is a language for visualizing, specifying, constructing, and documenting the artifacts of a software-intensive system, but it does draw the line as you move toward code. For example, complex branches and joins are better expressed in a textual programming language. The UML does have a tight mapping to a family of object languages so that you can get the best of both worlds. [1]

## 1.2    The Solution

Create an UML-Ada Profile to bridge the semantic gap between UML and Ada:

A *profile* is a stereotyped package that contains model elements that have been customized for a specific domain or purpose by extending the metamodel using stereotypes, tagged definitions, and constraints. A profile may specify model libraries on which it depends and the metamodel subset that it extends.

A *stereotype* is a model element that defines additional values (based on tag definitions), additional constraints, and optionally a new graphical representation. All model elements that are branded by one or more particular stereotypes receive these values and constraints in addition to the attributes, associations, and superclasses that the element has in the standard UML.

*Tag definitions* specify new kinds of properties that may be attached to model elements. The actual properties of individual model elements are specified using *Tagged Values*. These may either be simple datatype values or references to other model elements. [1]

Some readers may already be aware of ongoing work to define a UML Profile specifically for "the Development of Distributed Reactive Systems and Ada 95 code generation" [2]. This paper takes a broader look at the issue of modeling (and ultimately generating Ada code), using the UML.

# 2    The OMG's Model Driven Architecture (MDA)

The Object Management Group (OMG) are currently attempting to underpin their broad portfolio of efforts with their latest MDA initiative. Although still in its infancy, MDA encompasses many principles of best practice in the development of Software-intensive systems in terms of developing an abstract Platform Independent Model (PIM) and a Platform Specific Models (PSM). One justification for their initiative is the 'separation of concerns' related to the complexity of possible solution strategies available to engineers. At the same time as localizing each concern within a specific model, each model is expected to be inherently reusable at its own level of abstraction.

# 3    Anything New in MDA?

It could easily be argued that MDA is not a new initiative. The notion of 'separation of concerns' was successfully exploited by the likes of Larry Constantine, Tom DeMarco, Stephen Mellor, Paul Ward and Edward Yourdon in the 1970's and 1980's under the name of 'Structured Methods' and is attributed to previous work by Dykstra. Through a process-driven refinement of successively evolved models ('Requirements Model', 'Essential Model', 'Architectural Model', etc.), the designer would address each 'concern' in isolation within a single model and then move on to the next 'concern' by refinement of the current model – i.e. the 'Essential Model' was refined into the 'Implementation Model'. One word of caution about MDA, in the past, where models were refined (e.g. from the 'Essential Model' into the 'Implementation Model'), any subsequent *re-use* of the 'Essential Model' was rarely achieved – possibly because of the inflexibility of the notation used within the model (functional decomposition represented through data-flow diagrams) – but maybe because there was little attention paid to *potential future reuse* as the models evolved. Re-use does not come for free, not in the past and not today. We have to design with re-use in mind. Are the arguments for *re-use* (especially at the higher-levels of abstraction e.g. the PIM), more compelling today than they were some 20-30 years ago? We need to look at today's notation used whilst constructing a model the UML. The UML is a far better vehicle to provide the necessary flexibility (through modularity, encapsulation, polymorphism and abstraction) required to maximize reuse, but there are many issues still to be resolved about the 'mechanics' of managing multiple models (especially in the area of requirements traceability, and transitions from one model to another). In this paper we focus our attention on the transition from PIM to a notional 'Implementation Model' (IM) – the IM being, from practical project experience, considered a part of the model-set for the PSM. The IM addresses the specific issue of implementing a given PIM in a specific programming language. Even in today's '*need it yesterday*' world of systems development, software still needs to be 'engineered' (designed and written) – and in the area of safety-critical systems, '*must*' be engineered – the IM is proposed as a vehicle for addressing the specific concerns of software engineers allowing them to address language-specific issues whilst meeting the 'essential' requirements (from the PIM) and also bearing in mind the subsequent issues brought about by the 'execution environment' (hardware, firmware and operating systems etc.) captured in the (current definition of the) PSM. We all know that each programming language has its own 'execution model', syntax and semantics. We are also aware that each programming language can solve a specific problem in potentially many ways, where each solution has advantages and disadvantages – i.e. the 'quality of service' - over another. We also know that 'design patterns' exist to assist a designer in solving some of these recurrent problems in a 'commonly accepted' manner. As we transition from the PIM to an IM we need to bear all these solution strategies in mind and ensure that each strategy can be deployed with as much 'engineering pragmatism' and flexibility as possible – i.e. an engineer should not be constrained by a specific 'tool's' limitations e.g. restricting the application of a specific solution strategy like 'design patterns'. The existence of an Ada Profile enables the separate concern of an Ada implementation to be addressed in isolation from other concerns.

# 4 MDA & Reuse –A Modeling Nirvana?

Whilst MDA addresses (amongst others) the issue of retargeting legacy software onto new/different operating systems, firmware and hardware platforms, it also (by implication) addresses the issue of moving from an abstract 'essential' model (PIM) to an implementation model (IM) and finally into a platform specific model (PSM).

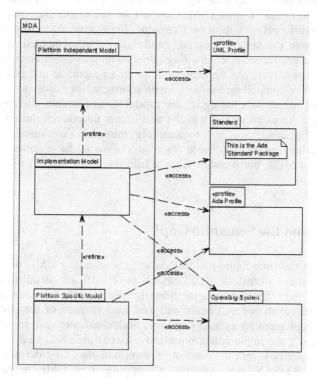

**Fig. 1.** UML Model Hierarchy for MDA

The necessity for an IM derives from the many implementation languages available, and the inherent solution strategies available within each language e.g. the Ada language inherently supports multi-tasking where as the C language does not, in one system this feature may want to be exploited. The ease through which a model can transition between an "Ada-IM" and a "C-IM" will be a significant factor in the adoption of MDA[3]. Figure 1 shows the refinement of successive models from PIM to PSM and highlights other areas of 'concern' required to support the refinement process.

---

[3] As MDA is still in its infancy, we at ARTiSAN do not necessarily restrict our view of MDA specifically to the 'forward development' of the PIM and PSM. We are also looking at offering maximum flexibility in allowing engineers to develop IM's in any supported language (Ada, C, C++ or Java), but also to be able to have a mixture of languages within a single IM (currently available). We are also considering 'rolling-back' and/or 'undo' capabilities in terms of applying Profiles and Patterns to ensure that a designer has access to all the possible solution strategies available.

The PIM, by definition, accesses no implementation-specific details. The IM accesses the pre-defined Ada Standard package (additionally, any other required Ada libraries) and the Ada Profile. The PSM also accesses the Ada Profile for platform-specific information (e.g. 'address' and 'length' representation clauses, etc.) and (if present) any parts of the system representing hardware, firmware and operating system. As the models evolve, addressing each concern in isolation, it should be apparent to the reader that a mechanism of 'undoing' refinements would be a positive advantage. Being able to 'roll-back' from a PSM to an IM would enable the same IM to be re-targeted onto a different execution environment. Being able to 'roll-back' from IM to PIM would enable the same 'essential' model to be implemented in another programming language. These are compelling ideas so long as the engineers, i.e. the end-users of these techniques, can implement them in a pragmatic and efficient manner. MDA forms a very compelling business-level argument, the ability to re-use as much of our corporate assets – including legacy models – as possible, thereby minimizing wasted effort (in 'reinventing the wheel') and errors introduced during the development process (because there is comparatively minimal 'development' required in assessing something as being suitable for re-use). We all have some way to go yet with the OMG's MDA but we should, at a minimum, all support its commendable aims.

## 5   Please Mind the Semantic Gap!

As explained in the Introduction to this paper, the standard UML does not contain sufficient semantics to create a model, which is of sufficient detail to ultimately be implemented in Ada. The Ada programming language has many features, which can in due course affect the overall behaviour and performance of the software system (for example, implementing a *timed* versus a *conditional* entry call to an Ada task). It is important that these implementation-specific issues are addressed in isolation (albeit not total isolation) from the 'execution environment'. One definition of an Ada Profile (from ARTiSAN) does provide a mapping from UML to Ada. However, knowing that each end-user implements Ada in a subtly different 'style' (generally captured in something like a 'Coding Standards' document), we also had to ensure the end-user had the ability to configure the ARTiSAN Ada Profile to meet their needs.

## 6   Models: Iterative Refinement

The different models within a MDA enable the separation of concerns: the PIM focusing on solving the end-user requirements; the Implementation Model focusing on programming language specific issues, and the PSM addressing issues related to both the API of the Operating system (if any) and issues of integrating the whole application to the target hardware platform – i.e. the execution environment. Reuse can potentially be achieved at any level. If it is required to implement the PIM in another programming language then the PIM can be reused (as the essential requirements have not changed) and another programming language Profile can be accessed to address the specific concerns of 'that' programming language. If the system is to be retargeted to either a new operating system or hardware platform – a different execution environment - then the Implementation Model (including its language-specific profile) can potentially be reused without modification. MDA places no constraints on how models are configured but it seems an intuitive assumption that several models of increasing refinement will be created (and more importantly) configured throughout a single project.

# 7  Proprietary versus Open Standards

Both the UML and the Ada Programming Language are classed as *open* standards. Automatic code generation for Ada exists. Some of these tools use proprietary techniques, which lock the end-user to a specific tool vendor. The philosophy implemented by ARTiSAN is to provide the end-user with a flexible and configurable technique to generate Ada source code whilst working, as much as possible, within the definitions of open standards. As a result of this, ARTiSAN are currently working with Ada-UK to define a standard UML-Ada Profile such that the Profile is 'owned' by the Ada-user community and will eventually exist as an open standard. When this work is completed, the Ada community will be the first programming language user group to have agreed and defined a standard for designing Ada using the UML.

# 8  Example Application of the UML-Ada Profile

The Ada Profile is an addition to the UML enabling Ada specific language features to enhance a basic mapping from the UML to Ada. The purpose of any profile is not to capture 'all' mapping issues as some of these issues are addressed by a 'basic mapping' of existing UML entities onto the Ada Language. Such mappings as: A UML Class maps onto an Ada package specification and body. An operation on a class (without a specified return parameter) maps onto an Ada procedure. An operation on a class (with a specified return parameter) maps onto an Ada function. The Profile is intended to address all the other mapping issues which cannot be mapped because the syntax and semantic of the standard UML are deficient for the Ada language. This should not be interpreted as a limitation of the UML, as the UML is not intended to capture language-specific issues directly and the Profile 'extension mechanism' was introduced to address this potential problem.

## 8.1  From the Abstract PIM to IM

It is beyond the scope of this paper to address the transition from IM to PSM. An analyst and/or designer will create a PIM, which reflects an abstract solution to the systems requirements. In general the PIM will be free from implementation-specific details. However when the designer moves to the implementation stage many other issues can come to light.

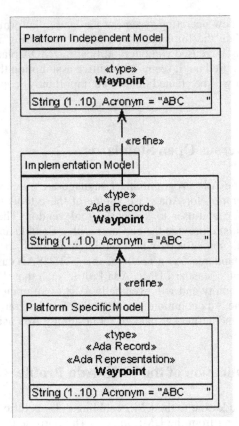

**Fig. 2.** From Abstract PIM to IM

## 8.2   Type Classes

Type definitions modeled using UML classes with the UML stereotype (<<Type>>) will need to be defined when the PIM is being constructed, to define the information being exchanged between model entities (e.g. between classes or sub-systems). However, it is unlikely that a System Engineer, developing the PIM, will be able to rigorously define these type definitions, as Ada Types and neither should they. Standard UML cannot be used, without extensions. The design must be refined from the PIM into the Implementation Model. In the example diagram, a *Waypoint* is successively refined as we transition from the PIM to the PSM. If the *Waypoint* were to be retargeted to a new hardware platform, only the <<Ada Record Representation Clause>> may need to be modified to reflect the addressing mechanism utilised by a target-specific compiler. If the *Waypoint* were to be implemented in C, then both the <<Ada Record Representation Clause>> and the <<Ada Record>> would need to be replaced with the stereotype <<C Struct>> from a UML-C Profile.

### 8.3    Modeling Application Classes in Ada

There exist Ada-specific modeling techniques that have to be adopted with the UML. An engineer must have an appreciation of the basic mapping from UML to Ada when constructing the IM. For example, we all know that the Ada language does not permit 'circular dependencies' between Ada packages. If we choose to map a UML Class onto an underlying Ada Package then any UML associations between UML Classes must always be specified using uni-directional (sometimes called 'navigational') association. Likewise, some UML entities are not strictly 'associated' in the UML definition of association, i.e. one entity is merely 'dependent' (not associated) with another. This requires the designer to specify what type of link exists between these entities. It is beyond the scope of this paper to detail the subtle application of the UML specifically for Ada. However, this diagram depicts a 'dependency' link between Task A and the Ada pre-defined Text_IO package.

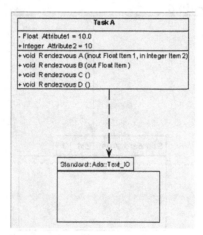

**Fig. 3.** Modeling Application Classes

### 8.4    Implementation Classes

Implementation classes, those that actually do something, can have various mappings into Ada. By default the attributes of a UML class map to some underlying record type (so that multiple instances can be created if required and further classes can inherit from this type). If however, a software engineer desired a specific class to be implemented as an Ada Task, Standard UML could capture this as an Active Class. However, the UML has no semantics to support the different *rendezvous* mechanisms available to the Ada programmer. As a result the UML must be extended (using the UML-Ada Profile) to support these semantics. Figure 2 and 3 (Figure 2 being part of the PIM, and Fig. 3 being part of the Implementation Model) show how the implementation of an Ada Task, including the identification of the different rendezvous mechanisms, can refine the information captured within the Implementation Model to represent the underlying semantics of Ada. Because the model is a 'visual' representation of the Ada constructs, it is apparent that the Implementation Model is incomplete. *Rendezvous D* has no specified rendezvous mechanism. It could be assumed that a rendezvous without a stereotype will be implemented as a standard entry call however, this denigrates the ability to create a 'precise' implementation model and the actual implementation of *Rendezvous D* would be open to interpretation by either a human engineer or 'hard-coded' into a proprietary code generator translation language.

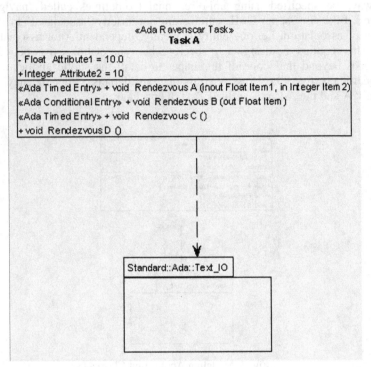

**Fig. 4.** Implementation Class

# 9   Automated Code Generation

It should now be apparent to the reader that once the UML-Ada Profile has been applied to the PIM to create the IM, there exists sufficient information within the model to generate the source code either manually or automatically. There are many issues in the adoption and adaptation of 'Automatic' code generation technologies, more so when the application is safety-critical.

I have deliberately used the word 'automated' and not 'automatic' code generation. Automatic code generation implies an autonomous process without the need for end-user intervention, something that proceeds on its own. 'Automated' implies the use of some 'automatic' capabilities but also relies on some level of end-user interaction. So, what level of end-user interaction are we speaking about? I am speaking here only of *end-user customization.* Typically (and certainly in the case of ARTiSAN's Real-time Studio's Ada Synchronizer), end-users will either:

1. Configure the Ada Synchronizer prior to starting a project to ensure it generates code that meets the organizational 'coding standards', or;
2. During a 'pilot' project, the Ada Synchronizer will be customized throughout the project, the aim here being to explore different UML-Ada mappings throughout the project so that a fully customized Ada Synchronizer is produced at the end of the 'pilot' study.

The ultimate aim of both approaches is to have the capability to 'automatically' generate code. The reason why I stress the 'automated' (i.e. with end-user interaction) is that no 'automatic' Ada code generator ('out of the box') will completely meet your organizations coding standards[4] and an Ada generator which does not permit end-user configuration/customization will cause more problems – in terms of your organization adopting the technology and any certification of the Ada generation capability.

Although it may be perceived, by some readers, that the existence of the Ada Profile is potentially removing the 'art' of engineering Ada software, this is not its intent. The purpose of a Standardized Ada Profile is to ensure consistency – either within an organization or amongst all the Ada users. There should be no limitations in a particular organization customizing, yet further, the Standard Ada Profile – indeed, such practices are actively encouraged to enable the Standard Ada Profile to evolve by capturing the needs of a specific organization or industrial sectors. Further to this, the implicit semantic extensions contained within the Standard Ada Profile (i.e. the 'meaning' of a stereotype such as <<Ada Task>>) will only truly be understood by Ada programmers; as such the Ada Profile is still the specialist reserve of Ada designers and Ada software engineers.

# 10   Conclusion

The application of an UML-Ada Profile supports the OMG's MDA initiative and clearly delineates the concerns of analysts and designers (PIM), software engineers (Implementation Model) and hardware engineers (PSM). The intent of the Ada-UK working group, in defining a Standard UML-Ada Profile, is to ensure consistency in the use of the UML in creating a model destined to be implemented in Ada. It is not true to think that we will all become 'designers' and never have to 'program' any more Ada code by hand - although this is still a worthwhile aim. What a Standard Ada Profile will give us is a consistent 'semantic bridge' between the UML and Ada, and hence a consistent approach to designing Ada code.

---

[4] How can it? Unless you work very closely with a tool-vendor as they are developing an Ada Code Generation capability, how will the tool vendor know what coding standards you use? Furthermore, if a tool vendor created an Automatic Code Generator that satisfied Company X, it is unlikely to satisfy Companies A, B and C at the same time... And finally, if you wanted to change your coding standards and these were 'hard-coded' into a tool, your tool vendor would be the only one to re-configure the generator... what we in the business call 'vendor lock-in'.

With the Ada Programming Language we have possibly the best language to address the complexity, safety, reliability, scalability, and the many more issues brought about by today's software-intensive systems. By 'marrying' (through the Ada Profile) this unique capability of a programming language with the de facto modeling language (the UML), we are facilitating faster (and ultimately cheaper) development of Ada source code (either 'automated' or 'automatic' generation) whilst encompassing the 'best practice' application of Ada - gained from many years experience – which is now embedded within the Ada Profile.

One ultimate aim of the Standard Ada Profile working group is to ensure that the Ada Language is the first to have a standardized UML Profile. Given the level of coordination within the Ada-users community for such activities as Ada-0X, the future of the Ada language – augmented by the ability to 'design' Ada using a standardized notation – will continue successfully for many years to come.

# References

[1]    Object Management Group (OMG), September 2001. *OMG Unified Modeling Language Specification*. [online] Available from: http://www.omg.org
[2]    Ada Users Journal, Volume 23, Number 3, September 2002. *Customizing UML for the Development of Distributed Reactive Systems and Ada 95 Code Generation*. Kersten et al.
[3]    *Structured Development for Real-Time Systems (Volumes 1, 2 & 3)*; P.T.Ward & S.J. Mellor; Yourdon Press; 1985.
[4]    *Strategies for Real-Time System Specification*; D.J. Hatley & I.A. Pirbhai; Dorset House Publications;1987.
[5]    *Modern Structured Analysis*; E. Yourdon; Prentice-Hall;1989.
[6]    *Use Cases combined with Booch, OMT and UML*; P.P. Texel & C.B. Williams; Prentice-Hall; 1997.
[7]    *Object Oriented Software Engineering*; I. Jacobson et al; Addison-Wesley; 1992.
[8]    *Object-Oriented Modelling and Design*; J. Rumbaugh et al; Prentice-Hall; 1991.
[9]    *Object-Oriented Design with Applications*; G. Booch; Benjamin/Cummings; 1991.
[10]   *Object-Oriented Development, The Fusion Method*; D. Coleman et al; Prentice-Hall; 1994.
[11]   UML Distilled, Applying the standard Object Modelling Language; M. Fowler & K. Scott; Addison-Wesley; 1997

# Authors Biography

**Francis Thom, Principal Consultant - Artisan Software Tools**
Francis Thom has been developing real-time systems for the past 14 years. He started developing C3I Systems for Surface and Sub-Surface vessels. He has also worked on multi-role military and commercial avionics systems, on safety-critical railway signalling systems and On-Board Computers for space vehicles. He has written software predominantly in Ada during this period. His roles have varied from team leader to developer. Francis has also worked on Process Improvement initiatives for a number of companies. His role at Artisan includes mentoring, sales presentations and training courses. He has also written a series of white papers on Process Improvement, Systems and Software Engineering, the use of UML for safety-relates systems and software and has also spoken at INCOSE several times and at Ada-Deutschland (2001), and at Ada-UK (2001, 2002).

# HRT-UML: Taking HRT-HOOD onto UML

Silvia Mazzini[1], Massimo D'Alessandro[1], Marco Di Natale[2], Andrea Domenici[2],
Giuseppe Lipari[2], and Tullio Vardanega[3]

[1] Intecs HRT, via L. Gereschi 32-34, I-56124 Pisa, Italy
`silvia.mazzini@pisa.intecs.it, max.dalex@libero.it`
[2] Scuola Superiore S. Anna, viale R. Piaggio 34, I-56025 Pontedera, Italy
`{marco,andrea,lipari}@sssup.it`
[3] Università di Padova, via G. Belzoni 7, I-35131 Padova, Italy
`tullio.vardanega@math.unipd.it`

**Abstract.** This paper discusses issues that arose in an effort to transpose the semantic and methodological value of HRT-HOOD into a carefully-crafted selection
of the UML meta-model. The main goal of the project was to prevent the massive
advent of the UML from obliterating the unique strengths of the HRT specialisation
of HOOD for the support of real-time design. Avoidance of closed conservatism
also afforded the project the opportunity to reflect on aspects of object-oriented
design that may arguably benefit the resulting HRT-UML method.

## 1 Introduction

In the niche market that the HOOD method addresses (initially aerospace and lately also
transport), its specialised derivative HRT-HOOD has increasingly earned the reputation
of being "the" answer to the demands from the development of time-critical systems.

From the methodological standpoint, therefore, it may well be argued that HRT-
HOOD was a definite success story, yet critically impaired by the lack of productive tool
support.

In this context, the massive advent of UML presents the HRT-HOOD community
with a threat and an opportunity: the threat that the irresistible thrust of the new trend
obliterate the methodological value of the HOOD stronghold; the opportunity of riding
the UML wave, with its large offer of support technology, while trying to preserve the
unique strengths of the HRT-HOOD method.

These considerations spawned a project, supported in part by the Italian Space
Agency, ASI, which produced the definition of a new, UML-based method, named HRT-
UML.

This paper discusses the technical and methodological choices that the project made
in devising the strategy for bringing together the best of both worlds: HRT-HOOD, as the
outgoing method; and UML, as the linguistic infrastructure on which domain-specific
methods and processes can be instantiated. This paper also positions HRT-UML with
respect to the on-going efforts for extending UML to the real-time domain.

Finally, which should be of major interest to the Ada community, this paper presents
its strategic choice to view the Ada Ravenscar Profile as the target language for the
automated code generation capability attached to the method.

J.-P. Rosen and A. Strohmeier (Eds.): Ada-Europe 2003, LNCS 2655, pp. 405–416, 2003.

## 2  The HOOD Dowry

In keeping with its denomination, the HOOD method [7] aims to to combine the advantages of object-oriented design (the OOD tail of its name) and hierarchical decomposition (whence the initial H) in the development of complex, composite projects.

With hindsight, it can perhaps be noted that the HOOD conception attempted to be object oriented well (possibly too much) before object orientation had found its way into virtually all aspects of software engineering, including the design of programming languages that qualify for the development of critical systems.

In fact, at the birth of HOOD in the early 80s, "the" programming language of choice for the method was Ada 83 [8], which effectively restrained it to being object based, thereby limiting the extent of object orientation of its design paradigm.

As a distinctive token of the restrained object-orientation of the method, a HOOD object is an object and a class at one time, where the class is in fact a *singleton* and the corresponding object statically instantiated at design time.

There is no arguing that this restrained character detracts from the object-oriented purity of the method. Yet, once coupled with the blend of deliberate and inherent limitations to its support for inheritance, it effectively exempts HOOD from the common drawbacks of "pure" object-oriented methods, which may complicate traceability, testability and maintenance. Paradoxically, therefore, what looks like a limitation in the eyes of the purist, makes HOOD (and its HRT-HOOD derivative) quite a good fit for the development of higher-integrity systems, where restraints to the allowable expressive power are often necessary and desirable.

If the OOD line of HOOD promised more than it actually delivered (not necessarily to the detriment of users, though, especially for what concerns discipline and rigor) its H component proved the dominant character of the method. Traditional HOOD designs often appear as *include* graphs and/or *parent-child* hierarchies of considerable depths, with quality metrics debating over limits, benefits and drawbacks of decomposition.

While hierarchical decomposition has not since made it into the mainstream of object-oriented design (which denounces a potential weakness of the latter) the ability to decompose a problem into finer-grained parts arguably remains a fundamental aid in handling complexity. In fact, HOOD adopts this view in a manner that serves the interface mastering, testing and integration needs of prime contractor and system integrator, who have to be able to control potentially deep levels of decomposition (and, thus, contractual delegation), more extensively than those of plain programmers.

As a further token of its restrained object orientation, HOOD is not equipped for requirements analysis proper, which instead seems to fit rather well in current object-oriented methods. The HOOD method covers from the design requirements analysis phase down to architectural design, detailed design, coding (which can be automated) and testing. The need to rely on foreign methods (and technology) for the requirements analysis phase exposes HOOD users to the intricacies of the bridging and importation process and to the variability of the allowable choices and techniques. Although some attempts were made to address this weakness, no structural solution was ever found, particularly because the energy that was put in the definition of the method was never equally available for its maintenance.

Overall, it can be argued that the HOOD dowry includes a considerable spectrum of useful features, along with a number of drawbacks and limitations, part of which are historical and part conceptual. If HOOD were to be replaced, care should therefore be taken to devise transition strategies that prize the wealth of its useful features; notably, a well-defined development process, strong support for hierarchical decomposition, static object-to-class binding.

## 3    The HRT-HOOD Specialisation

The very conception of HRT-HOOD [5] rested on two pillars: the mandate to follow suit from the HOOD method; and the goal to fulfill the demands of a modern design method suitable for hard real-time systems. The latter, which the resulting method thoroughly addresses, required:

- The explicit recognition of the types and the typical activities (i.e. cyclic and sporadic) of hard real-time systems.
- The integration of appropriate scheduling paradigms with the design process.
- The explicit definition of the application timing requirements for each activity.
- The definition of the relative importance (i.e. criticality) of each activity.
- The decomposition in a software architecture that facilitate processor allocation, schedulability and timing analysis.
- Facilities and tools to allow and advance static verification as early as possible and to facilitate consistent code generation.

The key to achieving these goals was the selection of a computational model that, while being amenable to static scheduling analysis, allowed the incorporation of sporadic activities, of interactions between concurrent activities, and of a blocking approach that provide an optimal bound to priority inversion.

Interestingly, these notions instigated a very strong synergy between promoting (methodologically) a computational model with the above characteristics and pursuing (by language design) the incorporation of the missing features for it in the Ada language revision process [9]. This process resulted in the definition of the Ravenscar Profile [4], which currently is the most accurate expression of the HRT-HOOD computational model and it definitely is one of the key assets to preserve in any departing route to a novel method.

HRT-HOOD was predominantly shaped by way of specialisation. Where the HOOD basic object types were just *passive* and *active*, HRT-HOOD added *cyclic*, *sporadic* and *protected*, with specific constraints on how parent objects could be decomposed without breaking the integrity of the parent type and the requirement that no terminal objects (i.e. those that are not further decomposed) could qualify as merely active but had to specialise as cyclic or sporadic.

HRT-HOOD further introduces a distinction between the synchronisation requirements on an object's operation and the internal concurrent activity within the (active) object itself. The method defines an agent for the synchronisation, which it calls the object control structure, whereas it uses the notion of thread to capture the concurrent activity within the object. The thread executes independently of the incoming operation requests (i.e. the invocation of the object's external operations), but when it honours them, it does so in the order dictated by the control object. In addition to being an especially useful specialisation of the unsatisfactory HOOD concept of 'active', these notions permit a very direct route to Ada code generation, with the thread mapping to a task and the object control structure mapping to a protected object.

The terminal level of an HRT-HOOD project (i.e. the set of all its terminal objects) ultimately consists of a collection of single-threaded cyclic and/or sporadic objects, along with as many protected and passive objects as expressed in the design. With certain constraints in place on the allowable object behaviour, which can and should in fact be asserted at design level (e.g.: no asynchronous transfer of control, no task termination, single activation event per task, non-suspending operations) the threads of cyclic and sporadic objects will just be Ravenscar tasks; protected objects will also be Ravenscar; and passive objects will be Ada packages whose operations abide by the Ravenscar restrictions [1,2]. Thanks to the same set of constraints, Ravenscar protected objects also become a natural fit for the implementation of object control structures.

# 4    A Mapping to the UML

The Unified Modeling Language, UML [10] is a language for specifying, visualizing, constructing, and documenting the artifacts of software systems, as well as for business modeling and other non-software systems. The UML represents a collection of the best engineering practices that have proven successful in the modeling of large and complex systems.

UML is a very general language that can be used to describe various systems. Unlike HOOD, it includes a graphical representation of the formal language. It intentionally re-frains from including any design process; rather, it expects that various design processes be defined using the base language. The general notation of UML can be specialized, by identifying certain uses of the constructs as bearing some special semantics; such specializations are called *stereotypes*, which in fact proved instrumental to our mapping.

At present, UML has emerged as the software industry's dominant modeling language. It has been successfully applied to a wide range of domains, ranging from health and finance to aerospace and e-commerce. At the time of this writing, over sixty UML CASE tools can be listed from the OMG resource page www.omg.technology/uml. The HOOD market never even approached this level of support, whose offer, at the peak of its success, counted perhaps a handful of tools.

Born at Rational in 1994, UML was taken over in 1997 at version 1.1 by the OMG Revision Task Force (RTF), which became responsible for its maintenance. The RTF released an editorial revision, UML 1.2 in June 1998, and two minor revisions, UML 1.3 in June 1999 and UML 1.4 in September 2001. The latter has been taken as reference for the definition of the HRT-UML method. A major revision, UML 2.0, which also aims to address the real-time dimension, is currently under the OMG specification process. Section 5 of this paper positions HRT-UML with respect to the on-going OMG effort in the domain [11].

Our approach to moving from HRT-HOOD to UML was founded on the determination to:

- Resist to the introduction of UML features that may clash with the HRT-HOOD way of thinking, even though they may improve the expressive power of the model.
- Refrain from liberally blending UML into HRT-HOOD so as to prevent creeping changes in the HRT-HOOD conceptual model.

Our proposal seeks a balance between novelty and tradition, with special care for the preservation of the HRT-HOOD methodological strengths.

In order to master the complexity of a direct mapping from HRT-HOOD to UML, we have opted to address the conceptual relationships between the pure HOOD and UML notations first, before delving into the real-time issues of the HRT-HOOD method proper.

## 4.1 Outlining the Mapping Approach

Arguably, regarding UML as a notation is not completely accurate. Some observers insist that the strength of UML lies not in its notation but in its meta-model, that is, in its selection and standardisation of concepts in support of object-oriented modeling. Furthermore, UML does not define a tight correspondence between models and notations. There can be different notational rendering of the same UML model (i.e. meta-model instance). In UML, meta-model and notation must thus be regarded as separate concerns.

A similar dichotomy is also present in HOOD: the so-called Object Description Skeleton (ODS) is a textual representation of the HOOD meta-model, with a diagram as its notation. Yet, a HOOD model can have only one representation in notational form, so that an ODS can be automatically rendered in terms of a diagram.

Specification (meta-model) and notational rendering thus are separate concepts in both HOOD and UML, in the latter more keenly so than in the former. This observation suggested that we should first focus on correspondences at meta-model level, and thereafter on correspondences between the HOOD ODS and the UML notations of interest.

In our view, the mapping between the HOOD and the UML meta-models should address the following concerns:

*Conceptual Correspondence between HOOD and UML.* Whenever one and the same concept is present in both HOOD and UML, or else when a HOOD concept can be viewed as a restriction of a UML one, a direct correspondence can be made between the two and the meta-model representation in UML is either trivial or easily achieved through the UML stereotype mechanism.

*Conceptual Substitution of HOOD Concepts with Standard UML Replacements.* Whenever any HOOD modeling mechanism appears to be obsolete with respect to the more modern and standardised approach in UML, we introduce a conceptual substitution instead of a correspondence, so long as the replacement suits the aspects of the overall design process philosophy that we want to preserve.

HOOD is quite deliberate in limiting the extent of its object orientation. In actual fact, HOOD [7] draws a clear separation between objects and data, whereby objects provide procedural services, whereas data are information exchanged between objects by these services. The rationale for this choice was one of separation of concerns: separating data analysis from structural and functional analysis was expected to reduce the complexity that has to be dealt with in the design process.

As we have seen earlier, a HOOD object is a UML object from a certain standpoint and a class from another: a HOOD object is thus an object and a class at the same time. More specifically, a HOOD object corresponds to a UML object and a descriptor for it (i.e. a UML class), provided that the class is never treated as a type: no class hierarchy; no attribute with it as a type and no parameter with it as a type in any operation. Consequently, descriptors of HOOD objects cannot be included in inheritance hierarchy structures. This definition results in a suitable «HOOD Object Descriptor» stereotype for a UML class.

HOOD forces a rigid top-down decompositional approach to the design process. In contrast with that, object orientation promotes a strong flavour of bottom-up reasoning. It is fairly infrequent that the external interface of the overall system be precisely defined at the start of an object-oriented architectural design phase, which instead is required for HOOD.

A HOOD non-terminal (parent) object is one which decomposes to a set of child objects. The non-terminal object is said to be in an *include relationship* with its child objects The descriptor (class) of a non-terminal object will therefore bear an equivalent relation with the classes of the internal objects.

A HOOD "include" relationship is complemented by an *implemented by* relationship, which sets the intended correspondence between the items provided by the interface of the non-terminal object and the items provided by the interface of its internals.

UML is not quite as sharp as HOOD in its support for hierarchical decomposition. Some tentative parallel can be drawn with the UML package and subsystem mechanisms, neither of which however proves satisfactory.

The *package* is only an envelope with some visibility filter (i.e. a name-space), which, unlike the HOOD parent object, has no semantics of its own and thus provides no attribute or operation. Note that, to further complicate matters, the semantics of HOOD decomposed objects, which is strongly enforced at design time, is fully replaced by that of their decomposition hierarchy at run time, when decomposed objects effectively disappear for good.

The *subsystem*, on the other hand, has a semantic identity, which, however, needs to be realised by its internals, which are in turn distinct from it. It is precisely this difference in nature that breaks the parallel with the HOOD notion of parent object, which is wholly indistinguishable from a child object until the decision is made to actually decompose. A UML subsystem may be viewed as an *interface* that will be subsequently replaced by a concrete realisation: where HOOD forces identity between an object interface (the parent) and its realisation (its children), UML introduces conceptual and methodological separation. In fact, there arguably is value in focusing on interfaces alone more than solely on the realisation view. Yet, this would be quite a diversion from the instantiation-centric HOOD development logic, which may be worth pursuing with a longer-term vision than our current project horizon. As a further restraint, we also note that the subsystem does not appear to be a particularly stable concept in UML yet.

Overall, the most effective object-oriented parallel to the "include" relationship that holds in our context is *composition*, whereas that of the complementary "implemented by" relationship is *delegation*. With this view in mind, we contend that:

- A HOOD non-terminal object is a UML object, which is composed of its child objects.
- "Implemented by" relationships between the parent-provided operations and child-provided operations are mapped into delegation dependencies.
- All the operations of the parent class must be delegated.

In HOOD, an object A may use the provided items of an object B as long as those are visible inside A. This visibility requirement holds transitively: a child object can only use provided items of uncle objects that are in an explicit *use relationship* with the parent object.

We insist that the HOOD "use" relationship between any two objects entails, for UML, a *link* between the corresponding instances and an *association* between their respective classes (descriptors). A link between objects is required for them to interact, but the existence of the link in turn requires an association between the respective classes.

Interestingly, the notion of required interface entailed by the HOOD "use" relationship offers another opportunity to explore the direction we outlined with regard to UML subsystems. The plain HOOD interpretation has the required interface merely define (i.e. select) which of the provided operations of the used object the other object will actually use. The strengthened interpretation takes that further to mean the required specification (i.e. the service contract) in place between the client object and any supplier object.

Using interfaces instead of classes as target entities of associations relaxes the need to attain exhaustive classification: the class of the actual object bound to a requiring object at design time is less important than the guarantee that the supplier object realise the required interface. This interpretation of required interface seems quite attractive for the realization of a component-based software development support for hard real-time systems in which the expected behaviour (i.e. the required interface) is more important than abstract classification. Recent work [12] provides some initial insights in this direction by showing how composability can be effectively achieved for non-terminal objects whose internals are strictly HRT-HOOD.

This required interface between objects directly maps to a dependency (an association) from the class of the source (client) object to a UML *interface* that defines the list and the signature of the operations required from the class of the target (supplier) object. Undoubtedly, further specific information (e.g.: the conformance of the operation instance to the non-suspension requirement and the worst-case execution time of it) will have to be tagged to such interface information to permit this concept to suit the hard real-time design philosophy.

On the basic conceptual infrastructure for the UML-to-HOOD mapping outlined above, and by use of the UML support for notation specialisation, we can easily produce the definition of the ≪HOOD Object≫ *stereotype*.

At this point, our next step was to reproduce on that stereotype the specialisations that seemed best suited to express the conceptual enhancements that HRT-HOOD added to HOOD. Before doing that, however, we had to reflect on what parts of the HRT-HOOD method descended more from inherent limitations of the HOOD expressive power than from intentional HRT characteristics.

## 4.2   HRT Design Issues

HRT systems typically exhibit shallow levels of hierarchical decomposition. This occurs because the verification needs of the HRT development process call for an early characterisation of the design components, which is obstructed from any too indefinite level of specification. The information distance that may incur between an active parent object and the set of terminal objects it decomposes to may become so huge to prevent early static analysis altogether.

This observation has two important consequences. First, that the HOOD notion of "root" object is so indeterminate that it has no useful value to an HRT design: an HRT system is thought of and viewed as a collaboration of concrete objects. Second, that a deep decomposition level may be viewed as the clue of a design to be refactored in some way.

In fact, the major semantic contribution of the notion of parent object in HRT-HOOD is that of placing strict constraints on the allowable decomposition choices, which permit the type characterisation of the parent to globally hold for its decomposed internals. Once that is granted, all attention is placed on the concrete attribute values (whether actual or estimates) of the terminal objects at the current level of decomposition.

HRT architects tend to use objects as first-level entities and regard classes as a secondary tool. This attitude, which is quite the reverse of classical object orientation, reflects the HRT requirement that all instances of all classes be statically known, and fully characterised, at design time.

In standard object-oriented methods, the architectural design of a system is mainly concerned with class modeling. Object modeling in this context configures as an additional activity intended to clarify behaviour and collaboration aspects through the use of "scenarios". Scenarios or collaborations do not tell the way the system is built; they merely are an instrument to better understand the system behaviour in some circumstances.

In contrast with that, the starting point for an HRT design is the topology of its objects. From the HRT perspective, a system is built directly defining its objects and their links, instead of its classes and their relationships.

This notwithstanding, HRT systems may well use multiple instances of objects, which all hold one and the same structure and properties, but differ in their instance attribute values and have an independent state evolution. In fact, a reuse-oriented HRT design is quite naturally conducive to multiple static instantiation. (Note that HRT-HOOD implicitly fosters low-level component reuse in the forced coupling between the task and the protected object that implement the thread and the object control structure of a cyclic and/or a sporadic object.)

This flavour of multiple static instances however can hardly be satisfied by the HOOD Generic Object concept, whose decomposition is external (and thus invisible) to the system under development and whose instances appear as terminal objects (and thus hiding their internals) and cannot interact with other objects in the system. In essence, classes can quite possibly exist in an HRT system as long as all of their instances are statically defined and expose their internal structure clearly.

Note that, under the guarantee of static resolution of the instance-to-class relationship (i.e. static binding), recent experiences show that even some degree of polymorphism can prove acceptable to the implementation of hard real-time systems in so far as beneficial to the elegance and, marginally, to the execution speed of the resulting source code (cf. e.g.: [3]).

We call *topology orientation* the characteristic of the HRT domain to require full knowledge about the way the objects in the system are interconnected (i.e. the system topology). We view this fact to cause any HRT-HOOD design to systematically map onto a pair of UML models: the class model and the object model. The class model alone would not fully describe the system, for it would miss its topology and the instance structure. The object model alone would also not suffice, for it would lack the specification of class properties. Object modeling will be in the foreground and class modeling in the background. In fact, under certain circumstances, it is possible to automatically derive a class model from an HRT object model, so that the object model actually attains total expressive power, thus giving rise to what we call *the underlying class model*.

In standard object orientation, a class model usually just describes the logical structure of a composite object (a non-terminal in HOOD), leaving to class constructors and factory operations the determination of most of the actual structure of individual instances of the composite class.

In the HRT realm, instead, the inner structure of a composite class must be precisely defined at design time, which all of its (static) instances must equally abide by. Variability is limited to the setting of specific attribute values (e.g.: worst-case execution time of a concrete operation, parameter initialisation, initial state). The structure of all instances of a class of HRT objects cannot be thus deferred to the logic of factory operations or class constructors, but it must be statically encoded in the (object) model.

An effective way to define this class-level structure would be *by example*, that is, by inferring it from the structure of a typical instance, which we call *the prototype instance*.

Albeit not a standard UML concept, the "OO prototype" design pattern proposed by the "Gang of Four" [6] is an attractive approach to realise this concept. A new object instance can be created by clonation of the structure defined by the prototype instance of a class. With this approach, the underlying class model would only contain the abstract inter-class associations, while the exact structure of the instances would be provided by the *prototype instance* of the object descriptor. The prototype instance would thus define part of the semantics of a class, expressed in terms of an object model, without however being part of the system. Note that this would be no different from when the dynamic semantics of a class are expressed in terms of a state transition model.

Topological issues also arise when we take the HRT view of the "use" relationship into account, which is only fully defined when set between specific instances. In order to preserve the information value of the prototype instance without breaking the integrity of the HRT dimension, we allow "use" relationships to be directed towards *placeholder objects*, which can be statically bound to specific instances at a later stage of design. In general, placeholder objects will be unbound when the prototype instance of an object descriptor is first created; subsequently, they will be bound to actual objects that are visible in the instantiation environment. The target object of the binding will have to be of the same class as determined by the placeholder, or, in fact, provide the required interface.

The natural evolution of this attractive view is the vision of HRT design as the static assembling of instances; in other words, the notion of component-based HRT development. The use of placeholder objects in prototype instances may give rise to the HRT equivalent of the topology-aware metaphor of GUI Builders.

## 4.3  Mapping the HRT Dimension

The realisation of the HRT-HOOD basic object types in HRT-UML quite simply results from the specialisation of the ≪HOOD object≫ stereotype. This specialisation adds semantics and attributes to the stereotype (e.g.: the period for periodic object; the deadline for all threaded objects; etc.), but it also introduces constraints on the allowable decomposition and visibility of objects. The HRT-HOOD reference manual [5] forms the basis for the definition of the required specialisations.

A more delicate issue arises with regard to operations. HOOD requires that *synchronisation constraints* be associated to the provided operations of objects, which specify how control should flow on both sides of the relevant execution request.

For example, HOOD uses a "highly synchronous execution request" constraint to model an Ada rendezvous with in and out parameters, which interrupts the control flow in the client until completion of the requested operation within the server.

HRT-HOOD retains this view and completes it with what required to model the invocation of protected operations. Of all allowable synchronisation constraints, the Ravenscar-minded version of HRT-HOOD then rules out those that might incur unbounded blocking of the caller, so as to ensure design-level compliance with the Ravenscar restrictions.

HOOD and HRT-HOOD view such synchronisation constraints as a static property of operations and associate them to the descriptor of the providing object. Conversely, UML views the synchronisation mode as a dynamic attribute of the invocation: different calls of one and the same operation can thus feature different synchronisation modes. Accordingly, UML defers this aspect of specification to the Interaction Diagrams, leaving it vacant in Class Diagrams.

This UML feature is clearly in contrast with the distinguishing methodological need of HRT design to allow and facilitate early static analysis of the system under construction. Consequently, the HOOD-sided view of synchronisation constraints has prevailed over the UML one in our definition of the HRT-UML method. Hence, the ≪HOOD Object≫ stereotype forms the basis for the HRT-UML specialisation of both the object characteristics and the corresponding operation constraints.

## 5  HRT-UML vs. the OMG Real-Time Profile

In our definition of an HRT-UML profile inspired to HRT-HOOD there were two distinct phases. The first one focused on the definition of a set of UML stereotypes that offered the best possible mapping for the HRT-HOOD meta-model. This paper has presented the resulting definition, which constitutes the core package and the core level of our HRT-UML profile. The stereotypes and tagged values defined in this phase allow the form of schedulability analysis supported by HRT-HOOD models, which fulfills the initial objectives of the project.

In March, 1999, the OMG issued a request of a profile for schedulability, performance and time (real-time analysis). The initial submission issued in August 2000 was followed by a revised version in July 2001, which had an entirely different structure and definitions. The revised submission of the OMG profile was then voted for adoption with minor changes only in November 2001.

Despite its belated arrival for the purposes of this project, the OMG profile for time and schedulability cannot be possibly ignored, for obvious pragmatic reasons. It is in fact expected that compliance with real-time analysis tools working on the OMG stereotypes and adoption (by specialization) of stereotypes and design patterns that are very likely to become a standard among real-time UML designers should enhance the portability and the readability of the UML schemes generated from our profile.

In order to attain the desired extent of compliance, we resolved to add two further levels to our HRT-UML profile.

The first additional level seeks compliance with the (simpler) static schedulability model defined by the OMG profile, which provides equivalent constructs for most of the HRT-HOOD meta-model entities. Unfortunately, however, the static model does not provide any expressive means to represent the synchronous/asynchronous semantics of execution requests (method calls) neither it allows the representation of precedence constraints. These limitations make most kinds of schedulability analysis (especially end-to-end analysis) impossible, unless the expressiveness of the static model were enhanced by specialising the stereotypes of the OMG static model.

The second additional level for our HRT-UML profile consists in the definition of an additional dynamic or behavioral UML model implied by the HRT-HOOD design. The definition of a dynamic model permits to sub-class the stereotypes defined in the dynamic model of the OMG profile, which are the basis for the most general forms of schedulability analysis. It is worth noting, however, that this enhanced compatibility with the OMG definitions comes at the price of a greater complexity, which may prove a substantial hurdle for those familiar (and satisfied) with HRT-HOOD.

Both of these two additional levels are currently built upon definitions and stereotypes that are not supported by any commercial tool yet and which are not validated by a sufficient span of design practice. It thus remains an on-going research effort the goal to build an HRT-UML profile that be both coherent with the HRT-HOOD meta-model semantics and yet compliant (by sub-classing) with OMG-profile stereotypes and definitions with actual use in the real-time design practice and support by static analysis tools.

# 6   Conclusions

The project provided an excellent opportunity of cross-contamination for the two starting points of our exercise.

In the HOOD-to-UML direction, focus was placed on transposing pivotal aspects of the base HOOD method and its HRT-HOOD derivative, which we deemed crucial to the fitness of the HRT-UML profile for the design of modern, high-integrity real-time systems. The highlight of this transaction was perhaps the preservation in UML of the inherent *topology orientation* of HRT design, which has resulted in the introduction of such innovative concepts as the *prototype instance* and the *placeholder object*.

The reverse UML-to-HOOD direction, in which the notion of UML *stereotype* proved instrumental, brought forward the attractive potential of the object-oriented concept of *interface*, an obvious candidate for supporting placeholder objects, and of *component-based HRT development*, as the means whereby domain-specific reuse may productively join forces with HRT design.

Finally, the transposition exercise also confirmed the strategic value and the methodological strength for HRT design to attain and preserve compliance with the Ravenscar profile.

# References

1. Ada Rapporteur Group: AI-249: Ravenscar profile for high-integrity systems. Technical report, ISO/IEC JTC1/SC22/WG9 (2002)
   http://www.ada-auth.org/cgi-bin/cvsweb.cgi/AIs/AI-00249.TXT
2. Ada Rapporteur Group: AI-305: New pragma and additional restriction identifiers for real-time systems. Technical report, ISO/IEC JTC1/SC22/WG9 (2002)
   http://www.ada-auth.org/cgi-bin/cvsweb.cgi/AIs/AI-00305.TXT
3. Alonso, A., López, R., Vardanega, T., de la Puente, J.A.: Using object orientation in high integrity applications: A case study. In J. Blieberger and A. Strohmeier, editors, Proceedings 7th Int'l Conference on Reliable Software Technologies | Ada Europe 2002, volume 2361 of LNCS. Springer-Verlag (2002)
4. Burns, A., Dobbing, B., Vardanega, T.: Guide for the use of the ada ravenscar profile in high integrity systems. Technical Report YCS-2003-348, University of York (2003)
   http://www.cs.york.ac.uk/ftpdir/reports/YCS-2003-348.pdf
5. Burns, A., Wellings, A.: HRT-HOOD: A Structured Design Method for Hard Real-Time Systems. Elsevier Science, Amsterdam, NL ISBN 0-444-82164-3 (1995)
6. Gamma, E., Helm, R., Johnson, R., Vlissideds, J.: Design Patterns: Elements of Reusable Object-Oriented Software. Addison-Wesley (1999)
7. HOOD Technical Group: HOOD Reference Manual 3.1. Prentice Hall, Englewood Cliffs, NJ (USA) (1993)
8. ISO: Ada Reference Manual. International Standardisation Organisation, Geneva, CH ISO/IEC 8652:1987 (1987)
9. ISO: Ada Reference Manual. International Standardisation Organisation, Geneva, CH ISO/IEC 8652:1995 (1995)
10. Revision Task Force: OMG Unified Modeling Language Specification, Version 1.4. Technical report, Object Management Group (OMG) OMG document formal/01-09-67 (2001)
11. Selic, B., Rumbaugh, J.: Using UML for Modeling Complex Real-Time Systems. Technical report, Rational (2002)
    http://www.rational.com/products/whitepapers/umlrt.pdf
12. Vardanega, T., Caspersen, G.: Engineering Reuse for On-board Embedded Real-Time Systems. Software - Practice and Experience, 32(3):233–264 John Wiley & Sons (2002)

# A Case Study in Performance Evaluation of Real-Time Teleoperation Software Architectures Using UML-MAST

Francisco Ortiz, Bárbara Álvarez, Juan Á. Pastor, and Pedro Sánchez

Universidad Politécnica de Cartagena (Spain)
francisco.ortiz@upct.es

**Abstract.** Reference architectures for specific domains can provide significant benefits in productivity and quality for real-time systems development. These systems require an exact characterisation based on quantitative evaluation of architectural features referred to timing properties, such as performance, reliability, etc. In this work, an UML-based tool has been used to obtain a measure of performance between two alternative architectures. These architectures share the same functional components, but with different interaction patterns. The used technique is illustrated with an industrial study in a well-known real-time domain: teleoperation systems. The obtained results show clear differences in performance between two architectures, giving a clear indication of which one is better from this point of view[1].

## 1   Introduction

Our experience in several research projects has allowed us to prove the interest of reusing a reference software architecture for teleoperation systems [2,5]. The use of such an architecture has allowed software components to be reused in different applications, making their development easier and decreasing their costs and development times.

The mentioned architecture was restricted to applications in which service robots work in structured environments, whose geometrical characteristics are perfectly known before the operation is performed, and in which reactive behaviour is usually limited to sensor-driven safety actions. The ROSA (*Remotely Operated System Arm*) and TRON (*Teleoperated and Robotized System for Maintenance Operation in Nuclear Power Plants Vessels*) systems are examples of this type of applications for maintenance tasks in nuclear power plants [4].

The development of a blasting robot for ship hulls [7,8], whose characteristics differ from the mentioned above, led us to evaluate whether the original architecture could be re-used for the new system. In this case, the system should work in unstructured environments and reactive behaviour is predominant during some operation phases. Given these requirements it is necessary either to include new services (i.e. a computer vision system) or use the existing ones in different ways.

---

[1] This work has been partially supported by FEDER (TAP-1FD97-0823) and GROWTH project (GRD2-2001-50004)

J.-P. Rosen and A. Strohmeier (Eds.): Ada-Europe 2003, LNCS 2655, pp. 417–428, 2003.

The result of the evaluation was that the performance requirements could not be met and it was necessary to propose modifications on the original architecture that combine the original components (trying to minimise the impact of the changes) with different interaction patterns. The original architecture and an alternative approach are compared in this paper.

The model used for representing the temporal and logical elements and real-time requirements of applications has been UML-MAST (*Modelling and Analysis Suite for Real-Time Applications*), developed by the University of Cantabria (Spain) [6]. UML-MAST comprises (1) an UML profile, where all the elements required to performed a RMA analysis [9] are modelled as stereotyped classes, (2) a set of analysis tools, and (3) a compiler that translates the UML model into a file, analysable by such analysis tools. The UML profile allows us to describe the system using an UML-based CASE tool [10]. The analysis model is directly obtainable from the UML description of the system design. UML-MAST can describe rather complex systems, including multi-processor and distributed architectures. We present our experience in using UML-MAST in the development of industrial applications.

The paper is organized as follows. In Sect. 2 the two alternative interaction patterns are introduced in order to compare the performance of both schemes. In Sect. 3 some relevant aspects of MAST are reviewed. Section 4 contains different models that represent different scenarios. In Sect. 5 the results of the analysis are described. Finally, Sect. 6 gives some conclusions and discusses further works.

## 2    Two Alternative Architectures Compared

Figures 1 and 2 show the considered components and interaction patterns taken into account for performance evaluation. Figure 1 describes the original architecture (*arch_1*) and Fig. 2 an alternative one (*arch_2*). Though the schemes showed in the figures are somewhat oversimplified, are enough to illustrate all the important issues. The boxes in Figs. 1 and 2 represent the main subsystems of the architectures and they could enclose tasks and shared resources. Note that the functional components of both schemes are the same, as well as their interfaces, but not their interaction patterns. The responsibilities of such subsystems are:

- *CinServer*. This server provides the system with operations for checking whether a given movement implies a collision between the robot and the environment.

    o   In *arch_1*, *CinServer* provides this service following an asynchronous client-server pattern. *HighLevelController* sends the current robot position to *CinServer* and retrieves asynchronously the answer that informs whether the current motion can cause a collision. In the original systems, this service was provided just *before* the execution of a given movement, but if the robot operates in unstructured environments then the service should be required and provided continuously during the motion. In these conditions, the interactions between *CinServer* and *HighLevelController* are a system bottleneck. For this reason the interaction pattern between them was modified according to the scheme showed in Fig. 2.

○ In *arch_2*, *CinServer* receives the robot position directly from *LowLevelController* and sends a message to *HighLevelController* if detects a collision. *HighLevelController* acts as an "Observer" subscribed to the information produced by *CinServer*.

- *UserInterface*. This subsystem is in charge of interacting with the user. It allows him to issue the desired command to the robot and to show the status of its execution.

- *LowLevelController*. This subsystem physically actuates the robot to move it and to get data from the sensors of the robot in order to evaluate its global state and position. This state information is sent to *HighLevelController* by means of *updateStatus*. In general this subsystem can be ported to any platform, nevertheless its execution has been considered in the same node of the other subsystems to simplify the analysis.

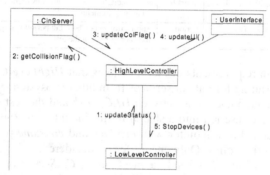

**Fig. 1.** *Arch_1*: Asynchronous client-server pattern between *CinServer* and *HighLevelController*

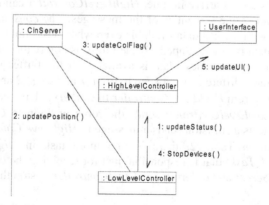

**Fig. 2.** *Arch_2*: Observer pattern between *CinServer* and *HighLevelController*

The model used for evaluating the timing requirements of *arch_1* was based on the characterization of the architecture timing behaviour described in [3]. Such a description, based on the *Rate Monotonic Analysis* method [10], is very exhaustive and allows the designers to reason with confidence about the timing correctness of the systems developed when using *arch_1*. But if the interaction patterns change, such model is difficult to use. For this reason the simpler model described in the following paragraphs has been used to compare *arch_1* and *arch_2*.

To simplify the analysis, the minimum significant number of tasks has been considered in each subsystem (boxes in Figs. 1 and 2). So, usually each subsystem has associated only one task. Following this approach the tasks considered to be executed in the subsystems of *arch_1* (Fig. 1) are described in Table 1.

**Table 1.** Task considered in *arch_1* (Fig. 1)

| Subsystem | Tasks | Activation | Minimum event interarrival time or period given by |
|---|---|---|---|
| **LowLevelController** | LLC_Task | Periodic | Motor controller loop. |
| **UserInterface** | UI_Task | Sporadic | The rate at which data should be updated in user interface. |
| **CinServer** | CinServer_Task | Sporadic | Short enough to avoid collisions (depends on robot speed and safety requirements) |
| **HighLevelController** | HLC_Task | Periodic | Application dependant, but period is larger than in *LLC_Task* and shorter than in the rest. |

One of the main requirements of the systems is that *HighLevelController* can not be blocked while waiting for data or services from other subsystems. In order to make asynchronous the communications between *HLC_Task* and the rest of tasks, the data produced by them are inserted into a shared buffer included in *HighLevelController*. This buffer is modelled as a monitor with *produce* and *consume* procedures and immediate ceiling priority policy. Only one buffer is considered.

When *HighLevelController* sends a message to *CinServer* or *UserInterface*, it does not wait for a response. Later, it will inspect the buffer looking for new data from *LowLevelController* and the responses to previous services requested to *CinServer*. If responses do not arrive in time, *HighLevelController* can take some action, but this is not a part of the semantics of the messages. The *consume* call always returns immediately with the last data available even when it has not been updated since the previous call (data is time-stamped). If the buffer is empty the *consume* procedure returns *null* meaning that no new data is available. If the buffer is full the data is overwritten[2]. The main differences between to *arch_1* and *arch_2* are:

- The queries from *HighLevelController* to *CinServer* have been eliminated.
- Instead, *LowLevelController* sends the robot position to *CinServer*. If *CinServer* detects a collision sends a message to *HighLevelController*.
- As shown in Table 2, there is one more task in *HighLevelController* (*HLC_Col_Task*) that is responsible just for receiving the collision messages from *CinServer* and ordering *LowLevelController* to stop the robot.

---

[2] The important issue is not to get as much as data arrives, but to get the data before deadline expires. If so, the good data is the last data.

**Table 2.** New tasks considered in *arch_2* (Fig. 2)

| Subsystem | Tasks | Activation | Minimum event interarrival time or period given by |
|---|---|---|---|
| **HighLevelController** | HLC_Task | Periodic | Application dependant, but period is larger than in *LLC_Task* and shorter than in the rest. |
| | HLC_Col_Task | Sporadic | Short enough to avoid collisions (depends on robot speed and safety requirements) |

## 3  Analysis of the Architectures Using UML-MAST

In UML-MAST the temporal behaviour is represented by means of three complementary views:

- **Platform Model.** It represents the processor, the tasks and the scheduling policies. The platform model corresponding to *arch_1* is shown in the Fig. 3. The platform model corresponding to *arch_2* is the same, but includes one more task (*HLC_Col_Task*). To simplify the model, the system considered is not distributed[3] and all the tasks are scheduled according to a fixed priority policy.

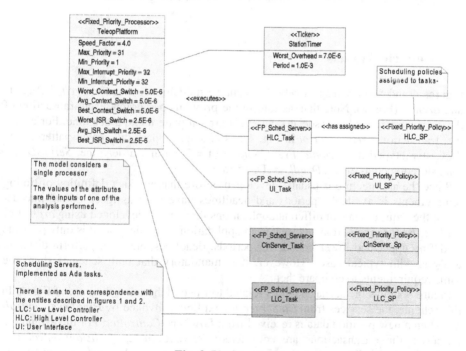

**Fig. 3.** Platform model

---

[3] For distributed systems, UML-MAST models the communication links as processing resources and the transmission times as execution times.

- **Logical Components Model.** The processing requirements of the operations can be modelled as well as the access to the resources shared among the different tasks. As said above, an immediate ceiling protocol has been used to limit the effects of priority inversion. In the implementation in Ada 95 [1], these resources are protected objects, otherwise the RMA analysis can not be applied.

- **Scenario Model.** This view allows us to model the dynamic behaviour of the system as a set of concurrent transactions. Each transaction represents a set of activities that will be executed sequentially by one or more tasks and it is activated by external events. A transaction is described attending to:

    1.  The events that cause its activation, its deadline and response time.
    2.  The way in which the different tasks interact to implement it.

The first point is described by means of associations that attach a given transaction to (1) a description of the arrival pattern of the events that activate it and (2) a description of its deadline. Transactions, event arrival patterns and deadlines are modelled by MAST using stereotyped classes. The second point is described by an activity diagram associated to the transaction. The activities generate events that are internal to the transaction in which they have been produced, but external to the rest and therefore may cause their activation. The scenario models corresponding to *arch_1* and *arch_2* are described in the following section.

## 4   Scenario Models

Figures 4 and 5 show respectively the scenario models associated to *arch_1* (Fig. 1) and *arch_2* (Fig. 2). Note that, as said in the previous paragraph, each transaction of the scenario models has associated an event arrival pattern and a deadline. For example, transaction *ServosControl* in Fig. 4 has attached a periodic arrival pattern, *Pid-Tick*; stereotyped as *Periodic_Event_Source* (T= 5ms) and a deadline, *ServosDeadline*, stereotyped as *ServosDeadline* (D = 5 ms).

Since the architecture is usually developed before that an exact definition of timing requirements is available, periods and deadlines have been assigned taking into account the requirements of different applications that can be developed using *arch_1* or *arch_2*. As these requirements vary from application to application, it is only possible to define upper and lower bounds for periods, deadlines, etc., and perform different analysis with typical cases. However, it is mandatory that both schemes share the same requirements to compare them.

Figure 4 shows the scenario model related to the collaboration diagram of Fig. 1. In this scheme, the services from *CinServer* are explicitly invoked by *HighLevelController* when a new position data is received from *LowLevelController*. For the evaluation purposes, three transactions are considered: *ServosControl, CollisionControl* and *UpdateDisStatus*. The transaction *ServosControl* represents the servo control of the teleoperated mechanisms. The transaction *UpdateDisStatus* gives the updating of data in the graphical user interface and the transaction *CollisionControl* represents how the collision detection is managed in *arch_1*. As said above, the interactions of the tasks that collaborates to perform a given transaction are described by an activity diagram. Figures 6, 7, and 8 show the activity diagrams corresponding to the transactions *ServosControl, UpdateDisStatus* and *CollisionControl*. Each element of the activity diagram has a stereotype that defines it in the UML-MAST model. It is beyond the scope of this paper to describe such a model and the reader is referred to [6] for a detailed information. But very roughly:

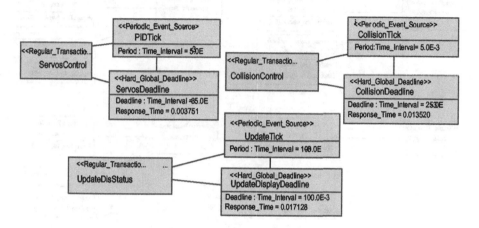

**Fig. 4.** Scenario model 1

- Each swimlane is labeled with the name of the task that performs the activities of such a swimlane. So, it is easy to identify the tasks involved in each transaction and their interactions.
- Each transaction starts with the arrival of a timed clock-driven event. So, initially the first task involved in a transaction is in a *Wait_State*. A *Wait_State* in the UML-MAST model represents that the task is waiting for an external event for its activation. In all the cases considered, the external events are clock events. Note the correspondence between the names of the *Periodic_Event_Source* in Figs. 4 and 5 and the names of the *Wait_States* of Figs. 6–10.
- All the activities are stereotyped as *Timed_Activity*. A *Timed_Activity* encloses an operation or a set of operations whose timing behaviour is well described in the model. When an operation implies the access to shared resources, the immediate ceiling protocol applies. It is important to remark that all the transactions can be performed concurrently, despite they are described in separate diagrams.
- Finally, a transaction ends when it reaches a timed state that describes the deadline of the transaction. Note the correspondence between the names of *Hard_Global_Deadline* of Fig. 4 and the names of the *Timed State* in Figs. 6–10.

The collaboration diagram of Fig. 2 is associated to the scenario model given in Fig. 5. This model considers the previous transactions *ServosControl* and *UpdateDisStatus*. However, the transaction *CollisionControl* has been replaced by two transactions. Note that in *arch_2* the processing of the collision in *HighLevelController* is separate from the rest and performed by a new task (*HLC_Col_Task*). So, the new transactions are *CollisionControl2*, that deals exclusively with the actions of *High-LevelController* associated with collisions, and *MonitorStatus* that represents the rest of processing. Figure 9 shows the activity diagram associated to the transaction *CollisionControl2* and Fig. 10 represents the transaction *MonitorState*.

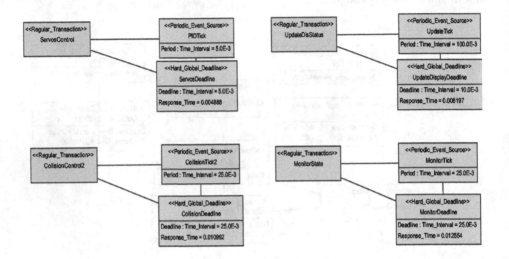

**Fig. 5.** Scenarios model 2

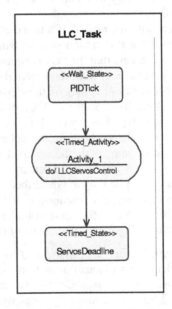

**Fig. 6.** Scenarios model 1 and 2. Transaction *ServosControl*

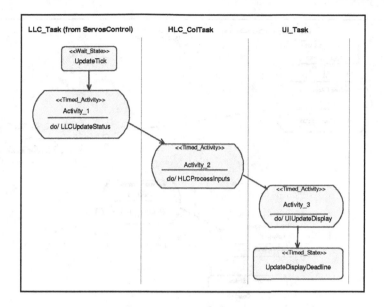

**Fig. 7.** Scenarios model 1 and 2. Transaction *UpdateDisStatus*

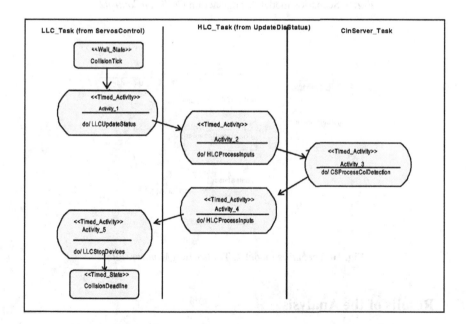

**Fig. 8.** Scenarios model 1. Transaction *CollisionControl*

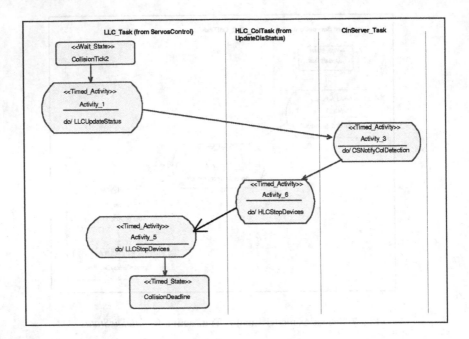

**Fig. 9.** Scenarios model 2. Transaction *CollisionControl2*

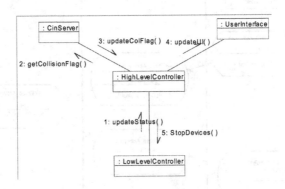

**Fig. 10.** Scenarios model 2. Transaction *MonitorState*

## 5    Results of the Analysis

Table 3 shows the response times of a classic RMA analysis of the previously described scenario models. The MAST tool automatically sets task priorities according to the *Rate Monotonic Scheduling* analysis technique. As the Table 3 shows, the response times for collision detection (*CollisionControl* in *arch_1* and *CollisionCon-*

*trol2* in *arch_2*) and the user interface update (*UpdateDisStatus*) are larger in the scenario model 1 (*arch_1*) than in the scenario model 2 (*arch_2*). However, the response time of the transaction *ServosControl* is a little larger in the scenario 2 (*arch_2*) than the scenario 1 (*arch_1*). This is due to the existence of one more task (*HLC_Col_Task*) and to the new responsibilities of *LowLevelController* (messages to *CinServer* not considered in *arch_1*) It must be highlighted that all the considered simplifications benefit the original scheme of interactions (*arch_1*). A more complex and realistic model (not included in this paper for space reasons), that includes more activities and distributes the *HLC_Task* responsibilities among several tasks, has produced the results showed in Table 4 (as more tasks were considered, CPU capacity was increased respect to the one used in Table 3 for avoiding deadlines expiration). The response times corresponding to *arch_1* worsen rather fast when new tasks are added, while the ones corresponding to the *arch_2* remain more stable. The unique advantage of the *arch_1* according to the tables is that the response time of *ServosControl* task is slightly shorter.

**Table 3.** Results of analysis (Only Tasks described in Tables 1 and 2)

| Transaction | Events arrival Pattern | Deadline | Scenario 1 Response time | Scenario 2 Response time |
|---|---|---|---|---|
| ServosControl | Periodic,  T = 5 ms | 5 ms | 3,8 ms | 4,8 ms |
| UpdateDisStatus | Periodic,  T = 100 ms | 100 ms | 17 ms | 6,1 ms |
| CollisionControl | Periodic,  T = 25 ms | 25 ms | 14 ms | |
| CollisionControl2 | Periodic,  T = 25 ms | 25 ms | | 11 ms |
| MonitorStatus | Periodic,  T = 25 ms | 25 ms | | 13 ms |

**Table 4.** Results of analysis. (Additional tasks in *HighLevelController*)

| Transaction | Events arrival Pattern | Deadline | Scenario 1 Response time | Scenario 2 Response time |
|---|---|---|---|---|
| ServosControl | Periodic,  T = 5 ms | 5 ms | 3,1 ms | 3.4 ms |
| UpdateDisStatus | Periodic,  T = 100 ms | 100 ms | 23 ms | 8,1 ms |
| CollisionControl | Periodic,  T = 25 ms | 25 ms | 23 ms | |
| CollisionControl2 | Periodic,  T = 25 ms | 25 ms | | 10 ms |
| MonitorStatus | Periodic,  T = 25 ms | 25 ms | | 16 ms |

# 6   Conclusions

The two interaction schemes introduced in this work share the same functional decomposition, but their timing behaviour are rather different. From the functional point of view both designs could be appropriate, but the first one (asynchronous client-server) is not acceptable with regard to its performance when *CinServer* has to be used on line.

Taking into account that the purposes of the analysis were to evaluate and to compare architectural schemes it is necessary to interpret these results under this point of view. In this way, the main conclusions are:

- Most of the relevant trade-offs are referred to the interaction patterns among components and not to their enclosed functionality.
- It would be possible to re-use a significant number of existing components if it would be possible to modify their interaction patterns maintaining their functionality and interfaces.

To summarise, it is much more interesting to define an architectural framework that defines a set of rules that allow the same components to be shared among systems with different architectures than trying to define a software architecture for large domains in which it is nearly impossible to reach the requirements of all of the potential applications. And one of the rules of such an architectural framework should be to consider the interaction patterns as design entities at the same level that the functional components. In this way, some original components could be reused to work in non-structured environments when other interaction patterns are selected, and other components could be replaced by other new ones (i.e. a collisions detection subsystem can be replaced by a computer vision subsystem).

The study of the performance during the first design phases is useful to compare different design solutions. At an architectural level such analysis can be of coarse grain and should be completed in later development phases. But even so, it is necessary an automated support of the timing evaluation process and a standard notation as UML, despite all its drawbacks for describing architectures.

# References

1. Ada 95 Reference Manual: Language and Standard Libraries. International Standard ANSI/ISO/IEC-8652, 1995.
2. A. Alonso, B.Alvarez, J.A. Pastor, J.A. de la Puente, A.Iborra. *"Software architecture for a robot teleoperation system"*. Proceedings of the 4th IFAC Workshop on Algorithms and Architectures for Real-Time Control. Portugal. April 1997.
3. Álvarez et al.. *"Timing Analysis of a Generic Robot Teleoperation Software Architecture"*, Control Engineering Practice, vol 6(6), pp. 409–416. June, 1998.
4. B. Álvarez et al.. *"Developing multi-application remote systems"* Nuclear Eng. International, vol. 45(548). March 2000.
5. L. Bass et al. *"Software Architecture in Practice"*. Addison-Wesley, 1998.
6. J.M. Drake et al. *"Mast Real-Time View: Graphic UML Tool for Modeling Object Oriented Real Time Systems"*. Group of Computers and Real-Time Systems. University de Cantabria (Internal Report), 2000. Spain. http://mast.unican.es/umlmast
7. Environmental Friendly and Cost-effective Technology for Coating Removal" (EFTCoR). UE- 5th FP. GROWTH project ref. GRD2-2001-50004, 2001.
8. A. Iborra et al. *"Service robot for hull blasting"*. The 27th Annual Conference of the IEEE Industrial Electronics Society (IECON'01), pp. 2178–2183. November, 2001.
9. M.H. Klein et al. *"A Practitioner's Handbook for Real-Time Analysis Guide to Rate Monotonic Analysis for Real-Time Systems"*. Kluwer Academic Publishers, 1993.
10. Reference manual. Rational Sw Corp, 2000. Available at http://www.rational.com.

# Author Index

# Lecture Notes in Computer Science

For information about Vols. 1–2601

please contact your bookseller or Springer-Verlag